This book introduces the tools of modern differential geometry – exterior calculus, manifolds, vector bundles, connections – to advanced undergraduates and beginning graduate students in mathematics, physics, and engineering. It covers both classical surface theory and the modern theory of connections and curvature, and includes a chapter on applications to theoretical physics. The only prerequisites are multivariate calculus and linear algebra; no knowledge of topology is assumed.

The powerful and concise calculus of differential forms is used throughout. Through the use of numerous concrete examples, the author develops computational skills in the familiar Euclidean context before exposing the reader to the more abstract setting of manifolds. There are nearly 200 exercises, making the book ideal for both classroom use and self-study.

$$\alpha \wedge \beta = \alpha \otimes \beta - \beta \otimes \alpha$$

(page 11)

Differential Forms and Connections

Differential Forms and Connections

R.W.R. Darling

University of South Florida

CAMBRIDGE
UNIVERSITY PRESS

Published by the Press Syndicate of the University of Cambridge
The Pitt Building, Trumpington Street, Cambridge CB2 1RP
40 West 20th Street, New York, NY 10011-4211, USA
10 Stamford Road, Oakleigh, Melbourne 3166, Australia

© Cambridge University Press 1994

First published 1994
Reprinted 1995, 1996

Printed in the United States of America

Library of Congress Cataloging-in-Publication Data is available.

A catalogue record for this book is available from the British Library.

ISBN 0-521-46259-2 hardback
ISBN 0-521-46800-0 paperback

Contents

Preface   ix

1   Exterior Algebra   1

1.1   Exterior Powers of a Vector Space   1
1.2   Multilinear Alternating Maps and Exterior Products   5
1.3   Exercises   7
1.4   Exterior Powers of a Linear Transformation   8
1.5   Exercises   12
1.6   Inner Products   13
1.7   The Hodge Star Operator   17
1.8   Exercises   20
1.9   Some Formal Algebraic Constructions   21
1.10 History and Bibliography   23

2   Exterior Calculus on Euclidean Space   24

2.1   Tangent Spaces – the Euclidean Case   24
2.2   Differential Forms on a Euclidean Space   28
2.3   Operations on Differential Forms   31
2.4   Exercises   33
2.5   Exterior Derivative   35
2.6   Exercises   39
2.7   The Differential of a Map   41
2.8   The Pullback of a Differential Form   43

2.9   Exercises   47

2.10 History and Bibliography   49

2.11 Appendix: Maxwell's Equations   50

3   Submanifolds of Euclidean Spaces   53

3.1   Immersions and Submersions   53

3.2   Definition and Examples of Submanifolds   55

3.3   Exercises   60

3.4   Parametrizations   61

3.5   Using the Implicit Function Theorem to Parametrize a Submanifold 64

3.6   Matrix Groups as Submanifolds   69

3.7   Groups of Complex Matrices   71

3.8   Exercises   72

3.9   Bibliography   75

4   Surface Theory Using Moving Frames   76

4.1   Moving Orthonormal Frames on Euclidean Space   76

4.2   The Structure Equations   78

4.3   Exercises   79

4.4   An Adapted Moving Orthonormal Frame on a Surface   81

4.5   The Area Form   85

4.6   Exercises   87

4.7   Curvature of a Surface   88

4.8   Explicit Calculation of Curvatures   91

4.9   Exercises   94

4.10 The Fundamental Forms: Exercises   95

4.11 History and Bibliography   97

5   Differential Manifolds   98

5.1   Definition of a Differential Manifold   98

5.2   Basic Topological Vocabulary   100

5.3   Differentiable Mappings between Manifolds   102

5.4   Exercises   104

5.5   Submanifolds   105

5.6   Embeddings   107

5.7	Constructing Submanifolds without Using Charts	110
5.8	Submanifolds-with-Boundary	111
5.9	Exercises	114
5.10	Appendix: Open Sets of a Submanifold	116
5.11	Appendix: Partitions of Unity	117
5.12	History and Bibliography	119

6	**Vector Bundles**	**120**
6.1	Local Vector Bundles	120
6.2	Constructions with Local Vector Bundles	122
6.3	General Vector Bundles	125
6.4	Constructing a Vector Bundle from Transition Functions	130
6.5	Exercises	132
6.6	The Tangent Bundle of a Manifold	134
6.7	Exercises	139
6.8	History and Bibliography	141
6.9	Appendix: Constructing Vector Bundles	141

7	**Frame Fields, Forms, and Metrics**	**144**
7.1	Frame Fields for Vector Bundles	144
7.2	Tangent Vectors as Equivalence Classes of Curves	147
7.3	Exterior Calculus on Manifolds	148
7.4	Exercises	151
7.5	Indefinite Riemannian Metrics	152
7.6	Examples of Riemannian Manifolds	153
7.7	Orthonormal Frame Fields	156
7.8	An Isomorphism between the Tangent and Cotangent Bundles	160
7.9	Exercises	161
7.10	History and Bibliography	163

8	**Integration on Oriented Manifolds**	**164**
8.1	Volume Forms and Orientation	164
8.2	Criterion for Orientability in Terms of an Atlas	167
8.3	Orientation of Boundaries	169
8.4	Exercises	172

8.5   Integration of an n-Form over a Single Chart   174

8.6   Global Integration of n-Forms   178

8.7   The Canonical Volume Form for a Metric   181

8.8   Stokes's Theorem   183

8.9   The Exterior Derivative Stands Revealed   184

8.10 Exercises   187

8.11 History and Bibliography   189

8.12 Appendix: Proof of Stokes's Theorem   189

9   **Connections on Vector Bundles**   **194**

9.1   Koszul Connections   194

9.2   Connections via Vector-Bundle-valued Forms   197

9.3   Curvature of a Connection   202

9.4   Exercises   206

9.5   Torsion-free Connections   212

9.6   Metric Connections   216

9.7   Exercises   219

9.8   History and Bibliography   222

10   **Applications to Gauge Field Theory**   **223**

10.1 The Role of Connections in Field Theory   223

10.2 Geometric Formulation of Gauge Field Theory   225

10.3 Special Unitary Groups and Quaternions   231

10.4 Quaternion Line Bundles   233

10.5 Exercises   238

10.6 The Yang–Mills Equations   242

10.7 Self-duality   244

10.8 Instantons   247

10.9 Exercises   249

10.10 History and Bibliography   250

      Bibliography   **251**

      Index   **253**

Preface

Purpose

This book represents an extended version of my lecture notes for a one-semester course on differential geometry, aimed at students without knowledge of topology. Indeed the only prerequisites are a solid grasp of multivariate calculus and of linear algebra. The goal is to train advanced undergraduates and beginning graduate students in exterior calculus (including integration), covariant differentiation (including curvature calculations), and the identification and uses of submanifolds and vector bundles. It is hoped that this will serve both the minority who proceed to study advanced texts in differential geometry, and the majority who specialize in other subjects, including physics and engineering.

Summary of the Contents

Every generation since Newton has seen a richer and deeper presentation of the differential and integral calculus. The nineteenth century gave us vector calculus and tensor analysis, and the twentieth century has produced, among other things, the exterior calculus and the theory of connections on vector bundles. As the title implies, this book is based on the premise that differential forms provide a concise and efficient approach to many constructions in geometry and in calculus on manifolds.

Chapter 1 is algebraic; Chapters 2, 4, 8, and 9 are mostly about differential forms; Chapters 4, 9, and 10 are about connections; and Chapters 3, 5, 6, and 7 are about underlying structures such as manifolds and vector bundles. The reader is not mistaken if he detects a strong influence of Harley Flanders's delightful 1989 text. I would also like to acknowledge that I have made heavy use of ideas from Berger and Gostiaux [1988], and (in Chapters 6 and 9) of my handwritten Warwick University 1981 lecture notes from John Rawnsley, as well as other standard differential geometry texts. Chapter 9 on connections is in the spirit of S. S. Chern [1989], p. ii, who remarks that "the notion of a connection in a vector bundle will soon find its way into a class on

advanced calculus, as it is a fundamental notion and its applications are widespread";
these applications include the field theories of physics (see Chapter 10), the study of
information loss in parametric statistics, and computer algorithms for recognizing
surface deformation. Regrettably the Frobenius Theorem and its applications, and de
Rham cohomology, are among many other topics which could not be included; see
Flanders [1989] for an excellent treatment of the former, and Berger and Gostiaux
[1988] for the latter.

Prerequisites

- **Linear Algebra:** finite-dimensional vector spaces and linear transformations,
  including the notions of image, kernel, rank, inner product, and determinant.
- **Vector Calculus:** derivative as a linear mapping; grad, div, and curl; line, surface,
  and volume integrals, including Green's Theorem and Stokes's Theorem; implicit
  function theorem; and the concept of an open set in Euclidean space.

Advice to the Instructor

In the diagram below, a solid arrow denotes dependency of chapters, and a fuzzy arrow
denotes a conceptual relationship. In one semester, an instructor would probably be hard
pressed to cover more than six chapters in depth. Chapters 1 and 2 are essential. Some
instructors may choose to emphasize the easier and more concrete material in Chapters
3 and 4, which is used in the sequel only as a source of examples, while others may
prefer to move rapidly into Chapters 5 and 6 so as to have time for Chapter 8 on
integration and/or Chapter 9 on connections. Alternatively one could deemphasize
abstract differential manifolds (i.e., skip over Chapter 5), cover only the "local vector
bundle" part of Chapter 6, and treat Chapters 7 to 10 in a similarly "local" fashion. As
always, students cannot expect to master the material without doing the exercises.

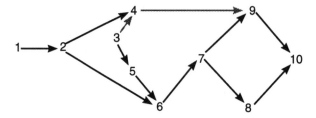

Acknowledgments and Comments

I wish to thank my Differential Geometry class of Spring 1992 for their patience, and
also Suzanne Joseph, Professor Ernest Thieleker, Greg Schreiber, and an anonymous
referee for their criticisms. The courteous guidance of editor Lauren Cowles of the
Cambridge University Press is gratefully acknowledged. The design is based on a
template from Frame Technology's program FrameMaker®. Lists of errors and
suggestions for improvement will be gratefully received at **rwrd@math.usf.edu** .

1   Exterior Algebra

Anyone who has studied linear algebra and vector calculus may have wondered whether the notion of cross product of vectors in 3-dimensional space generalizes to higher dimensions. Exterior algebra, which is a prerequisite for the study of differential forms, shows that the answer is yes. We shall adopt a constructive approach to exterior algebra, following closely the presentation given in Flanders [1989], and we will try to emphasize the connection with the vector algebra notions of cross product and triple product (see Table 1.2 on page 19).

1.1   Exterior Powers of a Vector Space

1.1.1   The Second Exterior Power

Let V be an n-dimensional vector space over R. Elements of V will be denoted u, v, w, u_i, v^j, etc., and real numbers will be denoted a, b, c, a_i, b_j, etc. For $p = 0, 1, ..., n$, the pth exterior power of V, denoted $\Lambda^p V$, is a real vector space, whose elements are referred to as "p-vectors." For $p = 0, 1$ the definition is straightforward: $\Lambda^0 V = R$, and $\Lambda^1 V = V$, respectively. $\Lambda^2 V$, consists of formal sums[1]

$$\sum_i a_i (u_i \wedge v_i),\tag{1.1}$$

where the "wedge product" $u \wedge v$ satisfies the following four rules:

$$(au + w) \wedge v = a(u \wedge v) + w \wedge v;\tag{1.2}$$

---

[1] A rigorous construction of the second exterior power is given in Section 1.9.

$$u \wedge (bv + w) = b(u \wedge v) + u \wedge w; \tag{1.3}$$

$$u \wedge u = 0; \tag{1.4}$$

$$\{v^1, \ldots, v^n\} \text{ is a basis for } V \implies \{v^i \wedge v^j : 1 \leq i < j \leq n\} \text{ is a basis for } \Lambda^2 V. \tag{1.5}$$

Postponing to the end of this chapter the question of whether a vector space with these properties exists, let us note two immediate consequences of (1. 2), (1. 3), and (1. 4). Apply (1. 4) to $(u + v) \wedge (u + v)$, and then express the latter as the sum of four terms using (1. 2) and (1. 3); two of these terms, namely, $u \wedge u$ and $v \wedge v$, are zero, and what remains shows that $u \wedge v + v \wedge u = 0$; hence

$$v \wedge u = -u \wedge v. \tag{1.6}$$

Second (1. 2), (1. 3), and (1. 4) by themselves imply that, for any basis $\{v^1, \ldots, v^n\}$ of V, the set of vectors $\{v^i \wedge v^j : 1 \leq i < j \leq n\}$ spans $\Lambda^2 V$, because it spans the set of "generators" $\{u \wedge w, u \text{ and } w \in V\}$; to check this, we express u and w in terms of the basis $\{v^1, \ldots, v^n\}$, and apply (1. 2), (1. 3), and (1. 6) to obtain:

$$u \wedge w = \left(\sum a_i v^i\right) \wedge \left(\sum b_j v^j\right) = \sum_{i,j} a_i b_j (v^i \wedge v^j)$$

$$= \sum_{i<j} (a_i b_j - a_j b_i) (v^i \wedge v^j).$$

The linear independence of $\{v^i \wedge v^j : 1 \leq i < j \leq n\}$ cannot, however, be deduced from (1. 2), (1. 3), and (1. 4), and is studied in Section 1.9.

1.1.2   Higher Exterior Powers

The description of $\Lambda^p V$ for any $2 \leq p \leq n$ follows the same lines; $\Lambda^p V$ is the set of formal sums[2]

$$\sum_{\gamma} a_{\gamma} (u_{\gamma(1)} \wedge \ldots \wedge u_{\gamma(p)}) \tag{1.7}$$

of "generators" $u_{\gamma(1)} \wedge \ldots \wedge u_{\gamma(p)}$, where each coefficient a_{γ} is indexed by a multi-index $\gamma = (\gamma(1), \ldots, \gamma(p))$; elements of $\Lambda^p V$ are called "*p*-vectors," and are subject to the rules (1. 8), (1. 9), and (1. 10):

$$(av + w) \wedge u_2 \wedge \ldots \wedge u_p = a(v \wedge u_2 \wedge \ldots \wedge u_p) + w \wedge u_2 \wedge \ldots \wedge u_p, \tag{1.8}$$

---

[2] A rigorous construction is given in Section 1.9.

and similarly if any of the u_i is replaced by such a linear combination;

$$u_i = u_j \text{ for some } i \neq j \Rightarrow u_1 \wedge \ldots \wedge u_p = 0; \tag{1.9}$$

and for any basis $\{v^1, \ldots, v^n\}$ of V, the following set of p-vectors forms a basis for $\Lambda^p V$:

$$\{v^{i(1)} \wedge \ldots \wedge v^{i(p)}, 1 \leq i(1) < \ldots < i(p) \leq n\} \tag{1.10}$$

The expression $u_1 \wedge \ldots \wedge u_{r-1} \wedge (v + w) \wedge u_{r+1} \wedge \ldots \wedge (v + w) \wedge \ldots \wedge u_p$, which is zero by (1.9), can be expanded using (1.8) into four terms, two of which are zero; what remains shows that

$$u_1 \wedge \ldots \wedge u_p \text{ changes sign if any two entries are transposed.} \tag{1.11}$$

Also it follows from (1.8) and (1.9) that, for any basis $\{v^1, \ldots, v^n\}$ of V, the set of vectors (1.10) spans $\Lambda^p V$; in order to demonstrate this, we shall need the language of permutations.

1.1.3 Permutations

Let Σ_p denote the set of permutations of the set $\{1, 2, \ldots, p\}$. For example, Σ_3 can be written as $\{e, (1, 2), (3, 1), (2, 3), (1, 3, 2), (1, 2, 3)\}$, where $\pi = (3, 1)$ means for example that $\pi(1) = 3, \pi(3) = 1$. A **transposition** is an element π of Σ_p that switches i and j for some $i \neq j$, but leaves k fixed for all $k \notin \{i, j\}$; thus in the list for Σ_3 above, the second, third, and fourth elements are transpositions. A result in algebra states that any permutation can be expressed as a composition of transpositions, and that the number m of transpositions is unique modulo 2; we define the **signature** $\text{sgn}(\pi)$ of the permutation π by

$$\text{sgn}(\pi) = (-1)^m. \tag{1.12}$$

It is also true, in the case of the composition $\pi \bullet \pi'$ of two permutations, that $\text{sgn}(\pi \bullet \pi') = \text{sgn}(\pi)\,\text{sgn}(\pi')$. It follows from (1.11) that

$$u_{\pi(1)} \wedge \ldots \wedge u_{\pi(p)} = \text{sgn}(\pi)(u_1 \wedge \ldots \wedge u_p). \tag{1.13}$$

Now we will show how to express an arbitrary generator of $\Lambda^p V$ as a linear combination of the set of vectors (1.10). We may write

$$u_1 \wedge \ldots \wedge u_p = (\sum_{j(1)} b_{1,j(1)} v^{j(1)}) \wedge \ldots \wedge (\sum_{j(p)} b_{p,j(p)} v^{j(p)})$$

$$= \sum_J c_J (v^{j(1)} \wedge \ldots \wedge v^{j(p)}),$$

where $J = (j(1), ..., j(p))$, and $c_J = b_{1,j(1)} ... b_{p,j(p)}$. For any J, there is a unique multi-index $I = (i(1), ..., i(p))$ such that $i(1) < ... < i(p)$, and a unique $\pi \in \Sigma_p$ such that $J = \pi(I)$, meaning that $(j(1), ..., j(p)) = (\pi(i(1)), ..., \pi(i(p)))$. Hence by (1. 13), we deduce

$$v^{j(1)} \wedge ... \wedge v^{j(p)} = \text{sgn}(\pi) (v^{i(1)} \wedge ... \wedge v^{i(p)}),$$

and therefore

$$u_1 \wedge ... \wedge u_p = \sum_I \left(\sum_\pi \text{sgn}(\pi) c_{\pi(I)} \right) (v^{i(1)} \wedge ... \wedge v^{i(p)}), \tag{1. 14}$$

where the first summation is over multi-indices I such that $i(1) < ... < i(p)$, and the second summation is over Σ_p. This completes the proof that the vectors (1. 10) span $\Lambda^p V$.

1.1.4   Calculating the Dimension of an Exterior Power

$$\dim(\Lambda^p V) = \frac{n!}{(n-p)! p!}, \ 0 \le p \le n. \tag{1. 15}$$

Proof: For any basis $\{v^1, ..., v^n\}$ of V, the set of p-vectors

$$\{v^{i(1)} \wedge ... \wedge v^{i(p)}, 1 \le i(1) < ... < i(p) \le n\}. \tag{1. 16}$$

forms a basis for $\Lambda^p V$, by (1. 10). The number of elements of this set is the number of ways of choosing p objects from n distinct objects, which is the expression shown.   ¤

Let us illustrate these ideas by writing down bases for the exterior powers of R^3.

p	Basis for $\Lambda^p V$	Dimension
0	$\{1\}$	1
1	$\{e_1, e_2, e_3\}$	3
2	$\{e_1 \wedge e_2, e_1 \wedge e_3, e_2 \wedge e_3\}$	3
3	$\{e_1 \wedge e_2 \wedge e_3\}$	1

Table 1.1   Exterior powers of Euclidean 3-space

1.2   Multilinear Alternating Maps and Exterior Products

For any set V, the set-theoretic product $V \times \ldots \times V$ (p copies) simply means the set of ordered p-tuples $(u_1, \ldots, u_p)$ where each $u_i \in V$. If V and W are vector spaces, a mapping $h: V \times \ldots \times V \to W$ is called:

- **Multilinear** if $h(au + bu', u_2, \ldots, u_p) = ah(u, u_2, \ldots, u_p) + bh(u', u_2, \ldots, u_p)$, and similarly for the other $(p-1)$ entries of h; h is called **bilinear** if p is 2;

- **Antisymmetric** (or **alternating**) if

$$h(u_{\pi(1)}, \ldots, u_{\pi(p)}) = \text{sgn}(\pi) h(u_1, \ldots, u_p), \pi \in \Sigma_p, \tag{1.17}$$

which implies $h(u_1, \ldots, u_p) = 0$ if $u_i = u_j$, some $i \neq j$; for when $u_i = u_j$, some $i \neq j$, transposing the ith and jth entries shows that $h(u_1, \ldots, u_p)$ is the same as its negative.

The student will have encountered the following examples of multilinear alternating maps in linear algebra or vector calculus courses:

$$(u,v) \to u \times v, R^3 \times R^3 \to R^3;$$

$$(u,v) \to \left| \begin{bmatrix} u_1 & u_2 \\ v_1 & v_2 \end{bmatrix} \right|, R^2 \times R^2 \to R;$$

$$(u, v, w) \to u \cdot (v \times w), R^3 \times R^3 \times R^3 \to R.$$

The linear maps from V to W will be denoted $L(V \to W)$, and the multilinear alternating maps will be denoted $A_p(V \to W)$. The following property of exterior powers will play a central role in the remainder of this chapter.

1.2.1 Universal Alternating Mapping Property

To every $g \in A_p(V \to W)$, there corresponds a unique $\hat{g} \in L(\Lambda^p V \to W)$ such that

$$\hat{g}(u_1 \wedge \ldots \wedge u_p) = g(u_1, \ldots, u_p), \forall u_1, \ldots, u_p; \tag{1.18}$$

in other words, a unique $\hat{g}$ such that the following diagram commutes.[3]

---

[3] A diagram is said to commute if following any sequence of arrows from one set to another yields the same mapping.

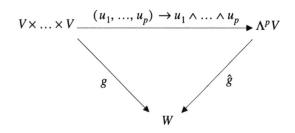

Proof: Deferred to Section 1.9.

1.2.2 Exterior Products

There exists a unique bilinear map $(\lambda, \mu) \to \lambda \wedge \mu$ from $\Lambda^p V \times \Lambda^q V$ to $\Lambda^{p+q} V$, whose effect on generators is that

$$(u_1 \wedge \ldots \wedge u_p) \wedge (w_1 \wedge \ldots \wedge w_q) = u_1 \wedge \ldots \wedge u_p \wedge w_1 \wedge \ldots \wedge w_q. \tag{1. 19}$$

To see that this is true, apply 1.2.1 twice: first to the multilinear, alternating map

$$(u_1, \ldots, u_p) \to u_1 \wedge \ldots \wedge u_p \wedge w_1 \wedge \ldots \wedge w_q,$$

for fixed $w_1 \wedge \ldots \wedge w_q$, so as to obtain a unique $f \in L(\Lambda^p V \to \Lambda^{p+q} V)$ such that

$$f(u_1 \wedge \ldots \wedge u_p) = u_1 \wedge \ldots \wedge u_p \wedge w_1 \wedge \ldots \wedge w_q,$$

so that we may define

$$\lambda \wedge (w_1 \wedge \ldots \wedge w_q) = f(\lambda);$$

and second to the multilinear, alternating map

$$(w_1, \ldots, w_q) \to \lambda \wedge (w_1 \wedge \ldots \wedge w_q),$$

for fixed λ, so as to obtain $g_\lambda \in L(\Lambda^q V \to \Lambda^{p+q} V)$ such that

$$g_\lambda (w_1 \wedge \ldots \wedge w_q) = \lambda \wedge (w_1 \wedge \ldots \wedge w_q). \tag{1. 20}$$

Finally the **exterior product** of $\lambda \in \Lambda^p V$ and $\mu \in \Lambda^q V$ is defined by $\lambda \wedge \mu = g_\lambda(\mu)$. The properties of the exterior product, the first two of which are immediate from the preceding construction, are:

- $(\lambda, \mu) \to \lambda \wedge \mu$ is **distributive** over addition and scalar multiplication;
- **associativity**: $(\lambda \wedge \mu) \wedge \nu = \lambda \wedge (\mu \wedge \nu)$;
- $\mu \wedge \lambda = (-1)^{pq} (\lambda \wedge \mu)$, so two vectors of odd degrees **anticommute**; otherwise the vectors commute.

The last property follows from Exercise 4 below, in the case where λ, μ are generators, and in general from linearity. In order to obtain a practical grasp of exterior products, try Exercises 5 and 6 below.

1.2.3 Example

Suppose V is 4-dimensional with a basis $\{v^1, v^2, v^3, v^4\}$. Then

$$(a\,(v^3 \wedge v^4) + b\,(v^1 \wedge v^3)) \wedge (c\,(v^1 \wedge v^2) + d\,(v^1 \wedge v^4)) \; = \; ac\,(v^3 \wedge v^4 \wedge v^1 \wedge v^2)$$

$$= \; (-1)^{2\,(2)}\, ac\,(v^1 \wedge v^2 \wedge v^3 \wedge v^4).$$

1.3   Exercises

1.   (a) Repeat Table 1.1 for the case of R^4, using the basis $\{e_1, e_2, e_3, e_4\}$.

     (b) Let $u = ae_1 + ce_3$, $v = be_2 + de_4$; express $u \wedge v$ in terms of your basis of $\Lambda^2 R^4$.

     (c) Let $w = a'e_1 + b'e_2$; express $u \wedge v \wedge w$ in terms of your basis of $\Lambda^3 R^4$.

     (d) Express $u \wedge v \wedge w \wedge e_3$ in terms of your basis of $\Lambda^4 R^4$.

2.   Verify that, when $V = R^3$, the cross product $(u,v) \to u \times v$, $R^3 \times R^3 \to R^3$, and the triple product $(u, v, w) \to u \cdot (v \times w)$, $R^3 \times R^3 \times R^3 \to R$, are multilinear, alternating maps.[4]

     Reminder: The cross product of $u = (a_1, a_2, a_3)$ and $v = (b_1, b_2, b_3)$ is

$$u \times v = \begin{vmatrix} a_2 & a_3 \\ b_2 & b_3 \end{vmatrix} e_1 - \begin{vmatrix} a_1 & a_3 \\ b_1 & b_3 \end{vmatrix} e_2 + \begin{vmatrix} a_1 & a_2 \\ b_1 & b_2 \end{vmatrix} e_3, \tag{1.21}$$

     and the triple product satisfies

$$u \cdot (v \times w) = v \cdot (w \times u) = w \cdot (u \times v) = -(w \cdot (v \times u)). \tag{1.22}$$

3.   Decompose the permutation $(6, 4, 3, 2, 1, 5) \in \Sigma_6$ into a product of transpositions in two different ways, and show that the number of transpositions used is the same modulo 2 in both cases.

4.   Prove, by induction or otherwise, that a permutation which sends $(1, 2, ..., p+q)$ into $(q+1, ..., q+p, 1, 2, ..., q)$ has signature

---

[4] By the end of this chapter, the reader will realize that, in terms of the star operator discussed in Section 1.7 below, $u \times v = *\,(u \wedge v)$, and $u \cdot (v \times w) = *\,(u \wedge v \wedge w)$.

$$\text{sgn}\,(\pi) \;=\; (-1)^{pq}. \tag{1.23}$$

Hint: A possible inductive hypothesis H_k is that whenever $p \geq 1,\, q \geq 1,\, p + q \leq k$, then the assertion above holds. To prove H_{k+1} from H_k, start by transposing q and $q + p$, and then rearrange $(q, 1, \ldots, q-1)$ so that H_k can be applied to the first $p + q - 1$ entries.

5. Let $V = R^3$, with any basis $\{v^1, v^2, v^3\}$; show that

$$(a\,(v^2 \wedge v^3) + b\,(v^3 \wedge v^1) + c\,(v^1 \wedge v^2)) \wedge (\tilde{a}v^1 + \tilde{b}v^2 + \tilde{c}v^3)$$

$$= (a\tilde{a} + b\tilde{b} + c\tilde{c})\, v^1 \wedge v^2 \wedge v^3. \tag{1.24}$$

6. Suppose V is 4-dimensional with a basis $\{v^1, v^2, v^3, v^4\}$. Express the following as multiples of $v^1 \wedge v^2 \wedge v^3 \wedge v^4$:

    (i) $(a\,(v^1 \wedge v^3) + b\,(v^2 \wedge v^4)) \wedge (c\,(v^1 \wedge v^3) + d\,(v^2 \wedge v^4))$;

    (ii) $(av^1 + bv^4) \wedge (c\,(v^1 \wedge v^2 \wedge v^3) + d\,(v^2 \wedge v^3 \wedge v^4))$.

7. The setting is the same as for Exercise 6. Suppose $\mu \in \Lambda^3 V,\, \mu \neq 0$. Characterize the vectors $u \in V$ such that $u \wedge \mu = 0$, and show that the vector space consisting of such u is of dimension 3.

    Hint: Write $u = u_1 v^1 + \ldots + u_4 v^4$, and express μ similarly in terms of the four basis elements of the third exterior power. Obtain a linear relation on the coefficients of u.

8. This is a generalization of Exercise 7. Suppose V is n-dimensional, and μ is an arbitrary nonzero element of $\Lambda^{n-1} V$. Prove that the subspace W^μ of elements u of V such that $u \wedge \mu = 0$ is of dimension $n - 1$, and deduce from this that there exist vectors $w^1, \ldots, w^{n-1}$ in V such that $\mu = w^1 \wedge \ldots \wedge w^{n-1}$.

    Hint: For the last part, take a basis for W^μ, extend it to a basis for V, and express μ in terms of the corresponding basis of $\Lambda^{n-1} V$. **Warning:** This kind of representation does not generally hold for elements of the other exterior powers.

1.4   Exterior Powers of a Linear Transformation

1.4.1 Determinants

Given $A \in L\,(V \to V)$, define $g_A \colon V^n \to \Lambda^n V \cong R$ by

$$g_A\,(u_1, \ldots, u_n) \;=\; (Au_1) \wedge \ldots \wedge (Au_n). \tag{1.25}$$

It follows immediately from the last equation that g_A is multilinear and antisymmetric, and so, by 1.2.1, there is a unique $f_A \in L\,(\Lambda^n V \to \Lambda^n V)$ such that

$$f_A(u_1 \wedge \ldots \wedge u_n) = (Au_1) \wedge \ldots \wedge (Au_n). \tag{1.26}$$

Since $\Lambda^n V$ is one-dimensional and f is linear, it follows that f is simply multiplication by a scalar, which we denote by $|A|$, the **determinant** of A. In other words,

$$|A|(u_1 \wedge \ldots \wedge u_n) = (Au_1) \wedge \ldots \wedge (Au_n). \tag{1.27}$$

It is somewhat surprising to discover that this abstract formulation refers to the same notion of determinant that the student has encountered in matrix algebra:

1.4.2 Formula for the Determinant of a Matrix

Suppose that, in terms of a basis $\{v^1, \ldots, v^n\}$ for V, A has the matrix representation $A = (a_{ij})_{1 \le i, j \le n}$ (a_{ij} may also be written $a_{i,j}$). Then taking

$$u_i = \sum_j a_{ij} v^j$$

gives, as in (1.14),

$$u_1 \wedge \ldots \wedge u_n = (\sum_{j(1)} a_{1,j(1)} v^{j(1)}) \wedge \ldots \wedge (\sum_{j(n)} a_{n,j(n)} v^{j(n)}),$$

$$= \sum_J a_{1,j(1)} \ldots a_{n,j(n)} (v^{j(1)} \wedge \ldots \wedge v^{j(n)}),$$

where $J = (j(1), \ldots, j(n))$. Any J with two entries the same makes no contribution to the sum, by (1.9). In all other cases there is a unique $\pi \in \Sigma_n$ such that $(j(1), \ldots, j(n)) = (\pi(1), \ldots, \pi(n))$. Hence by (1.13), we deduce

$$v^{j(1)} \wedge \ldots \wedge v^{j(n)} = \mathrm{sgn}(\pi)(v^1 \wedge \ldots \wedge v^n),$$

$$u_1 \wedge \ldots \wedge u_n = \left(\sum_{\pi \in \Sigma_n} \mathrm{sgn}(\pi) a_{1,\pi(1)} \ldots a_{n,\pi(n)}\right) v^1 \wedge \ldots \wedge v^n. \tag{1.28}$$

Thus the formula for the determinant of the matrix is

$$|A| = \sum_{\pi \in \Sigma_n} \mathrm{sgn}(\pi) a_{1,\pi(1)} \ldots a_{n,\pi(n)}. \tag{1.29}$$

For example, when $n = 2$,

$$\left|\begin{bmatrix} a_{11} & a_{12} \\ a_{21} & a_{22} \end{bmatrix}\right| = a_{11}a_{22} - a_{12}a_{21} = \sum_{\pi \in \Sigma_2} \text{sgn}(\pi)\, a_{1,\pi(1)} a_{2,\pi(2)}.$$

1.4.3   Other Exterior Powers of a Linear Transformation

A generalization of the notion of determinant is that of **exterior powers**[5] **of a linear transformation** $A \in L(V \to W)$. The map $V^p \to \Lambda^p W$ given by

$$(u_1, ..., u_p) \to (Au_1) \wedge ... \wedge (Au_p)$$

is multilinear and alternating, and so by 1.2.1 it defines an element of $L(\Lambda^p V \to \Lambda^p W)$ denoted $\Lambda^p A$, called the **exterior pth power** of A; in other words, $\Lambda^p A$ is specified by its action on generators as follows:

$$\Lambda^p A\,(u_1 \wedge ... \wedge u_p) = (Au_1) \wedge ... \wedge (Au_p). \tag{1.30}$$

The matrix representation of $\Lambda^p A$ may be obtained as follows. If $\{v^1, ..., v^n\}$ is a basis for V, and $\{w^1, ..., w^m\}$ for W, then $\{\sigma^I\}$ and $\{\tau^K\}$ are bases for $\Lambda^p V$ and $\Lambda^p W$, respectively, where

$$\sigma^I = v^{i(1)} \wedge ... \wedge v^{i(p)},\ 1 \le i(1) < ... < i(p) \le n; \tag{1.31}$$

$$\tau^K = w^{k(1)} \wedge ... \wedge w^{k(p)},\ 1 \le k(1) < ... < k(p) \le m. \tag{1.32}$$

If $Av^i = \sum_k a_k^i w^k$, then

$$(\Lambda^p A)\,\sigma^I = (Av^{i(1)}) \wedge ... \wedge (Av^{i(p)})$$

$$= \sum_J a_{j(1)}^{i(1)} ... a_{j(p)}^{i(p)}\, (w^{j(1)} \wedge ... \wedge w^{j(p)}),$$

where J runs through the set of all multi-indices. As usual, summands where $j(r) = j(s)$ for some $r \ne s$ are zero, and we express the other summands as in the steps preceding (1.14): there is a unique $K = (k(1), ..., k(p))$ such that $k(1) < ... < k(p)$, and a unique $\pi \in \Sigma_p$ such that $J = \pi(K)$, meaning that $(j(1), ..., j(p)) = (\pi(k(1)), ..., \pi(k(p)))$. Since

$$w^{j(1)} \wedge ... \wedge w^{j(p)} = \text{sgn}(\pi)\,(w^{k(1)} \wedge ... \wedge w^{k(p)}),$$

we obtain:

---

[5] This idea is needed in calculations related to the pullback of differential forms in Chapter 2, and is also relevant to Stokes's Theorem in Chapter 8.

$$(\Lambda^p A)\, \sigma^I = \sum_K \left(\sum_{\pi \in \Sigma_p} \text{sgn}\, (\pi)\, a_{\pi(k(1))}^{i(1)} \cdots a_{\pi(k(p))}^{i(p)} \right) (w^{k(1)} \wedge \ldots \wedge w^{k(p)}) \qquad \textbf{(1. 33)}$$

$$= \sum_K a_K^I \tau^K,$$

and so $\Lambda^p A$ is represented by the matrix (a_K^I) of all the $p \times p$ minors of A, where

$$a_K^I = \sum_{\pi \in \Sigma_p} \text{sgn}\, (\pi)\, a_{\pi(k(1))}^{i(1)} \cdots a_{\pi(k(p))}^{i(p)}. \qquad \textbf{(1. 34)}$$

An opportunity to evaluate this matrix when $m = n = 3, p = 2$, is provided in Exercise 10 below. This construction generalizes the notion of determinant because, when $V = W$ and $p = n$, then $\Lambda^n A$ has the effect of multiplication by $|A|$.

1.4.4 The Isomorphism $\Lambda^p (V^*) \cong (\Lambda^p V)^*$

Recall that the **dual space** $V^* = L(V \to R)$ of the n-dimensional vector space V is another n-dimensional vector space, consisting of the linear mappings from V to R, which are called **linear forms**. It is often helpful, though not necessary, to conceptualize elements of V as n-dimensional column vectors, and elements of V^* as n-dimensional row vectors which act on the column vectors by usual matrix multiplication.

Given linear forms $\psi_1, \ldots, \psi_p \in V^*$, where $p \leq n$, the isomorphism (constructed below) will show that $\psi_1 \wedge \ldots \wedge \psi_p \in \Lambda^p (V^*)$ acts linearly on $\Lambda^p V$ as follows:

$$(\psi_1 \wedge \ldots \wedge \psi_p) \cdot (u_1 \wedge \ldots \wedge u_p) = \sum_{\pi \in \Sigma_p} \text{sgn}\, (\pi)\, \psi_1(u_{\pi(1)}) \ldots \psi_p(u_{\pi(p)}). \qquad \textbf{(1. 35)}$$

1.4.4.1 Examples
When $p = 2$ and when $p = 3$, respectively,

$$(\varphi \wedge \psi) \cdot (u \wedge v) = \varphi(u)\, \psi(v) - \varphi(v)\, \psi(u); \qquad \textbf{(1. 36)}$$

$$(\psi_1 \wedge \psi_2 \wedge \psi_3) \cdot (u_1, u_2, u_3) = |(\psi_i(u_j))|. \qquad \textbf{(1. 37)}$$

1.4.4.2 Constructing the Isomorphism
Given linear forms $\psi_1, \ldots, \psi_p \in V^*$, where $p \leq n$, consider the mapping $A \in L(V \to R^p)$ given by

$$Au = \psi_1(u)\, e_1 + \ldots + \psi_p(u)\, e_p, \qquad \textbf{(1. 38)}$$

where $\{e_1, \ldots, e_p\}$ is the standard basis for R^p. Referring to (1. 30), we see that the range of $\Lambda^p A$ is the one-dimensional space $\Lambda^p R^p$ spanned by $e_1 \wedge \ldots \wedge e_p$; therefore there exists a unique linear form, temporarily denoted

$$\psi_1 \lozenge \dots \lozenge \psi_p \in (\Lambda^p V)^*, \tag{1.39}$$

given by the equation

$$\Lambda^p A (\lambda) = (\psi_1 \lozenge \dots \lozenge \psi_p)(\lambda)(e_1 \wedge \dots \wedge e_p), \lambda \in \Lambda^p V. \tag{1.40}$$

The reader may verify using (1.33) (see Exercise 11 below) that

$$(\psi_1 \lozenge \dots \lozenge \psi_p)(u_1 \wedge \dots \wedge u_p) = \sum_{\pi \in \Sigma_p} \text{sgn}(\pi)\, \psi_1(u_{\pi(1)}) \dots \psi_p(u_{\pi(p)}), \tag{1.41}$$

and also that the map $\psi_1 \wedge \dots \wedge \psi_p \to \psi_1 \lozenge \dots \lozenge \psi_p$ is linear and one-to-one (see Exercise 12). Since the dimension of $\Lambda^p(V^*)$ is the same as that of $(\Lambda^p V)^*$, this establishes an isomorphism from $\Lambda^p(V^*)$ to $L(\Lambda^p V \to R)$.   ¤

In subsequent chapters, we shall drop the $\lozenge$ notation, and identify $\psi_1 \lozenge \dots \lozenge \psi_p$ with $\psi_1 \wedge \dots \wedge \psi_p$. Thus equation (1.35) replaces (1.41).

1.5   Exercises

9. (a) Show using (1.27) that if $A, B \in L(V \to V)$, then $|AB| = |A||B|$.

(b) Show using (1.30) that if $B \in L(V \to W)$ and $A \in L(W \to Y)$, then

$$\Lambda^p(AB) = \Lambda^p(A)\,\Lambda^p(B).$$

10. Suppose $V = R^3$, and $A \in L(R^3 \to R^3)$ is expressible in terms of the usual basis $\{e_1, e_2, e_3\}$ as the matrix

$$\begin{bmatrix} \cos\varphi & \sin\varphi & 0 \\ -\sin\varphi & \cos\varphi & 0 \\ 0 & 0 & 1 \end{bmatrix}$$

for some real number φ. Express $\Lambda^2 A$ as a 3×3 matrix with respect to the basis $\{e_2 \wedge e_3, e_3 \wedge e_1, e_1 \wedge e_2\}$.

11. Verify the formula (1.41), using (1.40).

12. Show that the map $\psi_1 \wedge \dots \wedge \psi_p \to \psi_1 \lozenge \dots \lozenge \psi_p$, appearing in (1.40), is linear and one-to-one.

Hint: To show the map is one-to-one, note that by (1.41), $\psi_1 \lozenge \dots \lozenge \psi_p$ is zero if and only if $\{\psi_1, \dots, \psi_p\}$ is linearly independent; now appeal to (1.10).

13. Show that, for general finite-dimensional vector spaces V and W, the spaces $\Lambda^p L\,(V \to W)$ and $L\,(\Lambda^p V \to W)$ are not necessarily isomorphic.

14. Show that the exterior powers of a linear transformation $A \in L\,(V \to W)$ satisfy

$$(\Lambda^{p+q}A)\,(\lambda \wedge \mu) \;=\; \Lambda^p A\,(\lambda) \,\wedge\, \Lambda^q A\,(\mu) \tag{1.42}$$

for any $p + q \le n$, $\lambda \in \Lambda^p V$, $\mu \in \Lambda^q V$, by applying (1.30) to generators, and using the associativity of the exterior product.

15. For $p, q \ge 1$, let $\Sigma_{p,q}$ denote the set of permutations π of $(1, 2, \ldots, p + q)$ such that $\pi(1) < \ldots < \pi(p)$, $\pi(p+1) < \ldots < \pi(p+q)$ (think of splitting the top p cards from a deck of $p + q$ cards, and shuffling them in the usual way into the bottom q cards — there are $(p+q)!\,/\,(p!\,q!)$ such permutations). Notice that associativity of the exterior product implies that the image of the exterior product of

$$(\varphi_1 \wedge \ldots \wedge \varphi_p) \in \Lambda^p\,(V^*) \text{ and } (\psi_1 \wedge \ldots \wedge \psi_q) \in \Lambda^q\,(V^*)$$

under the isomorphism 1.4.4 must satisfy

$$(\varphi_1 \wedge \ldots \wedge \varphi_p) \wedge (\psi_1 \wedge \ldots \wedge \psi_q) \to (\varphi_1 \lozenge \ldots \lozenge \varphi_p) \lozenge (\psi_1 \lozenge \ldots \lozenge \psi_q)$$

$$= \varphi_1 \lozenge \ldots \lozenge \varphi_p \lozenge \psi_1 \lozenge \ldots \lozenge \psi_q$$

(see (1.39) for the notation). Prove that $\lozenge$ extends to a map

$$\lozenge : L\,(\Lambda^p V \to R) \times L\,(\Lambda^q V \to R) \to L\,(\Lambda^{p+q} V \to R);$$

$$(h \lozenge l)\,(u_1 \wedge \ldots \wedge u_{p+q}) \tag{1.43}$$

$$= \sum_{\pi \in \Sigma_{p,q}} \operatorname{sgn}(\pi)\, h\,(u_{\pi(1)} \wedge \ldots \wedge u_{\pi(p)})\, l\,(u_{\pi(p+1)} \wedge \ldots \wedge u_{\pi(p+q)}).$$

Hint: Use Exercise 14.

1.6   Inner Products

1.6.1 Definition of an Inner Product

An **inner product** on a vector space V is a map $V \times V \to R$, denoted $\langle \cdot | \cdot \rangle$, with:

* **Bilinearity:** $u \to \langle u | v \rangle$ is linear for every v, and $v \to \langle u | v \rangle$ is linear for every u;
* **Symmetry:** $\langle u | v \rangle = \langle v | u \rangle$;
* **Nondegeneracy:** If z satisfies $\langle z | u \rangle = 0$, $\forall u$, then $z = 0$.

Note that this definition is a little more general than the one often given in linear algebra courses, since it is **not** assumed that $\langle u|u \rangle \geq 0$.

1.6.1.1 Characterization of Nondegeneracy

If $\{v^1, \ldots, v^n\}$ is any basis for V, the nondegeneracy condition is equivalent to:

$$\left\| \begin{bmatrix} \langle v^1|v^1 \rangle & \ldots & \langle v^1|v^n \rangle \\ \ldots & \ldots & \ldots \\ \langle v^n|v^1 \rangle & \ldots & \langle v^n|v^n \rangle \end{bmatrix} \right\| \neq 0. \tag{1.44}$$

Proof: To check that this condition is sufficient, take any z which satisfies $\langle z|u \rangle = 0, \forall u$. Let us expand z in terms of the basis as $z = a_1 v^1 + \ldots + a_n v^n$. Taking inner products with each v^i in turn gives the system of linear equations:

$$\sum_j a_j \langle v^j|v^i \rangle = 0, \, i = 1, \ldots, n. \tag{1.45}$$

Condition (1.44) implies that the matrix $(\langle v^j|v^i \rangle)$ is invertible, and hence the only solution to (1.45) is for all the a_j to be zero, showing that $z = 0$. Proof of the converse is left as an exercise.   ¤

1.6.2 Examples

- The **dot product** in R^n.

$$\langle (a_1, \ldots, a_n)|(b_1, \ldots, b_n) \rangle = (a_1, \ldots, a_n) \cdot (b_1, \ldots, b_n) = a_1 b_1 + \ldots + a_n b_n.$$

- The **Lorentz inner product** in R^4: if c denotes the speed of light,

$$\langle (a_1, \ldots, a_4)|(b_1, \ldots, b_4) \rangle = a_1 b_1 + a_2 b_2 + a_3 b_3 - c^2 a_4 b_4. \tag{1.46}$$

1.6.3 Orthonormal Bases and Their Signatures

It follows from the axioms that every inner product space contains an element v such that

$$\langle v|v \rangle = \pm 1; \tag{1.47}$$

for if $\langle z|z \rangle = 0$ for all $z \in V$, then

$$2\langle u|w \rangle = \langle u + w|u + w \rangle - \langle u|u \rangle - \langle w|w \rangle = 0$$

for every u and w, which contradicts nondegeneracy; so take some z with $a = \langle z|z \rangle \neq 0$, and let $v = |a|^{-1/2} z$. A basis $\{v^1, \ldots, v^n\}$ for V is called an **orthonormal basis** if

$$\langle v^i | v^j \rangle = 0, \, i \neq j \; ; \; \langle v^i | v^i \rangle = \pm 1, \, i = 1, \ldots, n. \tag{1.48}$$

An induction argument, suggested in Exercise 23 below, shows that every inner product space has an orthonormal basis. Moreover if there are r plus signs and $s = n - r$ minus signs in (1.48), $t = r - s$ is called the **signature** of the inner product space; this does not depend on the choice of orthonormal basis (see Exercise 24).

A useful property of inner product spaces is the following.

1.6.4 Linear Forms on an Inner Product Space

Every $f \in L(V \to R)$ is of the form $f(\cdot) = \langle \cdot | u \rangle$ for some $u \in V$.

Proof: Take $u = f(v^1)\, v^1 + \ldots + f(v^n)\, v^n$, using the orthonormal basis in (1.48); then for any $w = a_1 v^1 + \ldots + a_n v^n$,

$$f(w) = \sum_j a_j f(v^j) = \sum_j (a_j) \langle v^j | u \rangle = \langle w | u \rangle. \qquad \qquad \text{¤}$$

1.6.5 Inner Products on Exterior Powers

Suppose V has an inner product $\langle \cdot | \cdot \rangle$. Then there exists a bilinear mapping $\langle \cdot | \cdot \rangle_p$ from $\Lambda^p V \times \Lambda^p V$ to R, characterized by the formula

$$\langle u_1 \wedge \ldots \wedge u_p | v_1 \wedge \ldots \wedge v_p \rangle_p = \left| \begin{bmatrix} \langle u_1 | v_1 \rangle & \ldots & \langle u_1 | v_p \rangle \\ \ldots & \ldots & \ldots \\ \langle u_p | v_1 \rangle & \ldots & \langle u_p | v_p \rangle \end{bmatrix} \right|. \tag{1.49}$$

To see that this is so, note that the determinant on the right is multilinear and alternating in $(u_1, \ldots, u_p)$ and in $(v_1, \ldots, v_p)$, respectively, and use 1.2.1 twice as in the construction of the exterior product in Section 1.2.2. Clearly $\langle \cdot | \cdot \rangle_p$ is symmetric, because transposing the matrix in (1.49) does not change the value of its determinant.

1.6.5.1 An Orthonormal Basis for an Exterior Power

$\langle \cdot | \cdot \rangle_p$ *is an inner product on $\Lambda^p V$. If $\{v^1, \ldots, v^n\}$ is an orthonormal basis for V, I is an ascending multi-index (i.e., $1 \leq i(1) < \ldots < i(p) \leq n$), and*

$$\sigma^I = v^{i(1)} \wedge \ldots \wedge v^{i(p)}, \tag{1.50}$$

then $\{\sigma^I\}$, as I ranges over ascending multi-indices, is an orthonormal basis of $\Lambda^p V$.

Proof: To show that $\langle \cdot | \cdot \rangle_p$ is an inner product on $\Lambda^p V$, it only remains to show that it is nondegenerate. We know from (1.10) that the $\{\sigma^I\}$ form a basis for $\Lambda^p V$, where now $\{v^1, \ldots, v^n\}$ is an orthonormal basis for V. Nondegeneracy follows from (1.44) once we show that the determinant of the matrix ($\langle \sigma^I | \sigma^H \rangle$), as I and H run through ascending sets of multi-indices, is nonzero. Now if $I \neq H$, then some entry $i(q)$ in I does not belong to the set $\{ h(1), \ldots, h(p) \}$. It follows that the qth row of the matrix

$$
\begin{bmatrix}
\langle v^{i(1)} | v^{h(1)} \rangle & \ldots & \langle v^{i(1)} | v^{h(p)} \rangle \\
\ldots & \ldots & \ldots \\
\langle v^{i(p)} | v^{h(1)} \rangle & \ldots & \langle v^{i(p)} | v^{h(p)} \rangle
\end{bmatrix}
$$

is zero, hence its determinant is zero. Thus

$$
\langle \sigma^I | \sigma^H \rangle = \pm \delta^{I,H}, ^6
\tag{1.51}
$$

and so $\langle \cdot | \cdot \rangle_p$ is nondegenerate as desired. This also demonstrates that the $\{ \sigma^I \}$ form an orthonormal basis for $\Lambda^p V$.   ¤

1.6.5.2   Example in Dimension 3
$V = R^3$ with the Euclidean inner product, and the standard orthonormal basis $\{ e_1, e_2, e_3 \}$. Then $\{ e_1 \wedge e_2, e_1 \wedge e_3, e_2 \wedge e_3 \}$ is an orthonormal basis of $\Lambda^2 R^3$, and $\{ e_1 \wedge e_2 \wedge e_3 \}$ is an orthonormal basis of $\Lambda^3 R^3$.

1.6.5.3   Example in Dimension 4
$V = R^4$ with the Lorentz inner product, and the standard orthonormal basis $\{ e_1, e_2, e_3, e_4 \}$, taking $c = 1$ for convenience. Then

$$
\{ e_1 \wedge e_2, e_1 \wedge e_3, e_1 \wedge e_4, e_2 \wedge e_3, e_2 \wedge e_4, e_3 \wedge e_4 \}
$$

is an orthonormal basis of $\Lambda^2 R^4$, with signature zero. To see that three of the basis elements give negative inner products with themselves, note that, for example,

$$
\langle e_1 \wedge e_4 | e_1 \wedge e_4 \rangle = \left| \begin{bmatrix} 1 & 0 \\ 0 & -1 \end{bmatrix} \right| = -1.
$$

1.6.5.4   Example in n Dimensions
For a general n-dimensional inner product space V, 1.6.5.1 shows that the n-vector

$$
\sigma = v^1 \wedge \ldots \wedge v^n
\tag{1.52}
$$

is by itself a basis for $\Lambda^n V$, and

$$
\langle \sigma | \sigma \rangle_n = \langle v^1 | v^1 \rangle \ldots \langle v^n | v^n \rangle = (1)^r (-1)^s = (-1)^{(n-t)/2}.
\tag{1.53}
$$

---

6 The "Kronecker delta" notation means that $\delta^{I,H} = 1$ if $I = H$, and $= 0$ otherwise.

1.7   The Hodge Star Operator

Let V be an n-dimensional vector space with an inner product $\langle \cdot | \cdot \rangle$. It follows already from (1.15) that $\dim (\Lambda^{n-p} V) = \dim (\Lambda^p V)$, $p = 0, 1, ..., n$. This section will provide a natural isomorphism, denoted $*$, from $\Lambda^p V$ to $\Lambda^{n-p} V$, which will finally clarify the relationship of the wedge product in R^3 with the familiar cross product.

An equivalence relation on the set of orthonormal bases of V can be defined as follows: $\{v^1, ..., v^n\}$ is said to have the same **orientation** as $\{\tilde{v}^1, ..., \tilde{v}^n\}$ if the linear transformation A, defined by $Av^i = \tilde{v}^i$, $i = 1, ..., n$, has positive determinant. This divides the set of orthonormal bases into two equivalence classes. The definition of the Hodge star operator depends, up to a sign, on which of these two orientations is selected. So we select an orientation, and then take an orthonormal basis $\{v^1, ..., v^n\}$ with this orientation; there is a corresponding basis vector σ for $\Lambda^n V$ as in (1.52).

For any $\lambda \in \Lambda^p V$, the map $\mu \rightarrow \lambda \wedge \mu$ from $\Lambda^{n-p} V$ to $\Lambda^n V$ is linear, so there exists a unique $f_\lambda \in L(\Lambda^{n-p} V \rightarrow R)$ such that

$$\lambda \wedge \mu = f_\lambda (\mu) \sigma.$$

Now it follows from 1.6.4 that there is a unique element of $\Lambda^{n-p} V$, denoted $*\lambda$, such that $f_\lambda (\mu) = \langle *\lambda | \mu \rangle_{n-p}$: in other words,

$$\lambda \wedge \mu = \langle *\lambda | \mu \rangle_{n-p} \sigma, \forall \mu \in \Lambda^{n-p} V. \tag{1.54}$$

The operation which sends λ to $*\lambda$ is called the **Hodge star operator**.

1.7.1   Example: The Hodge Star Operator in the 3-Dimensional Case

This is a continuation of Example 1.6.5.2; here $\sigma = e_1 \wedge e_2 \wedge e_3$ in the previous notation. We shall calculate $*\lambda$ for $\lambda = e_2 \wedge e_3$, which as we saw is one of the elements of an orthonormal basis for $\Lambda^2 R^3$. Clearly $*\lambda \in \Lambda^1 R^3 = R^3$, since $p = 2$ and $n - p = 1$, and so there are real numbers a, b, c, such that $*\lambda = ae_1 + be_2 + ce_3$. Equation (1.54) tells us that

$$e_2 \wedge e_3 \wedge \mu = \langle ae_1 + be_2 + ce_3 | \mu \rangle e_1 \wedge e_2 \wedge e_3, \forall \mu \in R^3.$$

Taking μ to be each basis vector in turn, we see that $b = c = 0$, while

$$e_2 \wedge e_3 \wedge e_1 = a(e_1 \wedge e_2 \wedge e_3),$$

and two transpositions on the left side show that $a = 1$. The same calculation can be carried out for the other elements of this orthonormal basis for $\Lambda^2 R^3$, showing that

$$* (e_2 \wedge e_3) = e_1, * (e_1 \wedge e_3) = -e_2, * (e_1 \wedge e_2) = e_3, \tag{1.55}$$

and so by linearity,

$$* (c_1 (e_2 \wedge e_3) + c_2 (e_1 \wedge e_3) + c_3 (e_1 \wedge e_2)) = c_1 e_1 - c_2 e_2 + c_3 e_3. \tag{1.56}$$

Since $e_2 \times e_3 = e_1$, etc., the last line shows that, if $u, v \in \{e_1, e_2, e_3\}$, then

$$u \times v = * (u \wedge v) \tag{1.57}$$

and by linearity, this extends to all $u, v \in R^3$, giving the exterior algebra interpretation of the cross product in vector algebra. Note that, if we had chosen a basis with the opposite orientation, such as $\{\tilde{e}_1, \tilde{e}_2, \tilde{e}_3\}$ where $\tilde{e}_1 = e_1, \tilde{e}_2 = e_3, \tilde{e}_3 = e_2$, then the right side of (1.57) would be minus the cross product.

1.7.2   Effect of the Hodge Star Operator on Basis Vectors

Given an orthonormal basis $\{v^1, ..., v^n\}$, we shall now derive a general formula for $* \lambda$ when $\lambda = v^1 \wedge ... \wedge v^p$. In other words $\lambda = \sigma^H$, where $H = (1, 2, ..., p)$. Using Section 1.6.5.1, we can specify $* \lambda$ by considering $\langle * \lambda | \sigma^K \rangle_{n-p}$ for $K = (k(1), ..., k(n-p))$, where $1 \le k(1) < ... < k(n-p) \le n$. The identity (1.54) gives

$$\lambda \wedge \sigma^K = \langle * \lambda | \sigma^K \rangle_{n-p} \sigma, \tag{1.58}$$

and the left side is zero unless $K = (p+1, ..., n) = H'$, in which case $\langle * \lambda | \sigma^K \rangle_{n-p} = 1$. It follows that $* \lambda = b \sigma^{H'}$ for some constant b, and (1.58) shows that $\sigma = b \langle \sigma^{H'} | \sigma^{H'} \rangle \sigma$, and therefore $b = \langle \sigma^{H'} | \sigma^{H'} \rangle = \pm 1$. In other words, for $H = (1, ..., p)$ and $H' = (p+1, ..., n)$,

$$* \sigma^H = \langle \sigma^{H'} | \sigma^{H'} \rangle_{n-p} \sigma^{H'}. \tag{1.59}$$

Referring back to the properties of the exterior product in Section 1.2.2, we observe that

$$\sigma^K \wedge \sigma^H = (-1)^{p(n-p)} (\sigma^H \wedge \sigma^K) = (-1)^{p(n-p)} \sigma = \langle * \sigma^K | \sigma^H \rangle_p \sigma,$$

which implies that

$$\langle * \sigma^K | \sigma^H \rangle_p = (-1)^{p(n-p)} \delta^{K, H'}, \tag{1.60}$$

in the notation of footnote 6, and the same reasoning as before shows that

$$* \sigma^{H'} = (-1)^{p(n-p)} \langle \sigma^H | \sigma^H \rangle_p \sigma^H. \tag{1.61}$$

Combining this with (1.59) gives

$$* \left(*\sigma^H \right) \ = \ \langle \sigma^{H'} | \sigma^{H'} \rangle_{n-p} \left(*\sigma^{H'} \right) \ = \ (-1)^{\,p\,(n-p)} \langle \sigma^H | \sigma^H \rangle_p \langle \sigma^{H'} | \sigma^{H'} \rangle_{n-p} \sigma^H. \quad \text{(1.62)}$$

However, (1.53) implies that

$$\langle \sigma | \sigma \rangle_n \ = \ \langle v^1 | v^1 \rangle \ldots \langle v^n | v^n \rangle \ = \ \langle \sigma^H | \sigma^H \rangle_p \langle \sigma^{H'} | \sigma^{H'} \rangle_{n-p} \ = \ (-1)^{\,(n-t)/2}, \quad \text{(1.63)}$$

and the last two identities combine to give:

$$* \left(*\lambda \right) \ = \ (-1)^{\,p\,(n-p)\,+\,(n-t)/2} \lambda. \quad \text{(1.64)}$$

in the case where $\lambda = \sigma^H$. This generalizes immediately to **any** p-element ascending multi-index set H, because we can simply relabel the basis so that H becomes $(1, 2, \ldots, p)$; this may cause a change of orientation when $p = n - 1$, but this does not affect (1.64). By linearity this formula extends to the whole of $\Lambda^p V$.

1.7.3 Examples

- For the 3-dimensional Euclidean case studied in Example 1.7.1, $n = t = 3$; so for every $p \in \{0, 1, 2, 3\}$, $p\,(n-p) + (n-t)/2$ is even, and

$$** \lambda \ = \ \lambda, \lambda \in \bigcup_{0 \leq p \leq 3} \Lambda^p R^3 \ . \quad \text{(1.65)}$$

- For the Lorentz inner product in Example 1.6.2, $n = 4$ and the signature t is 2, and so when $p = 2$ or $p = 4$, $p\,(n-p) + (n-t)/2$ is odd; thus

$$** \left(e_i \wedge e_j \right) \ = \ -\left(e_i \wedge e_j \right), 1 \leq i < j \leq 4; \quad \text{(1.66)}$$

$$** \left(e_1 \wedge e_2 \wedge e_3 \wedge e_4 \right) \ = \ -\left(e_1 \wedge e_2 \wedge e_3 \wedge e_4 \right). \quad \text{(1.67)}$$

1.7.4 Formula

For any $\lambda, \mu \in \Lambda^p V$, $\lambda \wedge *\mu = \mu \wedge *\lambda = (-1)^{\,(n-t)/2} \langle \lambda | \mu \rangle_p \sigma$.

Proof: Consider first the case where $\mu = \sigma^H$ as in (1.59); the only basis element λ for which $\lambda \wedge *\mu \neq 0$ is $\lambda = \sigma^H$, and in that case (1.59) and (1.63) give

Vector Algebra Expression	Exterior Algebra Version
cross product $u \times v$	$* \left(u \wedge v \right)$
triple product $u \cdot (v \times w)$	$* \left(u \wedge v \wedge w \right)$
$|u \times v|^2 = |u|^2 |v|^2 - (u \cdot v)^2$	$\langle u \wedge v | u \wedge v \rangle_2 = \langle u | u \rangle \langle v | v \rangle - \langle u | v \rangle^2$
$u \times (v \times w) = (u \cdot w)\,v - (u \cdot v)\,w$	$u \wedge * \left(v \wedge w \right) = \langle u | w \rangle \left(*v \right) - \langle u | v \rangle \left(*w \right)$

Table 1.2   Correspondence between exterior algebra and 3-dimensional vector algebra

$$\lambda \wedge {}^* \mu = \sigma^H \wedge (\langle \sigma^{H'} | \sigma^{H'} \rangle_{n-p} \sigma^{H'}) = \langle \sigma^{H'} | \sigma^{H'} \rangle_{n-p} \sigma$$

$$= (-1)^{(n-t)/2} \langle \sigma^H | \sigma^H \rangle_p \sigma = (-1)^{(n-t)/2} \langle \lambda | \mu \rangle_p \sigma.$$

The result extends by linearity to all $\lambda, \mu \in \Lambda^p V$.   ¤

With the help of this formula, it is possible to show that the constructions and formulas of 3-dimensional vector algebra are special cases of ones in exterior algebra. Note that in Table 1.2, the identities in the second column are valid in any inner product space (see Exercises 21 and 22 below).

1.8   Exercises

16. Calculate the signature of the induced inner products on $\Lambda^2 R^4$ and $\Lambda^3 R^4$ for the Lorentz inner product of Example 1.6.2, taking $c = 1$.

17. (Continuation) Find out the effect of the Hodge star operator (with respect to the Lorentz inner product) on each of the basis elements of the exterior powers of R^4 that you calculated in Exercise 1.

18. Let $\{e_1, \ldots, e_5\}$ be the standard basis of R^5, and give R^5 the inner product such that $\langle e_i | e_i \rangle = 1, 1 \le i \le 3, \langle e_j | e_j \rangle = -1, 4 \le j \le 5$.

(i) Write down an orthonormal basis for $\Lambda^2 R^5$, and calculate its signature.

(ii) Find $*(e_1 \wedge e_4)$ and $**(e_1 \wedge e_4)$.

19. Suppose $\{v^1, \ldots, v^n\}$ is an orthonormal basis of V, with signature t.

(a) Write down an orthonormal basis for $\Lambda^{n-1} V$, and calculate its signature.

(b) Do the same for $\Lambda^2 V$.

20. Prove that condition (1.44) is necessary for the nondegeneracy of an inner product.

Hint: Suppose the determinant in (1.44) is zero; show that there exists a nonzero z such that $\langle z | v^j \rangle = 0, \forall j$.

21. (a) Show that the formula

$$\langle u \wedge v | u \wedge v \rangle_2 = \langle u | u \rangle \langle v | v \rangle - \langle u | v \rangle^2, \tag{1.68}$$

which appears in Table 1.2, holds in any inner product space.

(b) Show that, in the case of R^3 with the Euclidean inner product, this formula is equivalent to the formula $|u \times v|^2 = |u|^2|v|^2 - (u \cdot v)^2$ in vector calculus.

Hint: For part (a), use (1. 49). For part (b), use 1.7.4, (1. 65), and 1.7.4 again.

22. Repeat Exercise 21 for the pair of formulas appearing on the last line of Table 1.2.

    Hint: Since $u \wedge *(v \wedge w) = \langle u|w \rangle (*v) - \langle u|v \rangle (*w)$ is linear in $u, v,$ and w, it suffices to verify it for elements of an orthonormal basis. Try taking the exterior product of both sides with another basis element, and use 1.7.4.

23. Prove, by induction on the dimension n, that every inner product space has an orthonormal basis.

    Hint: (1. 47) shows that the assertion is true when $n = 1$. In general, find a vector v such that (1. 47) holds, and apply the inductive hypothesis to the $(n-1)$-dimensional subspace consisting of vectors orthogonal to v.

24. Let $\{v^1, ..., v^n\}$ and $\{w^1, ..., w^n\}$ be two orthonormal bases of an inner product space V, arranged such that $\langle v^i|v^i \rangle = 1 = \langle w^j|w^j \rangle$ for $1 \leq i \leq q$, $1 \leq j \leq r$, but for no other indices. Let H denote the set $\{v \in V : \langle v|v \rangle \geq 0\}$. Show that H is a subspace, and it has both $\{v^1, ..., v^q\}$ and $\{w^1, ...w^r\}$ as bases. Conclude that $q = r$, and so the signature of an inner product space does not depend on the basis.

1.9 Some Formal Algebraic Constructions

This section is intended merely to fill in some of the logical gaps of earlier sections. The constructions given here will not play any part in later chapters, and may be omitted.

1.9.1 Formal Construction of the Second Exterior Power

There exists a vector space satisfying (1. 1), (1. 2), (1. 3), and (1. 4).

Proof: Let $V \times V$ denote the product set $\{(u,v): u, v \in V\}$. Let $F(V \times V)$ be the vector space consisting of all finite linear combinations of elements of $V \times V$, and let $S(V \times V)$ be the subspace of $F(V \times V)$ generated by the set of all elements of the following types:

$$(u + v, w) - (u, w) - (v, w), \quad (u, v + w) - (u, v) - (u, w), \tag{1. 69}$$

$$(au, v) - a(u, v), \quad (u, av) - a(u, v), \tag{1. 70}$$

$$(u, u). \tag{1. 71}$$

Define $\Lambda^2 V$ to be the quotient space $F(V \times V)/S(V \times V)$; in other words an element of $\Lambda^2 V$ is an equivalence class of vectors in $F(V \times V)$, where two vectors are called equivalent if their difference lies in $S(V \times V)$. We define

$$u \wedge v = [(u,v)] = \text{equivalence class containing } (u,v) \in V \times V. \tag{1. 72}$$

Properties (1. 2), (1. 3), and (1. 4) follow immediately from (1. 69), (1. 70), and (1. 71).   ¤

1.9.2   Formal Construction of the *p*th Exterior Power

There exists a vector space satisfying (1. 7), (1. 8), and (1. 9).

Proof: This is an obvious generalization of 1.9.1. The expressions $V \times \ldots \times V$, $F(V \times \ldots \times V)$, and $S(V \times \ldots \times V)$ refer to the obvious p-factor versions of those in the proof of 1.9.1. Note that in lines (1. 69) and (1. 70), we need p types of terms instead of just two, and (1. 71) becomes

$$(u_1, \ldots, u_p), \text{ where } u_i = u_j \text{ for some } i \neq j. \tag{1.73}$$

Thus $\Lambda^p V$ is defined to be the quotient space $F(V \times \ldots \times V)/S(V \times \ldots \times V)$, and we may define $u_1 \wedge \ldots \wedge u_p = [(u_1, \ldots, u_p)] =$ equivalence class of $(u_1, \ldots, u_p)$.   ¤

1.9.3   Proof of the Universal Alternating Mapping Property

To every $g \in A_p(V \to W)$, there corresponds a unique $\hat{g} \in L(\Lambda^p V \to W)$ such that

$$\hat{g}(u_1 \wedge \ldots \wedge u_p) = g(u_1, \ldots, u_p), \forall u_1, \ldots, u_p. \tag{1.74}$$

Proof: Any g which is multilinear and alternating may be uniquely extended to a map $\bar{g} \in L(F(V \times \ldots \times V) \to W)$ such that $S(V \times \ldots \times V)$ is contained in the kernel of $\bar{g}$, in the terminology of 1.9.1 and 1.9.2. Since $\Lambda^p V$ is defined to be the quotient space $F(V \times \ldots \times V)/S(V \times \ldots \times V)$, every $\phi \in F(V \times \ldots \times V)$ in the equivalence class $[\phi] \in \Lambda^p V$ is mapped to the same element $\bar{g}(\phi)$ in W. So define $\hat{g}([\phi]) = \bar{g}(\phi)$. In particular,

$$\hat{g}(u_1 \wedge \ldots \wedge u_p) = \hat{g}([(u_1, \ldots, u_p)]) = g(u_1, \ldots, u_p).$$

If $\tilde{g}$ were another such map, then

$$\tilde{g}(u_1 \wedge \ldots \wedge u_p) = \hat{g}(u_1 \wedge \ldots \wedge u_p), \forall (u_1, \ldots, u_p).$$

so $\tilde{g} - \hat{g}$ is zero on the generators of $\Lambda^p V$, and hence on all of $\Lambda^p V$.   ¤

1.9.4   Calculating a Basis for an Exterior Power

For any basis $\{v^1, \ldots, v^n\}$ of V,

$$\{v^{i(1)} \wedge \ldots \wedge v^{i(p)}, 1 \leq i(1) < \ldots < i(p) \leq n\} \tag{1.75}$$

forms a basis for $\Lambda^p V$.

Proof: We saw already in (1. 14) that these vectors span $\Lambda^p V$; only the linear independence remains to be proved. First take the case $p = n$; here it suffices to show that $v^1 \wedge \ldots \wedge v^n \neq 0$ for every basis $\{v^1, \ldots, v^n\}$ for V. Select such a basis, and for any vectors $\{u_1, \ldots, u_n\}$ in V, let $A = (a_{ij})_{1 \leq i,j \leq n}$ be the $n \times n$ matrix of coefficients given by

$$u_i = \sum_j a_{ij} v^j.$$

Define a multilinear alternating map $h: V^n \to R$ by the formula:

$$h(u_1, ..., u_n) = \sum_{\pi \in \Sigma_n} \text{sgn}(\pi) a_{1\pi(1)} \cdots a_{n\pi(n)}.$$

Note that $h(v^1, ..., v^n) = 1$, and hence the corresponding linear map $\hat{h}: \Lambda^n V \to R$, as in 1.2.1, satisfies $h(v^1 \wedge ... \wedge v^n) = 1$; therefore $v^1 \wedge ... \wedge v^n \neq 0$.

Now consider the case of an arbitrary p, $2 \leq p \leq n - 1$. Suppose we have some linear combination of the vectors (1. 75) which is equal to zero:

$$\sum_I a_I (v^{i(1)} \wedge ... \wedge v^{i(p)}) = 0, \tag{1.76}$$

where the summation is over multi-indices I such that $i(1) < ... < i(p)$. Pick a specific such I, and let $I' = (k(1) < ... < k(n-p))$ be the complementary set of indices. Taking the exterior product of the vector $v^{k(1)} \wedge ... \wedge v^{k(n-p)}$ with the left side of (1. 14) makes all entries vanish except

$$a_I (v^{k(1)} \wedge ... \wedge v^{k(n-p)}) \wedge (v^{i(1)} \wedge ... \wedge v^{i(p)}),$$

$$= \pm a_I (v^1 \wedge ... \wedge v^n),$$

which must equal zero by (1. 14). Since $v^1 \wedge ... \wedge v^n \neq 0$, it follows that $a_I = 0$. Thus the linear independence of the vectors (1. 75) is proved. ¤

1.10   History and Bibliography

Exterior algebra is attributed to Hermann Grassmann (1809–77). Many books on manifolds and geometry give a brief exposition of exterior algebra, usually in the more general context of tensor algebra, and usually with greater emphasis on multilinear mappings; see, for example, Chapter 6 of Abraham, Marsden, and Ratiu [1988]. For a fuller treatment of the subject, see Greub [1978].

2   Exterior Calculus on Euclidean Space

The calculus of differential forms, known as exterior calculus, offers another approach to multivariable calculus, including line, surface, and volume integrals, which is ultimately more powerful than vector calculus, and, being coordinate-free, is ideally suited to the context of the "differential manifolds" we shall encounter later. In this chapter we shall adopt an algebraic approach to exterior calculus, which is efficient in terms of proofs but lacking in intuitive content; Chapter 8 will attempt to remedy this deficiency by showing the role of differential forms in multidimensional integration.

2.1   Tangent Spaces – the Euclidean Case

In vector calculus, the distinction between the position vector of a point in space, and directions of motion from a point in space, is not clearly drawn. Before commencing the study of differential geometry, it is necessary to formalize this distinction by putting these two kinds of vectors into different vector spaces.

A function on an open[1] subset U of R^n into R^p is said to be **smooth** if its partial derivatives of all orders exist and are continuous. Let $C^\infty(U)$ denote the set of smooth functions from U to R.

The **tangent space to R^n at y**, denoted $T_y R^n$, is simply a copy of R^n labeled with the element $y \in U$. An element $\xi \in T_y R^n$ is identified with the mapping which takes every

---

[1] We say that U is open when, for every $y \in U$, the "open ball" $B(y, \varepsilon) = \{x : \|x - y\| < \varepsilon\}$ is contained in U for all sufficiently small ε; here $\| \bullet \|$ denotes Euclidean length. An arbitrary union of open sets is open, and a finite intersection of open sets is open.

smooth function f, defined on any neighborhood[2] of y, to its directional derivative at y along ξ, denoted ξf; in symbols,

$$\xi f = \frac{d}{dt} f(y + t\xi) \Big|_{t=0}. \tag{2.1}$$

Take a basis for R^n, and let $\{x^1, ..., x^n\}$ denote the corresponding set of coordinate functions. Formally speaking, this means that $x^i : R^n \to R$ is the function such that $x^i(a_1, ..., a_n) = a_i$. Then we may express (2.1) as

$$\xi f = \xi^1 \frac{\partial f}{\partial x^1}(y) + ... + \xi^n \frac{\partial f}{\partial x^n}(y),$$

$$= (\xi^1 \frac{\partial}{\partial x^1}\Big|_y + ... + \xi^n \frac{\partial}{\partial x^n}\Big|_y) f.$$

In view of the last line, we see that $T_y R^n$ can be regarded as the vector space spanned by the differential operators

$$\{\frac{\partial}{\partial x^1}\Big|_y, ..., \frac{\partial}{\partial x^n}\Big|_y\}.$$

Elements of $T_y R^n$ are called **tangent vectors** at y. Let U be an open subset of R^n, and consider a function X which assigns an element $X(y) \in T_y R^n$ to each $y \in U$; define a mapping Xf from U to R, by taking

$$Xf(y) = X(y)f. \tag{2.2}$$

X is said to be a (smooth) **vector field** on U if $Xf \in C^\infty(U)$.

The set of smooth vector fields on U will be denoted $\mathfrak{I}(U)$. The representation of $X(y)$ in terms of the basis above determines functions $\xi^i : U \to R$, called the **component functions** with respect to this coordinate system, by the formula:

$$X(y) = \xi^1(y) \frac{\partial}{\partial x^1}\Big|_y + ... + \xi^n(y) \frac{\partial}{\partial x^n}\Big|_y. \tag{2.3}$$

By taking f in (2.2) to be each of the coordinate functions in turn, we see that X is a smooth vector field if and only if all the component functions are smooth. In that case we may abbreviate (2.3) to

---

[2] A "neighborhood" of y means an open subset of R^n containing y.

$$X = \xi^1 \frac{\partial}{\partial x^1} + \dots + \xi^n \frac{\partial}{\partial x^n}. \qquad (2.4)$$

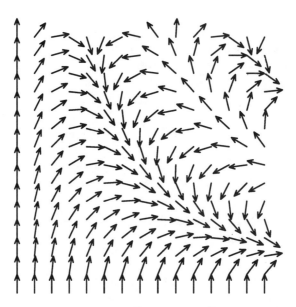

Figure 2. 1   The vector field $X = \sin(xy)\dfrac{\partial}{\partial x} + \cos(xy)\dfrac{\partial}{\partial y}$ on $(0, \pi) \times (0, \pi)$

2.1.1 Derivations

It is useful to have a more abstract characterization of vector fields. We say that $Z: C^\infty(U) \to C^\infty(U)$ is a **derivation** of $C^\infty(U)$ if, for all $f, g \in C^\infty(U), a, b \in R$, the following two properties hold:

- $Z(af + bg) = aZf + bZg$;
- $Z(fg) = f \cdot Zg + Zf \cdot g$, where $(f \cdot Zg)(x) = f(x) Zg(x)$.

2.1.1.1 The Set of Vector Fields May Be Identified with the Set of Derivations

$\mathfrak{I}(U)$ *is identical with the set of derivations of* $C^\infty(U)$.

Proof (may be omitted): It follows immediately from the product rule for differentiation that every vector field is a derivation; conversely if Z is a derivation, we claim that Z can be expressed, as in (2. 4), as the vector field:

$$Z = Zx^1 \frac{\partial}{\partial x^1} + \dots + Zx^n \frac{\partial}{\partial x^n}.$$

This representation is obtained as follows; for ease of notation let $y = 0$, and notice that a smooth function f, defined on a neighborhood of zero, has the following Taylor expansion:

$$f(x) - f(0) = \int_0^1 \frac{d}{dt} f(tx) \, dt = \sum_{i=1}^n x^i \int_0^1 \frac{\partial f}{\partial x^i}(tx) \, (dt) = \sum_{i=1}^n x^i h_i(x).$$

For any derivation Z, $Z(1 \cdot 1) = Z(1) + Z(1) \Rightarrow Z(1) = 0$, and so Z annihilates constants; applying Z to both sides of the equation above, and using the second of the two rules for derivations, shows that

$$Zf(x) = \sum_{i=1}^n Zx^i(x) h_i(x) + \sum_{i=1}^n x^i(Zh_i)(x).$$

Taking $x = 0$ removes the second term, and so

$$Zf(0) = \sum_{i=1}^n Zx^i(0) h_i(0) = \sum_{i=1}^n Zx^i(0) \frac{\partial f}{\partial x^i}(0).$$

This shows that $\mathfrak{I}(U)$ is identical with the set of derivations of $C^\infty(U)$.　　　　¤

2.1.2 Lie Derivative with Respect to a Vector Field

Given vector fields X and Y on U, the mapping $[X,Y]: C^\infty(U) \to C^\infty(U)$, defined by

$$[X,Y]f = X(Yf) - Y(Xf), \tag{2.5}$$

is indeed a derivation, as the reader may verify in Exercise 1, and hence $[X,Y]$ is a vector field by 2.1.1.1; we call this the **bracket** of X and Y, or the **Lie derivative** of Y along (or with respect to) X, also denoted $L_X Y$.

The bracket of vector fields is anticommutative, that is, $[Y,X] = -[X,Y]$, and satisfies the **Jacobi identity**, verified in Exercise 2 below:

$$[X,[Y,Z]] + [Y,[Z,X]] + [Z,[X,Y]] = 0. \tag{2.6}$$

The Lie derivative can also be applied to differential forms (see Exercises 15, 16, 17, 22, and 23), and has an important dynamical interpretation in terms of the "flow" of a vector field (see Exercise 23). It is useful in dynamical systems and in Riemannian geometry. The Jacobi identity crops up in the study of curvature, and elsewhere.

2.1.3 Example of a Lie Derivative

Let $U = R^3 \setminus \{0\}$, and take the usual Cartesian coordinate system $\{x, y, z\}$. The "gravitational field" associated with a point mass at 0 is (up to a constant multiple):

$$X = -\{x^2 + y^2 + z^2\}^{-3/2} \{x \frac{\partial}{\partial x} + y \frac{\partial}{\partial y} + z \frac{\partial}{\partial z}\}.$$

One may visualize the value of X at a point (x, y, z) in U as a vector pointing in the direction $(-x, -y, -z)$, with a Euclidean length proportional to $1/\{x^2 + y^2 + z^2\}$. The vector field associated with linear flow parallel to the y-axis is:

$$Y = \frac{\partial}{\partial y}.$$

If $r = \{x^2 + y^2 + z^2\}^{1/2}$, then the Lie derivative of Y along X is given by:

$$(L_X Y)f = -r^{-3} (x\frac{\partial}{\partial x} + y\frac{\partial}{\partial y} + z\frac{\partial}{\partial z}) \frac{\partial f}{\partial y} + \frac{\partial}{\partial y} \{ r^{-3} (x\frac{\partial}{\partial x} + y\frac{\partial}{\partial y} + z\frac{\partial}{\partial z}) \} f,$$

$$= r^{-5} \{ -3xy\frac{\partial}{\partial x} + y(r^2 - 3y^2)\frac{\partial}{\partial y} - 3zy\frac{\partial}{\partial z} \} f.$$

2.2   Differential Forms on a Euclidean Space

In the last section we introduced the set $\mathfrak{I}\,(U)$ of vector fields on an open set $U \subseteq R^n$, and showed that this notion is "intrinsic," that is, does not depend on the basis used for R^n. Before giving an intrinsic (but cryptic) definition of a differential form of order p on U, we shall first give a description, using the coordinate system $\{x^1, ..., x^n\}$, which may be more illuminating for the reader.

Let us recall the notion of the **dual space** $V^* = L(V \to R)$ of an n-dimensional vector space V. Given a basis $\{v^1, ..., v^n\}$ for V, the **dual basis**[3] $\{\lambda_1, ..., \lambda_n\}$ of V^* consists of the linear mappings defined by

$$\lambda_j (v^k) = \delta_j^k \equiv 1 \text{ if } k = j, 0 \text{ otherwise.}$$

Therefore

$$\lambda_j (a_1 v^1 + ... + a_n v^n) = a_j. \tag{2.7}$$

The **cotangent space** at $y \in U$ is defined to be $(T_y R^n)^*$, that is, the n-dimensional vector space of linear forms on the tangent space at y. Elements of $(T_y R^n)^*$ are called **cotangent vectors** at y. Define a basis for $(T_y R^n)^*$ as follows:

---

[3] To see that $\{\lambda_1, ..., \lambda_n\}$ is linearly independent, observe that if $b_1 \lambda_1 + ... + b_n \lambda_n = 0$, then applying the left side to each v_j in turn shows that each $b_j = 0$; to see that it spans, note that any $\psi \in V^*$ can be expressed as $\psi = \psi(v^1) \lambda_1 + ... + \psi(v^n) \lambda_n$.

$$\{dx^1(y), ..., dx^n(y)\} \text{ is the dual basis to } \{\frac{\partial}{\partial x^1}\Big|_y, ..., \frac{\partial}{\partial x^n}\Big|_y\}.$$

Using (2.7) gives that

$$dx^j(y) \; (\xi^1\frac{\partial}{\partial x^1}\Big|_y + ... + \xi^n\frac{\partial}{\partial x^n}\Big|_y) \; = \; \xi^j. \tag{2.8}$$

Suppose ω is an assignment of an element $\omega(y) \in (T_yR^n)^*$ to each $y \in U$; expressing $\omega(y)$ in terms of the basis $\{dx^1(y), ..., dx^n(y)\}$ defines components $h_i: U \to R$:

$$\omega(y) \; = \; h_1(y) \, dx^1(y) + ... + h_n(y) \, dx^n(y). \tag{2.9}$$

A (smooth) differential form of degree 1 on U, or 1-form, is such a mapping ω with the property that every $h_i \in C^\infty(U)$. We shall abbreviate (2.9) to:

$$\omega \; = \; h_1 dx^1 + ... + h_n dx^n.$$

Likewise a 2-form ω is an assignment of an element $\omega(y) \in \Lambda^2((T_yR^n)^*)$ to each $y \in U$, so that in terms of the basis $\{dx^i(y) \wedge dx^j(y) : 1 \le i < j \le n\}$ for $\Lambda^2((T_yR^n)^*)$, we have the abbreviated representation:

$$\omega \; = \; \sum_{i<j} h_{ij}(dx^i \wedge dx^j),$$

where each $h_{ij}: U \to R$ is smooth. The reader may look ahead to (2.28) for the representation of a p-form; but before the notation becomes too bloated, let's move into something a little more abstract.

2.2.1 The $\omega \cdot X$ Notation

A much neater way of expressing the smoothness condition for 1-forms, without reference to a coordinate system, is that

$$y \to \omega(y) \, (X(y)) \in C^\infty(U), \forall X \in \mathfrak{I}(U).$$

We shall henceforward use the notation

$$(\omega \cdot X) \, (y) \; = \; \omega(y) \, (X(y)) \tag{2.10}$$

to express the duality between differential forms and vector fields. Let us emphasize that $(\omega \cdot X)(y)$ is a real number. To see why this notation makes sense, suppose X is a vector field as in (2.4), and ω is a 1-form as in (2.9); then (2.8) shows that the notation $\omega \cdot X$ can be interpreted as follows:

$$(\omega \cdot X)\,(y) \;=\; (h_1 dx^1 + \ldots + h_n dx^n) \cdot (\xi^1 \frac{\partial}{\partial x^1} + \ldots + \xi^n \frac{\partial}{\partial x^n})\,(y), \qquad (2.11)$$

$$= h_1\,(y)\,\xi^1\,(y) + \ldots + h_n\,(y)\,\xi^n\,(y),$$

$$= \vec{h}\,(y) \cdot \vec{\xi}\,(y);$$

that is, the dot product of the vectors whose entries are the components of ω and X, respectively.

To express the smoothness condition for p-forms, we shall use the fact, discussed in Chapter 1, that an element $\eta\,(y) \in \Lambda^p\,((T_y R^n)\,*)$ can be considered as a linear form on $\Lambda^p\,(T_y R^n)$, that is, on the pth exterior power of the tangent space. Following the pattern of (2.10), we shall use the notation

$$\eta \cdot (X_1 \wedge \ldots \wedge X_p)\,(y) \;=\; \eta\,(y)\,(X_1\,(y) \wedge \ldots \wedge X_p\,(y)) \in R \qquad (2.12)$$

for $X_1, \ldots, X_p \in \mathfrak{I}\,(U)$. For the local coordinate version of (2.12) when $p = 2$, see Exercise 7; in general, in the case of a monomial $\eta = h dx^{i\,(1)} \wedge \ldots \wedge dx^{i\,(p)}$ (as always, each $x^{i\,(j)}$ is a coordinate function) the calculations presented in Chapter 1 show that

$$\eta \cdot (X_1 \wedge \ldots \wedge X_p) \;=\; \sum_{\pi \in \Sigma_p} (\operatorname{sgn}\pi)\, h\,(X_{\pi\,(1)} x^{i\,(1)}) \ldots (X_{\pi\,(p)} x^{i\,(p)}). \qquad (2.13)$$

Now we are ready for the formal, coordinate-free definition of a p-form, after which we shall exhibit some concrete examples in dimension 3.

2.2.2   Definition of Differential Forms

*A mapping ω which assigns to each $y \in U$ an element $\omega\,(y) \in \Lambda^p\,((T_y R^n)\,*)$, where $p \in \{1, \ldots, n\}$, is called a (smooth) **differential form** of degree p, or "p-form," if, for all $X_1, \ldots, X_p \in \mathfrak{I}\,(U)$, the mapping (see (2.12))*

$$y \to \omega \cdot (X_1 \wedge \ldots \wedge X_p)\,(y) \qquad (2.14)$$

belongs to $C^\infty\,(U)$. The set of p-forms on U will be denoted $\Omega^p U$. In the case $p = 0$, define $\Omega^0 U = C^\infty\,(U)$.

Note that it follows immediately from this that $\Omega^p U$ is nontrivial only for $0 \le p \le n$.

2.2.3   The Space of *p*-forms

An element of $\Omega^p U$, when evaluated at a **specific** $y \in U$, takes values in an $n!/p!\,(n-p)\,!$-dimensional space of linear forms. However, $\Omega^p U$ itself is not finite-dimensional, because $C^{\infty}(U)$ is an infinite-dimensional vector space over the reals. Its vector space structure is defined in the obvious way: For p-forms ω, η and real numbers a, b, the p-form $a\omega + b\eta$ is defined by

$$(a\omega + b\eta)\,(y)\ =\ a\omega\,(y) + b\eta\,(y).$$

Moreover[4] for every p-form ω and smooth function f on U, a p-form $f\omega$ is defined by $(f\omega)\,(y)\ =\ f(y)\,\omega\,(y)$.

For example, take the usual Cartesian coordinate system $\{x, y, z\}$ on an open subset $U \subseteq R^3$. General expressions for differential forms of order 0, 1, 2, and 3 are shown in Table 2.1. Symbols F, A, B, C, P, Q, R denote smooth functions $F = F(x, y, z)$, etc.

Name	Differential Form	dim $(\Lambda^p\,((T_yR^3)\,{}^*))$
0-form	$F = F(x, y, z)$	1
1-form	$A\,dx + B\,dy + C\,dz$	3
2-form	$P\,(dy \wedge dz) + Q\,(dz \wedge dx) + R\,(dx \wedge dy)$	3
3-form	$F\,(dx \wedge dy \wedge dz)$	1

Table 2.1   General expressions for differential forms in dimension 3

2.3   Operations on Differential Forms

2.3.1 Exterior Product of Differential Forms

For any $p, q \in \{0, ..., n\}$ with $p + q \leq n$, and for any $\omega \in \Omega^p U$ and $\eta \in \Omega^q U$, we may define the **exterior product** $\omega \wedge \eta \in \Omega^{p+q}U$,

$$(\omega \wedge \eta)\,(y)\ =\ \omega\,(y) \wedge \eta\,(y),$$

where the expression on the right is the exterior product developed in Chapter 1; the rules for exterior product apply immediately here. In particular,

$$\eta \wedge \omega\ =\ (-1)^{pq}\,(\omega \wedge \eta).$$

A special case of this formula is

$$(dx^{h\,(1)} \wedge ... \wedge dx^{h\,(p)}) \wedge (dx^{i\,(1)} \wedge ... \wedge dx^{i\,(q)})$$

---

[4] The sophisticated reader will recognize that $\Omega^p U$ is a "$C^{\infty}(U)$-module." This point of view is presented in Helgason [1978].

$$= (-1)^{pq} (dx^{i(1)} \wedge \ldots \wedge dx^{i(q)}) \wedge (dx^{h(1)} \wedge \ldots \wedge dx^{h(p)}).$$

For example, using the differential forms in Table 2.1,

$$(Adx + Bdy + Cdz) \wedge (P(dy \wedge dz) + Q(dz \wedge dx) + R(dx \wedge dy))$$

$$= (AP + BQ + CR)(dx \wedge dy \wedge dz),$$

and in R^4, with coordinates $\{x, y, z, t\}$,

$$(A(dx \wedge dz)) \wedge (B(dy \wedge dt)) = -AB(dx \wedge dy \wedge dz \wedge dt).$$

2.3.2 Hodge Star Operator on Differential Forms

If each cotangent space $(T_y R^n)^*$ is equipped with an inner product $\langle \cdot | \cdot \rangle$ (which may depend on y), then the Hodge star operator gives a mapping

$$* : \Omega^p U \rightarrow \Omega^{n-p} U,$$

where $*\omega$ is the $(n-p)$-form given by $(*\omega)(y) = *(\omega(y))$. For example, the Lorentz inner product with $c = 1$ may be applied to every cotangent space to R^4, and may be expressed in terms of coordinate functions $\{x, y, z, t\}$ by saying that $\{dx, dy, dz, dt\}$ is an orthonormal basis for each cotangent space, with

$$\langle dx|dx \rangle = 1, \langle dy|dy \rangle = 1, \langle dz|dz \rangle = 1, \langle dt|dt \rangle = -1. \tag{2.15}$$

The calculations performed at the end of Chapter 1 can be duplicated in the differential form notation: for example,

$$*dx = -dy \wedge dz \wedge dt, *dy = dx \wedge dz \wedge dt, \tag{2.16}$$

$$*dz = -dx \wedge dy \wedge dt, *dt = -dx \wedge dy \wedge dz. \tag{2.17}$$

2.3.3 Tensor Product of Differential Forms

The well-informed reader will know that the exterior algebra described in Chapter 1 is usually presented within the broader context of tensor algebra. In this book we shall only need the following simple case of a tensor product. If V and W are real finite-dimensional vector spaces, $V \otimes W$ is a vector space consisting of formal sums

$$\sum_i a_i (v_i \otimes w_i)$$

obeying the linearity conditions

$$(av_1 + v_2) \otimes w = a(v_1 \otimes w) + v_2 \otimes w, v \otimes (bw_1 + w_2) = b(v \otimes w_1) + v \otimes w_2,$$

and the condition that if $\{v^1, ..., v^m\}$ is a basis for V and $\{w^1, ..., w^n\}$ is a basis for W, then $\{v^i \otimes w^j : 1 \leq i \leq m, 1 \leq j \leq n\}$ is a basis for $V \otimes W$. The construction of such a space follows in the same way as that of the second exterior power at the end of Chapter 1. Analogous to the isomorphism $\Lambda^p(V^*) \cong (\Lambda^p V)^*$ in Chapter 1, the duals V^* and W^* satisfy:

$$V^* \otimes W^* \cong \{\text{bilinear maps } V \times W \to R\};$$

$$V^* \otimes W \cong L(V \to W);$$

for the second isomorphism, identify $\lambda \otimes w \in V^* \otimes W$ with the linear transformation $v \to \lambda(v) w$ in $L(V \to W)$, and extend to all of $V^* \otimes W$ by linearity.

The **tensor product** of 1-forms ω^1, ω^2, denoted $\omega^1 \otimes \omega^2$, is the map which assigns to each $y \in U$ the bilinear map

$$\omega^1 \otimes \omega^2(y) \in (T_y R^n)^* \otimes (T_y R^n)^*,$$

$$\omega^1 \otimes \omega^2(y)(\xi, \zeta) = \omega^1(y)(\xi)\, \omega^2(y)(\zeta).$$

Thus for vector fields X, Y, we may write:

$$\omega^1 \otimes \omega^2(X, Y) = (\omega^1 \cdot X)(\omega^2 \cdot Y) \in C^\infty(U).$$

For example, $dx \otimes x dy\, (e^x \frac{\partial}{\partial x}, \frac{\partial}{\partial y} + y \frac{\partial}{\partial z}) = xe^x$. The relationship with the exterior product is:

$$\omega^1 \wedge \omega^2 = \omega^1 \otimes \omega^2 - \omega^2 \otimes \omega^1.$$

A map which assigns to each $y \in U$ an element of $(T_y R^n)^* \otimes (T_y R^n)^*$ is called a **(0,2)-tensor**; a general (0,2)-tensor can be expressed in the form

$$\sum_{i,j} h_{ij}(dx^i \otimes dx^j)$$

where the $\{h_{ij}\}$ are smooth functions on U.

2.4 Exercises

1. Verify that for any vector fields X and Y on an open set $U \subseteq R^n$, $[X,Y]$ defined by (2.5) is a derivation, and hence a vector field.

Hint: In verifying that $[X,Y]$ is a derivation, the fact that (2. 5) is linear in f is clear, and to check the second property, expand the expression $X(Y(fg)) - Y(X(fg))$ into eight terms, four of which cancel out, leaving $f \cdot [X,Y]g + [X,Y]f \cdot g$.

2. Verify directly (i.e., without using the notion of a derivation), that for any vector fields X and Y on an open set $U \subseteq R^n$, $[X,Y]$ defined by (2. 5) is indeed a vector field.

    Hint: Since $[X,Y]$ is clearly linear in X and Y, it suffices to check this when

    $$X = \xi \frac{\partial}{\partial x^i}, Y = \zeta \frac{\partial}{\partial x^j},$$

    where $\xi = \xi(x^1, ..., x^n), \zeta = \zeta(x^1, ..., x^n)$. Show that the two second-derivative terms cancel out.

3. Verify the Jacobi identity (2. 6) for vector fields.

    Hint: Work in terms of derivations; don't differentiate anything!

4. Show that if X and Y are the vector fields on $U = R^3 \setminus \{0\}$ given below, then $L_X Y = 0$:

    $$X = -\{x^2 + y^2 + z^2\}^{-3/2} \{x \frac{\partial}{\partial x} + y \frac{\partial}{\partial y} + z \frac{\partial}{\partial z}\}; Y = -y \frac{\partial}{\partial x} + x \frac{\partial}{\partial y}.$$

    Hint: For brevity, take $r = \{x^2 + y^2 + z^2\}^{1/2}$, and note that $(x \frac{\partial r^{-3}}{\partial y} - y \frac{\partial r^{-3}}{\partial x}) = 0$.

5. Repeat Table 2.1 for the case of R^4 with coordinates $\{x, y, z, t\}$.

6. Express the 2-form $(Adx + Bdy + Cdz) \wedge (Fdx + Gdy + Hdz)$ as a linear combination of three basis elements, as in the third row of Table 2.1.

7. Show that if $\eta = P_1(dy \wedge dz) + P_2(dz \wedge dx) + P_3(dx \wedge dy)$ is a 2-form on an open set $U \subseteq R^3$, as in Table 2.1, and if

    $$X = \xi^1 \frac{\partial}{\partial x} + \xi^2 \frac{\partial}{\partial y} + \xi^3 \frac{\partial}{\partial z}, Y = \zeta^1 \frac{\partial}{\partial x} + \zeta^2 \frac{\partial}{\partial y} + \zeta^3 \frac{\partial}{\partial z},$$

    then the coordinate expression for (2. 14) when $p = 2$ is

    $$\eta \cdot (X \wedge Y) = \begin{Vmatrix} P_1 & P_2 & P_3 \\ \xi^1 & \xi^2 & \xi^3 \\ \zeta^1 & \zeta^2 & \zeta^3 \end{Vmatrix}.$$

    Hint: Calculate $(dy \wedge dz) \cdot (X \wedge Y)$ using methods of Chapter 1, and then use linearity.

8. Suppose we apply the Lorentz inner product to every cotangent space to R^4, as in (2. 15).

(a) Verify the formulas (2. 16) and (2. 17).

(b) Calculate $* \, (E \, (dx \wedge dt) + F \, (dy \wedge dt) + G \, (dz \wedge dt))$.

2.5   Exterior Derivative

The exterior derivative is an operation on differential forms which does not make sense for exterior algebra in general. It is a concept of enormous power and wide application, and it plays a central role throughout the rest of the book. As we see in Table 2.1, all the "differentiation" concepts that a student encounters in calculus, such as grad, div, and curl, are special cases of this one. Here we shall follow an "algebraic" approach to the construction of the exterior derivative; this means that we stipulate in advance the algebraic rules that it must obey, and show that there is a unique operation satisfying those rules. The "geometric" significance of the exterior derivative will not become apparent until the end of Chapter 8.

For U open in R^n, there is a natural mapping $d : \Omega^0 U \to \Omega^1 U$ which associates to the smooth function f the 1-form df, also called the **differential** of f, whose value at y is defined by

$$df(y) \, (\zeta) \; = \; \zeta f, \zeta \in T_y U, \tag{2. 18}$$

or, in the notation of (2. 10),

$$df \cdot X \; = \; Xf, X \in \mathfrak{I} \, (U) \, . \tag{2. 19}$$

If df is expressed in terms of coordinates by $df = h_1 dx^1 + \ldots + h_n dx^n$, then (2. 11) and (2. 19) combine to show that

$$(h_1 dx^1 + \ldots + h_n dx^n) \cdot \frac{\partial}{\partial x^j} = h_j = \frac{\partial f}{\partial x^j},$$

and therefore

$$df = \frac{\partial f}{\partial x^1} dx^1 + \ldots + \frac{\partial f}{\partial x^n} dx^n. \tag{2. 20}$$

2.5.1   Theorem: Existence and Uniqueness of the Exterior Derivative

There is a unique R-linear mapping $d : \Omega^p U \to \Omega^{p+1} U$ for $p = 0, 1, \ldots, n-1$, which agrees with (2. 19) when $p = 0$, such that for all differential forms ω, η:

$$d(\omega \wedge \eta) = d\omega \wedge \eta + (-1)^{\deg \omega}(\omega \wedge d\eta), \tag{2.21}$$

$$d(d\omega) = 0, \tag{2.22}$$

where $\deg \omega$ *denotes the degree of* ω, *and where* $f \wedge \eta$ *means the same as* $f\eta$ *if* $f \in \Omega^0 U$.

This mapping d is called the **exterior derivative**. The proof of the theorem, which depends on nothing more than combining the rules of multivariable calculus with those of exterior algebra, will be given a little later. Condition (2. 21) is a sophisticated form of the "product rule" in calculus, designed to specify the relationship between the exterior derivative and the exterior product of differential forms, while condition (2. 22) is tantamount to the equality of the mixed second partial derivatives, and could be called the **Iteration Rule**.

We are now going to relate exterior differentiation of differential forms in R^3 to vector calculus. Given a vector field $X = \xi^1 \frac{\partial}{\partial x} + \xi^2 \frac{\partial}{\partial y} + \xi^3 \frac{\partial}{\partial z}$ on U, there exist differential forms

$$\varpi_X \in \Omega^1 U, \varpi_X = \xi^1 dx + \xi^2 dy + \xi^3 dz, \tag{2.23}$$

called the **work form**[5] of X, and

$$\phi_X \in \Omega^2 U, \phi_X = \xi^1 (dy \wedge dz) + \xi^2 (dz \wedge dx) + \xi^3 (dx \wedge dy), \tag{2.24}$$

called the **flux form** of X. The logic of these terms will become more apparent in Chapter 8, when we shall see that the "line integral" of the work form along a path measures the work done by the vector field X in moving a particle along the path, and the "surface integral" of the flux form across a surface measures the flux of the field through the surface.

Finally, given a smooth function F on U, define a differential form

$$\rho_F \in \Omega^3 U, \rho_F = F(dx \wedge dy \wedge dz) \tag{2.25}$$

called the **density form** of F. To see how exterior differentiation relates to grad, div, and curl of a vector field, consider the following extension of Table 2.1.

Let us indicate how the rules (2. 21) and (2. 22) above are used to obtain the second row of the table: Taking only the first summand, and using the fact that $d(dx) = 0$, and writing A for ξ^1,

---

[5] I am indebted to John Hubbard (Cornell) for this terminology.

Differential Form ω	Exterior Derivative $d\omega$	
$F = F(x, y, z)$	$dF = \varpi_{\text{grad } F} = \dfrac{\partial F}{\partial x} dx + \dfrac{\partial F}{\partial y} dy + \dfrac{\partial F}{\partial z} dz$	"grad"
"work form" $\varpi_X = \xi^1 dx + \xi^2 dy + \xi^3 dz$	$d\varpi_X = \phi_{\text{curl } X} = (\dfrac{\partial \xi^3}{\partial y} - \dfrac{\partial \xi^2}{\partial z})\,(dy \wedge dz) +$ $(\dfrac{\partial \xi^1}{\partial z} - \dfrac{\partial \xi^3}{\partial x})\,(dz \wedge dx) + (\dfrac{\partial \xi^2}{\partial x} - \dfrac{\partial \xi^1}{\partial y})\,(dx \wedge dy)$	"curl"
"flux form" $\phi_X = \xi^1 (dy \wedge dz) + \xi^2 (dz \wedge dx)$ $+ \xi^3 (dx \wedge dy)$	$d\phi_X = \rho_{\text{div } X}$ $= (\dfrac{\partial \xi^1}{\partial x} + \dfrac{\partial \xi^2}{\partial y} + \dfrac{\partial \xi^3}{\partial z})\,(dx \wedge dy \wedge dz)$	"div"
$F(dx \wedge dy \wedge dz)$	0	

Table 2.2   Exterior derivatives of differential forms in dimension 3

$$d(A dx) = d(A \wedge dx) = dA \wedge dx + (-1)^0 (A \wedge d(dx))$$

$$= (\frac{\partial A}{\partial x} dx + \frac{\partial A}{\partial y} dy + \frac{\partial A}{\partial z} dz) \wedge dx$$

$$= \frac{\partial A}{\partial y}(dy \wedge dx) + \frac{\partial A}{\partial z}(dz \wedge dx) = -\frac{\partial A}{\partial y}(dx \wedge dy) + \frac{\partial A}{\partial z}(dz \wedge dx).$$

The other four terms are obtained similarly.

2.5.1.1   Uniqueness Part of Theorem 2.5.1

As a consequence of the discussion in Chapter 1, we know that the p-forms

$$\{dx^{i(1)} \wedge \ldots \wedge dx^{i(p)} : 1 \le i(1) < \ldots < i(p) \le n\} \tag{2.26}$$

provide a basis of $\Lambda^p (T_y R^n)^*$ at each $y \in U$. We shall begin by showing that if a mapping d with the properties described in the theorem exists, then it must satisfy:

$$d(dx^{i(1)} \wedge \ldots \wedge dx^{i(p)}) = 0. \tag{2.27}$$

We shall use induction on p. The assertion holds when $p = 1$ by (2. 22), since the coordinate function $x^{i(1)}$ is a 0–form. If it is true for $p - 1$, then by (2. 21),

$$d(x^{i(1)} \wedge dx^{i(2)} \wedge \ldots \wedge dx^{i(p)}) = dx^{i(1)} \wedge \ldots \wedge dx^{i(p)};$$

$$\Rightarrow d(dx^{i(1)} \wedge \ldots \wedge dx^{i(p)}) = d(d(x^{i(1)} \wedge dx^{i(2)} \wedge \ldots \wedge dx^{i(p)})) = 0$$

by (2. 22), which completes the induction. Now we shall demonstrate the uniqueness part of the theorem by showing that, for every $\omega \in \Omega^p U$, $d\omega$ is uniquely specified by the rules above. For, summing over ascending multi-indices $I = (i(1), ..., i(p))$, we may express $\omega \in \Omega^p U$ in terms of (2. 26) as:

$$\omega = \sum_I g_I (dx^{i(1)} \wedge ... \wedge dx^{i(p)}), \tag{2. 28}$$

where each g_I is a smooth function from U to R. Using (2. 21), (2. 27), and (2. 20):

$$d\omega = \sum_I d(g_I (dx^{i(1)} \wedge ... \wedge dx^{i(p)}))$$

$$= \sum_I dg_I \wedge dx^{i(1)} \wedge ... \wedge dx^{i(p)}$$

$$= \sum_I (\frac{\partial g_I}{\partial x^1} dx^1 + ... + \frac{\partial g_I}{\partial x^n} dx^n) \wedge dx^{i(1)} \wedge ... \wedge dx^{i(p)}, \tag{2. 29}$$

so now we see that $d\omega$ is uniquely specified.   ¤

2.5.1.2   Existence Part of Theorem 2.5.1

Note that (2. 19) specifies the action of d on 0-forms (and shows that df is indeed a 1-form); let us define d on the p-forms for $p \geq 1$ by (2. 29), which is clearly linear in ω. We must now show that (2. 21) and (2. 22) hold for all p-forms. By linearity over R, it is sufficient to demonstrate these for "monomials"

$$\omega = gdx^{i(1)} \wedge ... \wedge dx^{i(p)}, \eta = hdx^{j(1)} \wedge ... \wedge dx^{j(q)}, \tag{2. 30}$$

where I and J are ascending multi-indices, as in (2. 28). Using only (2. 20), (2. 29), and the rules for the exterior product from Chapter 1, we obtain:

$$d(\omega \wedge \eta) = d(gh(dx^{i(1)} \wedge ... \wedge dx^{i(p)}) \wedge (dx^{j(1)} \wedge ... \wedge dx^{j(q)}))$$

$$= ((\frac{\partial}{\partial x^1} gh) dx^1 + ... + (\frac{\partial}{\partial x^n} gh) dx^n) \wedge (dx^{i(1)} \wedge ... \wedge dx^{j(q)})$$

$$= \sum_k \{ \frac{\partial g}{\partial x^k} h(dx^k \wedge dx^{i(1)} \wedge ... \wedge dx^{j(q)})$$

$$+ \frac{\partial h}{\partial x^k} g(dx^k \wedge dx^{i(1)} \wedge ... \wedge dx^{j(q)}) \}$$

$$= \sum_k (\frac{\partial g}{\partial x^k}) dx^k \wedge (dx^{i(1)} \wedge ... \wedge dx^{i(p)}) \wedge \eta$$

$$+ \sum_k (-1)^p (\omega \wedge dx^k \wedge (dx^{j\,(1)} \wedge \ldots \wedge dx^{j\,(q)})) \frac{\partial h}{\partial x^k};$$

this says that $d(\omega \wedge \eta) = d\omega \wedge \eta + (-1)^{\deg \omega} \omega \wedge d\eta$, which establishes (2. 21); as for the "Iteration Rule,"

$$d(d\omega) = d\left(\sum_k \frac{\partial g}{\partial x^k} dx^k \wedge (dx^{i\,(1)} \wedge \ldots \wedge dx^{i\,(p)}) \right)$$

$$= \sum_{k,\,m} \frac{\partial^2 g}{\partial x^m \partial x^k} (dx^m \wedge dx^k \wedge (dx^{i\,(1)} \wedge \ldots \wedge dx^{i\,(p)}))$$

$$= \sum_{k < m} \left(\frac{\partial^2 g}{\partial x^m \partial x^k} - \frac{\partial^2 g}{\partial x^k \partial x^m} \right) (dx^m \wedge dx^k \wedge (dx^{i\,(1)} \wedge \ldots \wedge dx^{i\,(p)})),$$

which is zero, by equality of the mixed second partial derivatives.   ¤

2.6   Exercises

9. (i) Verify the formula given in Table 2.1 for the exterior derivative of the flux form
    $\phi_X = \xi^1 (dy \wedge dz) + \xi^2 (dz \wedge dx) + \xi^3 (dx \wedge dy)$, where X is the vector field

    $$X = \xi^1 \frac{\partial}{\partial x} + \xi^2 \frac{\partial}{\partial y} + \xi^3 \frac{\partial}{\partial z}.$$

    (ii) If $F \in C^\infty (R^3)$, calculate $d(F\phi_X)$ using (2. 21), and use the result to verify the following vector calculus formula:

    $$\mathrm{div}\,(FX) = \mathrm{grad}\,F \cdot X + F\,\mathrm{div}\,X.$$

10. When $\varpi_X = \xi^1 dx + \xi^2 dy + \xi^3 dz$ as in Table 2.1, and $F \in C^\infty (R^3)$, calculate $d(F\varpi_X)$ using (2. 21), and use the result to verify the following vector calculus formula:

    $$\mathrm{curl}\,(FX) = (\mathrm{grad}\,F) \times X + F\,\mathrm{curl}\,X$$

11. Using the Iteration Rule (2. 22) and the ideas of Table 2.1, deduce the vector calculus identities:

    $$\mathrm{curl}\,(\mathrm{grad}\,F) = 0; \quad \mathrm{div}\,(\mathrm{curl}\,X) = 0.$$

12. (a) In R^4, with coordinates $\{x, y, z, t\}$, calculate the exterior derivative of the following expressions; here $P, Q, F, G,$ and H are functions of (x, y, z, t).

$$Pdx + Qdt;$$

$$F(dy \wedge dz) + G(dz \wedge dx) + H(dx \wedge dy).$$

(b) Calculate $* (df \wedge dg)$ if $f = f(x, z)$, $g = g(y, t)$, with respect to the Lorentz inner product on every cotangent space to R^4, as in (2. 15).

13. Calculate $* (df)$ for $f \in C^\infty (R^3)$, with respect to the Euclidean inner product.

14. (i) Prove the following identity holds for any 1-form ω and vector fields X and Y:

$$d\omega \cdot (X \wedge Y) = X(\omega \cdot Y) - Y(\omega \cdot X) - \omega \cdot [X,Y]. \tag{2.31}$$

Hint: Since the equation is linear in ω, the case $\omega = hdx^k$ suffices.

(ii) Using induction starting from (2. 31) and formula (2. 13), verify that for any p-form ω and vector fields $X_0, \ldots, X_p$,

$$d\omega \cdot X_0 \wedge \ldots \wedge X_p = \sum_{i=0}^{p} (-1)^i X_i (\omega \cdot X_0 \wedge \ldots \wedge \hat{X}_i \wedge \ldots \wedge X_p)$$

$$+ \sum_{i<j} (-1)^{i+j} \omega \cdot [X_i, X_j] \wedge X_0 \wedge \ldots \wedge \hat{X}_i \wedge \ldots \wedge \hat{X}_j \wedge \ldots \wedge X_p,$$

where the hat $\wedge$ denotes a missing entry.

15. Let $U \subseteq R^n$ be an open set, and let X be a vector field on U. A set of linear mappings $L_X : \Omega^p U \to \Omega^p U$ for each $p \in \{0, 1, \ldots, n\}$ will be called a **Lie derivative of differential forms** along X if it satisfies the following three rules:

$$L_X f = Xf, f \in \Omega^0 U; \tag{2.32}$$

$$L_X (df) = d(L_X f), f \in \Omega^0 U; \tag{2.33}$$

$$L_X (\omega \wedge \eta) = L_X \omega \wedge \eta + \omega \wedge L_X \eta, \omega \in \Omega^p U, \eta \in \Omega^q U. \tag{2.34}$$

(i) Assuming that such a mapping exists, derive from (2. 32), (2. 33), and (2. 34) the formula $L_X (gdf) = (Xg) df + gd(Xf)$, where $f, g \in \Omega^0 U$.

(ii) Using induction on p, or otherwise, show that $L_X \omega$, if it exists, is uniquely defined for all $\omega \in \Omega^p U$, and show that the formula for the Lie derivative is

$$L_X (gdx^{i(1)} \wedge \ldots \wedge dx^{i(p)}) = \tag{2.35}$$

$$(Xg) dx^{i(1)} \wedge \ldots \wedge dx^{i(p)} + g \sum_{1 \le m \le p} dx^{i(1)} \wedge \ldots \wedge d(Xx^{i(m)}) \wedge \ldots \wedge dx^{i(p)}.$$

Hint: It suffices, by linearity, to consider p-forms as in (2. 30), that is, monomials.

(iii) Now prove the existence of the Lie derivative of differential forms as follows; define $L_X \omega$ for $\omega \in \Omega^p U$, $p \geq 1$, using (2. 35), and show that it satisfies (2. 34).

(iv) Finally, using induction or otherwise, prove that

$$L_X(d\omega) = d(L_X\omega), \omega \in \Omega^p U.$$

16. (i) Show that the Lie derivative of a 1-form, specified by (2. 32), (2. 33), and (2. 34), has the following representation when $\omega = h_1 dx^1 + \ldots + h_n dx^n$:

$$L_X\omega = \sum_j (Xh_j) \, dx^j + \sum_k h_k d\xi^k, \qquad (2.36)$$

where $X = \xi^1 \dfrac{\partial}{\partial x^1} + \ldots + \xi^n \dfrac{\partial}{\partial x^n}$.

(ii) Let U be R^2 without the x-axis, with coordinates $\{x, y\}$; calculate $L_X(y^{-1}dx - y^{-1}dy)$ where $X = y^2 \partial/\partial y$.

17. Show that the Lie derivative of a 1-form, specified by (2. 32), (2. 33), and (2. 34), satisfies the following identity, for all vector fields X and Y:

$$L_X\omega \cdot Y = L_X(\omega \cdot Y) - (\omega \cdot L_X Y). \qquad (2.37)$$

Hint: Since (2. 37) is linear in ω, the case $\omega = hdx^k$ suffices.

2.7 The Differential of a Map

In this section we extend the notion of the differential of a function to the case of functions with values in another Euclidean space. Suppose $U \subseteq R^n$ and $V \subseteq R^m$ are open sets, and $\varphi: U \to V$ is a smooth map. If $f \in C^\infty(V)$, then we have the following diagram:

$$U \subseteq R^n \xrightarrow{\quad \varphi \quad} V \subseteq R^m \xrightarrow{\quad f \quad} R$$

The **differential** of φ at $x \in U$ is the linear map $d\varphi(x) \in L(T_x R^n \to T_{\varphi(x)} R^m)$ given by

$$(d\varphi(x)\xi)f = \xi(f \bullet \varphi), \xi \in T_x R^n, f \in C^\infty(V). \qquad (2.38)$$

In other words, the effect of the differential operator $d\varphi(x)\xi$ on f is given by applying the differential operator ξ to the composite function $f \bullet \varphi \in C^\infty(U)$. This is consistent with (2. 18) when $m = 1$, provided we regard the real number $df(x)\xi = \xi f$ as an element of the tangent space $T_{f(x)}R$.

An equivalent description of the differential, once a basis is chosen for V, is to say that $d\varphi$ is the m-vector of 1-forms $(d\varphi^1, ..., d\varphi^m)$, where $\varphi = (\varphi^1, ..., \varphi^m)$. However, the beauty of (2. 38) is that the chain rule is already encoded in it.

2.7.1   The Chain Rule

If $\psi: V \to W \subseteq R^q$ is another smooth map, then for $y = \varphi(x)$,

$$d(\psi \bullet \varphi)(x) = d\psi(y) \bullet d\varphi(x). \tag{2. 39}$$

Proof: Two applications of (2. 38) show that, for $\xi \in T_xR^n, f \in C^\infty(W)$,

$$(d(\psi \bullet \varphi)(x)\xi)f = \xi(f \bullet \psi \bullet \varphi) = (d\varphi(x)\xi)f \bullet \psi = (d\psi(y) \bullet d\varphi(x)\xi)f. \text{¤}$$

2.7.2   Interpretation of the Differential in Terms of Multivariable Calculus

Using coordinate systems $\{x^1, ..., x^n\}$ on U and $\{y^1, ..., y^m\}$ on V, the smooth map $\varphi = (\varphi^1, ..., \varphi^m)$ has a derivative $D\varphi(x) \in L(R^n \to R^m)$ at each $x \in U$, which may be written as:

$$D\varphi = \begin{bmatrix} D_1\varphi^1 & ... & D_n\varphi^1 \\ ... & ... & ... \\ D_1\varphi^m & ... & D_n\varphi^m \end{bmatrix}, D_i\varphi^j = \frac{\partial \varphi^j}{\partial x^i}.$$

We shall compute in terms of $D\varphi$ the effect of applying $d\varphi(x)$ to the tangent vector

$$\xi = \xi^1 \frac{\partial}{\partial x^1}\bigg|_x + ... + \xi^n \frac{\partial}{\partial x^n}\bigg|_x.$$

2.7.2.1   The Derivative Matrix Is the Local Expression for the Differential

If $\vec{\xi} = [\xi^1, ..., \xi^n]^T, \vec{\zeta} = [\zeta^1, ..., \zeta^m]^T$, and $\zeta = \zeta^1 \frac{\partial}{\partial y^1}\bigg|_y + ... + \zeta^m \frac{\partial}{\partial y^m}\bigg|_y$, then

$$\zeta = d\varphi(x)\xi \Leftrightarrow \vec{\zeta} = D\varphi(x)\vec{\xi}$$

Proof: By the usual chain rule for differentiation,

$$(d\varphi(x)\xi)f = (\xi^1 \frac{\partial}{\partial x^1}\bigg|_x + ... + \xi^n \frac{\partial}{\partial x^n}\bigg|_x)(f \bullet \varphi)$$

$$= \sum_{i=1}^{n} \xi^i \sum_{j=1}^{m} (D_i \varphi^j)\,(x)\,\frac{\partial f}{\partial y^j}$$

$$= \left(\sum_{j=1}^{m} (D\varphi\,(x)\,\vec{\xi})^j \frac{\partial}{\partial y^j} \right) f.$$

So $\vec{\zeta} = D\varphi\,(x)\,\vec{\xi}$ is necessary and sufficient for $d\varphi\,(x)\,\xi = \zeta^1 \dfrac{\partial}{\partial y^1}\Big|_y + \ldots + \zeta^m \dfrac{\partial}{\partial y^m}\Big|_y$. ¤

2.8 The Pullback of a Differential Form

In 2.8.1, we shall give the definition of an operation on differential forms called the pullback, which is really nothing more than a change of variable formula. This is a very important construction, because in later chapters it will allow the apparatus of differential forms on Euclidean spaces to be transferred to differentiable manifolds. After the definition we give a slick characterization in a frequently encountered special case, and then the messy local coordinate version. Our suggestion to the novice is to check that one understands the syntax of the definition (i.e., check that a function, applied to an argument, really does lie in the intended domain), to use the local coordinate version for purposes of reassurance, but to memorize only the slick characterization. By the way, the "*" in this section has nothing to do with the Hodge Star Operator of Chapter 1!

The notation is the same as in Section 2.7. If $\lambda \in L\,(T_{\varphi\,(x)} R^m \to R)$ for some $x \in U$, then the composite linear map $\lambda \bullet d\varphi\,(x) : T_x R^n \to R$ is well defined, as shown in the following diagram:

$$T_x R^n \xrightarrow{\ \ d\varphi\,(x)\ \ } T_{\varphi\,(x)} R^m \xrightarrow{\ \ \lambda\ \ } R.$$

Now define $\varphi_x^* \in L\,(\,(T_{\varphi\,(x)} R^m)\,{}^* \to (T_x R^n)\,{}^*)$ (note the direction!) by

$$\varphi_x^* \lambda = \lambda \bullet d\varphi\,(x),\ \lambda \in (T_{\varphi\,(x)} R^m)\,{}^*.$$

2.8.1 Definition of the Pullback

The **pullback** $\varphi^* : \Omega^p V \to \Omega^p U$, for each $p \in \{0, 1, \ldots, m\}$, is a linear map created out of the linear maps $\{\varphi_x^* : x \in U\}$ as follows:

* $p = 0$: for $f \in \Omega^0 V = C^\infty\,(V)$, define $\varphi^* f = f \bullet \varphi$;
* $p \geq 1$: for $\omega \in \Omega^p V$, define $(\varphi^* \omega)\,(x) = (\Lambda^p \varphi_x^*)\,(\omega\,(y))$, where $y = \varphi\,(x)$.

Here we have used the *p*- exterior power of a linear transformation, introduced in Chapter 1.

2.8.2   The Slick Characterization

The following special case will be encountered frequently in later chapters. When $\varphi: U \subseteq R^n \to V \subseteq R^n$ is a smooth bijection with a smooth inverse (a "diffeomorphism"), then every vector field X on U induces a vector field $\varphi_* X$ on V, called the **push-forward** of X under φ, by the formula

$$\varphi_* X(y) \;=\; d\varphi(x) \,(X(\varphi^{-1}(y))). \tag{2.40}$$

The formula is elucidated by the following diagram.

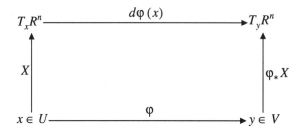

The definition of pullback says that if ω is a 1-form, then for every vector field X on U,

$$(\varphi^* \omega) \cdot X(x) \;=\; \omega \cdot \varphi_* X(y).$$

In this notation, the pullback $\varphi^* \eta$ of a *p*-form η on V is uniquely characterized when $p \geq 1$ by

$$\varphi^* \eta \cdot (X_1 \wedge \dots \wedge X_p) \;=\; \eta \cdot (\varphi_* X_1 \wedge \dots \wedge \varphi_* X_p),$$

for all $X_1, \dots, X_p \in \mathfrak{S}(U)$.

2.8.3   Interpretation of the Pullback in Terms of Multivariable Calculus

Suppose $\omega = A_1 dy^1 + \dots + A_m dy^m$ is a 1-form on V, and $\eta = \varphi^* \omega$ has an expression $\eta = B_1 dx^1 + \dots + B_n dx^n$ as a 1-form on U. Let us represent ω and η as the row vectors

$$\vec{\omega} \;=\; [A_1, \dots, A_m], \vec{\eta} \;=\; [B_1, \dots, B_n],$$

respectively.

2.8.3.1 Formula for the Pullback in Terms of the Derivative Matrix

$\eta = \varphi^* \omega$ *if and only if* $\vec{\eta}(x) = \vec{\omega}(y) D\varphi(x)$ *for all* $y = \varphi(x)$. *Written out in full, the condition says*

$$[B_1, ..., B_n] = [A_1, ..., A_m] \begin{bmatrix} D_1\varphi^1 & ... & D_n\varphi^1 \\ ... & ... & ... \\ D_1\varphi^m & ... & D_n\varphi^m \end{bmatrix}. \tag{2.41}$$

Proof: From the formula $\varphi^* \omega(x)(\xi) = \omega(y)(d\varphi(x)\xi)$ and from 2.7.2.1, we see that $\eta = \varphi^* \omega$ means that $\vec{\eta}\vec{\xi} = \vec{\omega}(D\varphi)\vec{\xi}$ for every ξ.   ¤

Thus the differential and the pullback correspond to premultiplication and postmultiplication by the $m \times n$ matrix $D\varphi$, respectively. Clearly the pullback of a p-form can be expressed in terms of the pth exterior power of the matrix $D\varphi$, as was discussed in Chapter 1. A specific example which will be important for the theory of integration is the following: When $p = m = n$

$$\omega = A(dy^1 \wedge ... \wedge dy^n) \Rightarrow \varphi^* \omega = |D\varphi| A(dx^1 \wedge ... \wedge dx^n). \tag{2.42}$$

2.8.4 Example: Spherical Coordinates

Let $U = \{(r, \theta, \phi) \in R^3 : r > 0, 0 < \theta < 2\pi, 0 < \phi < \pi\}$, which is open in R^3, and define $\Phi: U \to R^3$ to be the usual change of variable map from spherical coordinates to Cartesian ones, namely,

$$\Phi(r, \theta, \phi) = (r\cos\theta\sin\phi, r\sin\theta\sin\phi, r\cos\phi) = (x, y, z).$$

This map is not onto (for example, $(0, 0, 1) \notin \Phi(U)$), but it is a diffeomorphism onto its image V. As one may easily check,

$$D\Phi = \begin{bmatrix} \cos\theta\sin\phi & -r\sin\theta\sin\phi & r\cos\theta\cos\phi \\ \sin\theta\sin\phi & r\cos\theta\sin\phi & r\sin\theta\cos\phi \\ \cos\phi & 0 & -r\sin\phi \end{bmatrix},$$

and now it follows from (2.41) that

$$\Phi^* dx = \cos\theta\sin\phi\, dr - r\sin\theta\sin\phi\, d\theta + r\cos\theta\cos\phi\, d\phi; \tag{2.43}$$

$$\Phi^* dy = \sin\theta\sin\phi\, dr + r\cos\theta\sin\phi\, d\theta + r\sin\theta\cos\phi\, d\phi; \tag{2.44}$$

$$\Phi^* dz = \cos\phi\, dr - r\sin\phi\, d\phi. \tag{2.45}$$

It follows from (2.42) that

$$\Phi^* \, (dx \wedge dy \wedge dz) \ = \ -r^2 \sin\phi \, (dr \wedge d\theta \wedge d\phi), \tag{2.46}$$

which is supposed to remind you of the Jacobian determinant which you use when evaluating a volume integral in spherical coordinates; more of this in Chapter 8.

2.8.5 Formal Properties of the Pullback

The pullback defined in 2.8.1 has the following properties, $\forall \omega, \eta \in \Omega^p V, \, \rho \in \Omega^q V$:

$$\varphi^* \, (\omega + \eta) \ = \ \varphi^* \, \omega + \varphi^* \, \eta; \tag{2.47}$$

$$\varphi^* \, (\eta \wedge \rho) \ = \ \varphi^* \, \eta \wedge \varphi^* \, \rho, \tag{2.48}$$

and in particular $\varphi^* \, (f\rho) \ = \ (f \bullet \varphi) \, \varphi^* \, \rho$ *for* $f \in C^\infty (V)$;

$$d \, (\varphi^* \, \omega) \ = \ \varphi^* \, (d\omega); \tag{2.49}$$

and if $\psi: V \to W \subseteq R^k$ *is also smooth, then*

$$(\psi \bullet \varphi)^* \ = \ \varphi^* \bullet \psi^*. \tag{2.50}$$

Proof: (2.47) is immediate from the definition, and (2.50) is immediate from the chain rule 2.7.1. Property (2.48) is simply an algebraic property of the exterior powers of the linear map $\varphi_x^{\ *}$, mentioned in Chapter 1. To prove (2.49), note first that for $f \in C^\infty (V)$ and $X \in \mathfrak{I} \, (U)$,

$$\varphi^* \, df \cdot X \ = \ df \cdot \varphi_* X \ = \ (\varphi_* X) f \ = \ X (f \bullet \varphi) \ = \ X (\varphi^* f) \ = \ d \, (\varphi^* f) \cdot X;$$

in other words

$$\varphi^* \, df \ = \ d \, (\varphi^* f), \tag{2.51}$$

which proves that (2.49) holds for 0-forms. It suffices by (2.47) to prove (2.49) when ω is a monomial, that is, to show that

$$d \, (\varphi^* \, (h dy^{i\,(1)} \wedge \dots \wedge dy^{i\,(p)})) \ = \ \varphi^* \, (d \, (h dy^{i\,(1)} \wedge \dots \wedge dy^{i\,(p)})). \tag{2.52}$$

Now it follows from (2.21) and (2.22) that

$$d \, (h dy^{i\,(1)} \wedge \dots \wedge dy^{i\,(p)}) \ = \ dh \wedge dy^{i\,(1)} \wedge \dots \wedge dy^{i\,(p)}, \tag{2.53}$$

and so by (2.48) and (2.51), the right side of (2.52) is

$$= \ d \, (\varphi^* \, h) \wedge d \, (\varphi^* \, y^{i\,(1)}) \wedge \dots \wedge d \, (\varphi^* \, y^{i\,(p)})$$

$$= d\left(\left(\varphi^* h\right) d\left(\varphi^* y^{i\,(1)}\right) \wedge \ldots \wedge d\left(\varphi^* y^{i\,(p)}\right)\right)$$

$$= d\left(\left(\varphi^* h\right)\left(\varphi^* dy^{i\,(1)} \wedge \ldots \wedge \varphi^* dy^{i\,(p)}\right)\right)$$

$$= d\left(\varphi^*\left(h dy^{i\,(1)} \wedge \ldots \wedge dy^{i\,(p)}\right)\right),$$

where the second line uses (2. 53), the third uses (2. 48) and (2. 51), and the fourth uses (2. 48). This verifies (2. 52) and completes the proof.   ¤

2.8.6 Spherical Coordinates Example; Continued

It follows from (2. 48), for example, that

$$\Phi^*\left(dx \wedge dy\right) = \left(\Phi^* dx\right) \wedge \left(\Phi^* dy\right),$$

and by substituting from equations (2. 43) and (2. 44), and performing the usual exterior algebra operations on the resulting nine terms, we obtain:

$$r\left(\sin\phi\right)^2\left(dr \wedge d\theta\right) - r^2 \sin\phi\cos\phi\left(d\theta \wedge d\phi\right).$$

2.9   Exercises

18. In Example 2.8.4, calculate $\Phi^*\left(dy \wedge dz\right)$, and verify directly that

$$\Phi^*\left(dx\right) \wedge \Phi^*\left(dy \wedge dz\right)$$

coincides with the formula (2. 46) for $\Phi^*\left(dx \wedge dy \wedge dz\right)$.

19. Let $U = \{(r, \theta) : r > 0, 0 < \theta < 2\pi\} \subseteq R^2$, and take the usual radial coordinate map $\Phi(r, \theta) = (r\cos\theta, r\sin\theta) = (x, y)$. Calculate $\Phi^*\left(dx\right)$, $\Phi^*\left(dy\right)$, and $\Phi^*\left(dx \wedge dy\right)$.

20. Let $U = \{(t, \theta) : -\infty < t < \infty, 0 < \theta < 2\pi\} \subseteq R^2$, and map U into R^3 using:

$$\Phi(t, \theta) = (\cosh t \cos\theta, \cosh t \sin\theta, \sinh t) = (x, y, z).$$

Calculate $\Phi^*\left(dx\right)$, $\Phi^*\left(dy\right)$, and $\Phi^*\left(dx \wedge dy\right)$.

Remark: This map is a parametrization of the "hyperboloid" $x^2 + y^2 - z^2 = 1$.

21. Consider the map $\Phi: R^3 \to R^4$ given by

$$(x, y, z, w) = \Phi(s, t, u) = (u(s+t), u(s-t), s(t+u), s(t-u)).$$

Calculate $\Phi^*(dx)$, $\Phi^*(dy)$, $\Phi^*(dz)$, and $\Phi^*(dx \wedge dy \wedge dz)$.

Remark: The image of this map is the set $\{(x, y, z, w) : x^2 - y^2 = z^2 - w^2\}$.

22. Suppose $\varphi : U \subseteq R^n \to V \subseteq R^n$ is a diffeomorphism between open sets U and V (i.e., a smooth bijection with a smooth inverse).

(i) Show that the push-forward of vector fields, defined in (2. 40), commutes with the Lie derivative of vector fields and differential forms (see Exercise 15), in the following sense: For all $X, Y \in \mathfrak{I}(U)$ and $\omega \in \Omega^p V$,

$$\varphi_*(L_X Y) = L_{\varphi_* X}(\varphi_* Y), \tag{2. 54}$$

$$\varphi^*(L_{\varphi_* X} \omega) = L_X(\varphi^* \omega). \tag{2. 55}$$

(ii) Calculate the push-forward of the vector fields $\partial/\partial r$ and $\partial/\partial\theta$ under the spherical coordinate map Φ of Example 2.8.4, and verify directly that formula (2. 54) holds.

(iii) Using formula (2. 36), calculate the Lie derivative of $\Phi^*(dx)$ along $\partial/\partial r$ in Example 2.8.4, and thus verify (2. 55) directly for $\omega = dx$ and $X = \partial/\partial r$.

23. (Dynamical interpretation of the Lie derivative) A family of diffeomorphisms[6] $\{\phi_t\}$ of R^n, where t takes values in some open interval of R including 0, is called the flow[7] of a vector field X on R^n if, for all $x \in R^n$,

$$X(\phi_t(x)) = \frac{\partial}{\partial t}\phi_t(x).$$

(i) Using the fact that $\phi_s \bullet \phi_t = \phi_{s+t}$, prove that if Y is another vector field, then

$$L_X Y = \frac{\partial}{\partial t}((\phi_{-t})_* Y)\bigg|_{t=0}, \tag{2. 56}$$

where the symbolism on the right side uses the push-forward defined in (2. 40).

(ii) By writing $\phi_t^* df$ as $\sum_{i=1}^{n} \frac{\partial}{\partial x^i}(f \bullet \phi_t)(x) dx^i$, prove that for every $f \in \Omega^0 R^n$,

$$\frac{\partial}{\partial t}(\phi_t^* df)\bigg|_{t=0} = d(Xf) = L_X df. \tag{2. 57}$$

(iii) Using the properties of the pullback proved in 2.8.5, derive the formula for the Lie derivative of an arbitrary differential form in terms of the flow:

---

[6] A diffeomorphism means a smooth map with a smooth inverse.
[7] For the existence and uniqueness of the flow, and other technical matters, see books on differential equations, or Abraham, Marsden, and Ratiu [1988].

$$L_X\omega = \frac{\partial}{\partial t}(\phi_t^* \,\omega)\Big|_{t=0}, \ \forall \omega \in \Omega^p R^n. \tag{2.58}$$

24. Given a p-form ω and a vector field X, the **interior product** or **contraction** of X and ω is the $(p-1)$-form $\iota_X\omega$ defined as follows: $\iota_X\omega = 0$ if $p = 0$, $\iota_X\omega = \omega \cdot X$ if $p = 1$, and

$$\iota_X\omega \cdot X_1 \wedge \dots \wedge X_{p-1} = \omega \cdot X \wedge X_1 \wedge \dots \wedge X_{p-1}, \ p \ge 2. \tag{2.59}$$

(i) Explain why the identities $\iota_{hX}\omega = h\iota_X\omega$ and $\iota_X dh = L_X h = Xh$ are valid for every smooth function h.

(ii) Show, using (2.13), that for any p-form λ and any q-form μ,

$$\iota_X(\lambda \wedge \mu) = \iota_X\lambda \wedge \mu + (-1)^{\deg \lambda}\lambda \wedge \iota_X\mu. \tag{2.60}$$

(iii) Prove, by induction on the degree of ω, the identity

$$L_X\omega = \iota_X(d\omega) + d(\iota_X\omega), \tag{2.61}$$

where the left side is the Lie derivative of the form ω as discussed in Exercise 15.

Hint: It suffices to prove the result for monomials. Write a monomial p-form ω as $\omega = df \wedge \mu$ for a $(p-1)$-form μ and use formulas (2.21), (2.34), and (2.60).

(iv) Prove that for a smooth function h, $L_{hX}\omega = hL_X\omega + dh \wedge \iota_X\omega$.

25. (Continuation) Suppose $U \subseteq R^n$ and $V \subseteq R^n$ are open sets, and $\varphi: U \to V$ is a diffeomorphism. Prove that, for every vector field X on U,

$$\iota_X(\varphi^* \,\omega) = \varphi^*(\iota_{\varphi_* X}\omega). \tag{2.62}$$

2.10   History and Bibliography

Differential forms in the sense discussed here originated in 1899 in an article by Élie Cartan (1869–1951) and in the third volume of *Les Méthodes Nouvelles de la Mécanique Céleste* by Henri Poincaré (1854–1912). The program of restating the laws of physics in an invariant notation, including the use of differential forms, was initiated by Gregorio Ricci-Curbastro (1853–1925) and his student Tullio Levi-Civita (1873–1941), and helped to provide the framework in which Albert Einstein (1879–1955) developed the theory of relativity.

The account given here follows along the same lines as Flanders [1989]. For much more detailed information about exterior calculus and its use in mathematical physics, see Edelen [1985] and Curtis and Miller [1985].

2.11   Appendix: Maxwell's Equations

The purpose of this section is to illustrate the power of exterior calculus to express physical laws in a coordinate-free manner. Flanders [1989] gives many other examples.

As in the earlier part of this chapter, the Lorentz inner product may be applied to every cotangent space to R^4, and may be expressed in terms of coordinate functions $\{x, y, z, t\}$ by saying that $\{dx, dy, dz, dt\}$ is an orthonormal basis for each cotangent space, with

$$\langle dx|dx\rangle = 1, \langle dy|dy\rangle = 1, \langle dz|dz\rangle = 1, c^2\langle dt|dt\rangle = -1, \qquad \text{(2. 63)}$$

where c denotes the speed of light. Assume that the following "fields" are smooth maps from R^4 to R^3; according to electromagnetic theory, they are related by Maxwell's equations in the table below:

- **Electric Field** $E = E_1\dfrac{\partial}{\partial x} + E_2\dfrac{\partial}{\partial y} + E_3\dfrac{\partial}{\partial z}$;

- **Magnetic Field** $B = B_1\dfrac{\partial}{\partial x} + B_2\dfrac{\partial}{\partial y} + B_3\dfrac{\partial}{\partial z}$;

- **Electric Current Density** $J = J_1\dfrac{\partial}{\partial x} + J_2\dfrac{\partial}{\partial y} + J_3\dfrac{\partial}{\partial z}$.

The symbol ρ above refers to a smooth map from R^4 to R called the **charge density**. For ease of comparison with our earlier calculations, let us subsume the speed of light into the t variable, which has the effect of setting $c = 1$. The electric and magnetic fields may be encoded in "work forms"

$\operatorname{curl} E = -\dfrac{1}{c}\dfrac{\partial B}{\partial t}$	Faraday's Law (Electric field produced by a changing magnetic field)
$\operatorname{curl} B = \dfrac{4\pi}{c}J + \dfrac{1}{c}\dfrac{\partial E}{\partial t}$	Ampère's Law[a] (Magnetic field produced by a changing electric field)
$\operatorname{div} E = 4\pi\rho$	Gauss's Law
$\operatorname{div} B = 0$	Nonexistence of true magnetism

Table 2.3   Maxwell's equations

[a] We omit the dielectric constant and the permeability, and assume we deal with bodies at rest.

$$\varpi_E = E_1 dx + E_2 dy + E_3 dz, \varpi_B = B_1 dx + B_2 dy + B_3 dz \tag{2.64}$$

and also in "flux forms"

$$\phi_E = E_1 dy \wedge dz + E_2 dz \wedge dx + E_3 dx \wedge dy, \tag{2.65}$$

$$\phi_B = B_1 dy \wedge dz + B_2 dz \wedge dx + B_3 dx \wedge dy, \tag{2.66}$$

while the "4-current" is the 1-form:

$$\gamma = J_1 dx + J_2 dy + J_3 dz - \rho dt. \tag{2.67}$$

It follows from formulas above that

$$*\gamma = -J_1 dy \wedge dz \wedge dt - J_2 dx \wedge dz \wedge dt - J_3 dx \wedge dy \wedge dt + \rho dx \wedge dy \wedge dz. \tag{2.68}$$

A restatement of the results of an exercise in Chapter 1 gives:

$$* (dx \wedge dt) = dy \wedge dz, * (dy \wedge dt) = dz \wedge dx, * (dz \wedge dt) = dx \wedge dy,$$

$$* (dy \wedge dz) = -dx \wedge dt, * (dz \wedge dx) = -dy \wedge dt, * (dx \wedge dy) = -dz \wedge dt;$$

and from these formulas, the reader may check that

$$* (\varpi_E \wedge dt) = \phi_E, *\phi_B = dt \wedge \varpi_B. \tag{2.69}$$

Define a 2-form

$$\eta = \varpi_E \wedge dt + \phi_B; \tag{2.70}$$

$$\therefore d\eta = d\varpi_E \wedge dt + d\phi_B. \tag{2.71}$$

Noting that $d (E_1 dx) = \partial E_1/\partial y (dy \wedge dx) + \partial E_1/\partial z (dz \wedge dx) + \partial E_1/\partial t (dt \wedge dx)$, etc., the reader is invited to verify, by the methods of this chapter, that the right side of (2.71) is equal to

$$d\varpi_E \wedge dt + d\phi_B = \phi_{\text{curl } E} \wedge dt + \rho_{\text{div } B} + dt \wedge \phi_{\dot{B}}, \tag{2.72}$$

where $\dot{B}_i = \partial B_i/\partial t$, using the work-form, flux-form, and density-form notation. Thus the first and fourth of Maxwell's equations together are equivalent to

$$d\eta = 0. \tag{2.73}$$

On the other hand, (2.69) implies that $*\eta = \phi_E + dt \wedge \varpi_B$, and so the rules for exterior differentiation give:

$$d(*\eta) = d\phi_E - dt \wedge d\varpi_B,$$

and a calculation similar to the one above shows that the second and third of Maxwell's equations together are equivalent to

$$*d(*\eta) = 4\pi\gamma,$$

using the fact, deduced from the methods of Chapter 1, that $* \ (*\gamma) = \gamma$. It turns out that a deeper understanding of the theory is obtained by replacing (2. 73) by a certain equation which implies it; Maxwell's equations can be solved by finding an **electromagnetic potential** α, which means a 1–form $A_1 dx + A_2 dy + A_3 dz + A_0 dt$, such that

$$d\alpha = \eta, \ *d(*\eta) = 4\pi\gamma. \tag{2.74}$$

Note that, by properties of the exterior derivative, $d\alpha = \eta \Rightarrow d\eta = d(d\alpha) = 0$. The concise formulation (2. 74) is more than just a clever trick. It is an intrinsic formulation of natural law, independent of any coordinate system, whereas the original formulation of Maxwell's equations is specific to the $\{x, y, z, t\}$ coordinate system. Moreover, as we shall see in Chapter 10, (2. 74) arises naturally as the solution of a variational problem, phrased in geometric language; indeed, it is a special case of so-called gauge field theories in mathematical physics, which will be described in Chapter 10.

2.11.1 Exercise

26. (i) Verify in detail that (2. 72) is a correct expression for $d\eta$.

(ii) Calculate in detail $d\phi_E - dt \wedge d\varpi_B$, and check that this is $4\pi \ (*\gamma)$ under Maxwell's equations.

3 Submanifolds of Euclidean Spaces

The main purpose of this chapter is to introduce a class of concrete examples – the submanifolds of Euclidean space – in order to motivate the definitions of abstract manifold, vector bundle, etc., in later chapters. The primary technical tool in much of this work will be the Implicit Function Theorem, which will be restated in geometric language.

3.1 Immersions and Submersions

Before we can define submanifolds of R^{n+k}, we need to consider carefully the notion of the rank of the derivative of a mapping; to avoid constant mention of degrees of differentiability, we work with smooth mappings throughout.

For W open in R^n, a C^∞ map $\Psi : W \to R^{n+k}$ is called a (smooth) **immersion** if its differential $d\Psi(u) \in L(T_u R^n \to T_{\Psi(u)} R^{n+k})$ is a one-to-one map at every $u \in W$. Here are two statements, each of which, according to linear algebra, is equivalent to the definition of immersion:

- for every $u \in W$, multiplication by the $(n+k) \times n$ derivative matrix

$$D\Psi(u) = \begin{bmatrix} \partial\Psi^1/\partial u_1 & \ldots & \partial\Psi^1/\partial u_n \\ \ldots & \ldots & \ldots \\ \partial\Psi^{n+k}/\partial u_1 & \ldots & \partial\Psi^{n+k}/\partial u_n \end{bmatrix} \qquad (3.1)$$

  induces a one-to-one linear map from R^n to R^{n+k}.

- for every $u \in W$, $D\Psi(u)$ has rank n, which here means that it has n linearly independent rows (or equivalently, n linearly independent columns).

Now we introduce the dual notion of submersion. Let U be open in R^{n+k}, and suppose $f: U \to R^k$ is a smooth map. To keep things in their proper place, think of the following diagram:

$$W \subseteq R^n \xrightarrow{\quad \Psi \quad} R^{n+k} \supseteq U \xrightarrow{\quad f \quad} R^k$$

For the case $n = 2$ and $k = 1$, Figure 3.8 on page 64 may be suggestive.

We say that f is a **submersion** if its differential $df(x) \in L(T_x R^{n+k} \to T_{f(x)} R^k)$ is an onto map; according to linear algebra, two equivalent statements are:

*   multiplication by the $k \times (n+k)$ derivative matrix $Df(x)$, where

$$Df(x) = \begin{bmatrix} \partial f^1 / \partial x_1 & \dots & \partial f^1 / \partial x_{n+k} \\ \dots & \dots & \dots \\ \partial f^k / \partial x_1 & \dots & \partial f^k / \partial x_{n+k} \end{bmatrix}, \tag{3.2}$$

    induces an onto map from R^{n+k} to R^k for every $x \in U$.
*   for every $x \in U$, the derivative matrix $Df(x)$ has rank k, which here means that it has k linearly independent columns (or equivalently, k linearly independent rows).

3.1.1   Examples of Immersions

Let us consider the case where $n = 2$ and $k = 1$. Let $h \in C^\infty(R)$ be a strictly positive function, and consider the map $\Psi: W = R^2 \to R^3$ which rotates the curve $x = h(z)$ infinitely many times around the z-axis, namely,

$$\Psi(u, \theta) = (h(u)\cos\theta, h(u)\sin\theta, u). \tag{3.3}$$

This gives what is known as a (parametrized) **surface of revolution**. In this case

$$D\Psi(u, \theta) = \begin{bmatrix} h'(u)\cos\theta & -h(u)\sin\theta \\ h'(u)\sin\theta & h(u)\cos\theta \\ 1 & 0 \end{bmatrix}.$$

By inspection of the last row, the only way that the two columns of this matrix could be linearly dependent is for the second column to be zero; this is impossible because $\| [-h(u)\sin\theta, h(u)\cos\theta, 0]^T \| = h(u) > 0$. Hence Ψ is an immersion. Note, incidentally, that neither the map Ψ nor the map $(u, \theta) \to D\Psi(u, \theta)$ is one-to-one; it is merely that, for fixed (u, θ), the linear transformation $D\Psi(u, \theta)$ is one-to-one.

On the other hand the derivative of the map $\Phi(s, t) = (s, s, st)$ from R^2 to R^3 is a 3×2 matrix with second column $[0, 0, s]^T$, which is zero on $\{(s, t) : s = 0\}$. Thus Φ is not an immersion.

3.1.2 Examples of Submersions

When we consider the case where $n = 2$ and $k = 1$, and U is open in R^3, then $f: U \to R$ is a submersion if its derivative does not vanish on U. For example, let U be the set $\{(x_1, x_2, x_3) : x_1^2 + x_2^2 + x_3^2 > 0\}$, and let $f(x_1, x_2, x_3) = x_1^2 + x_2^2 + x_3^2$. Then $Df(x_1, x_2, x_3) = 2[x_1, x_2, x_3]$, which is never zero on U, and so f is a submersion. On the other hand the map $g(x_1, x_2, x_3) = x_1 x_2 x_3$ has derivative $[x_2 x_3, x_1 x_3, x_1 x_2]$, which is zero, for example, at $(0, 0, 1) \in U$, and so g is not a submersion on U.

The complementary notions of immersion and submersion are so important that we summarize the information above in the following table, using the terminology of differentials introduced in Chapter 2.

	Immersions	**Submersions**
Domain and range	$\Psi: W \subseteq R^n \to R^{n+k}$	$f: U \subseteq R^{n+k} \to R^k$
Defining property	$d\Psi(u)$ is 1-1 $\forall u \in W$	$df(x)$ is onto $\forall x \in U$
Derivative matrix is	$(n+k) \times n$	$k \times (n+k)$
... whose rank should be	rank $= n$	rank $= k$
... or in other words	columns are independent	rows are independent

Table 3.1   Immersions and submersions

3.2   Definition and Examples of Submanifolds

Our primary goal is to define a class of subsets of R^{n+k} which are "smooth" and "locally n-dimensional," in some sense which we shall try to make precise. When $n = 2$ and $k = 1$, this class should include certain parametrized surfaces such as the sphere, torus, and ellipsoid; it should also include open subsets of the subspace R^2, such as the $\{x, y\}$-plane with $(0, 0)$ deleted. On the other hand this class should exclude anything with sharp corners, such as the surface of a cube or of a cone, and anything whose "dimension" could be said to vary, such as

$$\{(x, y, 0) : -x^2 < y < x^2\} \cup \{(0, 0, 0)\} \subset R^3,$$

which collapses from being "2-dimensional" to "1-dimensional" at $(0, 0, 0)$. There is also a topological subtlety in the definition which will be apparent a little later.

We have two main choices when it comes to giving the definition of an n-dimensional submanifold M of R^{n+k}. The first choice is to try to generalize the vector calculus notions of "parametrized curve" and "parametrized surface," and say that M has to be the union of images of certain one-to-one immersions $\Psi_i: W_i \subseteq R^n \to R^{n+k}$, called n-dimensional parametrizations, which are discussed in Section 3.4; this would give the same end result as the definition below, but it requires a lot more work, because

- it would put us under the obligation to construct the parametrizations in order to check that M is a submanifold;

- we would have to include a topological condition on each of the immersions, which is tedious to state and to verify.

Instead we choose a second way using submersions, which is slick but initially incomprehensible; we shall try to explain heuristically why it works for surfaces; then we shall show that it is equivalent to the first kind of definition, using the Implicit Function Theorem.

3.2.1   Definition of an n-Dimensional Submanifold of R^{n+k}

A subset M of R^{n+k} is called an n-dimensional submanifold of R^{n+k} if, for each $x \in M$, there is a neighborhood U of x in R^{n+k} and a submersion $f: U \to R^k$ such that

$$U \cap M = f^{-1}(0). \tag{3.4}$$

Note incidentally that in the trivial case $k = 0$, the condition that f be a submersion is vacuous, and so any open subset of R^n is an n-dimensional submanifold of R^n.

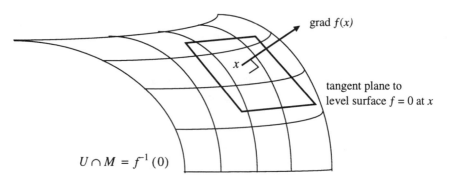

Figure 3. 1   The level surface of a submersion

The figure above is supposed to help in understanding this definition, at least when $n = 2$ and $k = 1$. For f to be a submersion amounts to saying that $\operatorname{grad} f(x)$ is a nonzero vector at every x in U. The reader may recall from multivariate calculus, or derive directly, that existence of $\operatorname{grad} f(x)$ implies that there exists a tangent plane to the level surface $f = 0$ at x, namely, the plane through x normal to $\operatorname{grad} f(x)$. Thus the level surface $f = 0$, which is already "smooth" because f is a smooth function, has in some sense a "2-dimensional structure."

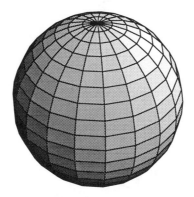

Figure 3. 2   The 2-sphere

3.2.2 Example: The *n*-Sphere

The ***n*-sphere** $S^n \subset R^{n+1}$ is defined as

$$\{x = (x_0, ..., x_n) : f(x) = x_0^2 + ... + x_n^2 - 1 = 0\}. \tag{3.5}$$

If U is any open set containing S^n but not containing 0, then f restricted to U is a submersion, so S^n is an n-dimensional submanifold of R^{n+1}. A special case is the circle S^1. The sphere S^2 is pictured.

3.2.3 Example: The *n*-Torus

The ***n*-torus** $T^n \subset R^{2n}$ is defined as

$$\{x = (x_1, ..., x_{2n}) : g(x) = (x_1^2 + x_2^2 - 1, ..., x_{2n-1}^2 + x_{2n}^2 - 1) = (0, ..., 0)\}. \tag{3.6}$$

Similarly, T^n is an n-dimensional submanifold of R^{2n} (see Exercise 4). Note that this is not the only way to describe the torus; for example, T^2 can be viewed as a 2-dimensional submanifold of R^3, as is shown in Figure 3.7 and described in Section 3.4. Another way to describe the torus is as

$$T^n = S^1 \times ... \times S^1 \ (n \text{ factors}). \tag{3.7}$$

3.2.4   Example: The n-Hyperboloid

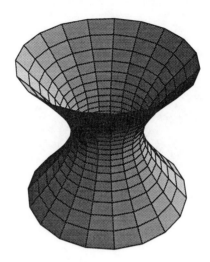

Figure 3. 3   Portion of the hyperboloid H^2_{-1}

The n-**hyperboloid** $H^n_c \subset R^{n+1}$ is defined as

$$\{x = (x_0, \ldots, x_n) : h(x) = x_0^2 - \ldots - x_n^2 - c = 0\}. \tag{3. 8}$$

For $c \neq 0$, H^n_c is an n-dimensional submanifold of R^{n+1}, as the reader may verify. A truncated picture of H^2_{-1} is shown above; this shows

$$\{(x_0, x_1, x_2) : x_0^2 - x_1^2 - x_2^2 + 1 = 0, -1.5 \leq x_0 \leq 1.5\}.$$

3.2.5   Example: Graph of a Smooth Function

The **graph of a smooth function** f from an open set W in R^2 into R gives a 2-dimensional submanifold $G = \{(x, y, f(x, y)) : (x, y) \in W\} \subseteq R^3$; this is because the function $g(x, y, z) = z - f(x, y)$ is a submersion on $W \times R$. In the following picture, W is the annulus $\{(x, y) \in R^2 : 1 < x^2 + y^2 < 2\}$, which is shown along with G.

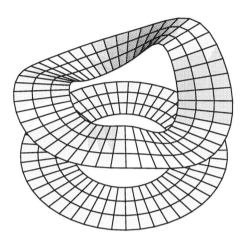

Figure 3. 4   Graph of a function on an annulus in R^2

3.2.6 A Nonexample of a Submanifold

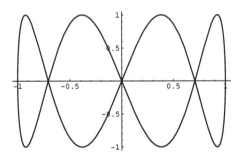

Figure 3. 5   A curve that is not a submanifold of R^2

Let $P = \{ (\sin t, \sin 4t) \in R^2 : 0 \le t \le 2\pi \}$, which can also be expressed using the addition formulas as $P = g^{-1}(0)$, where

$$g(x, y) = y^2 - 16x^2 (1 - x^2) (1 - 2x^2)^2. \tag{3.9}$$

A picture of P is presented Figure 3.5.

Warning! Showing that g fails to be a submersion at $(0, 0)$ is not sufficient to prove that P is not a submanifold, because it does not address the possibility that there might be some other submersion f on a neighborhood U of $(0, 0)$ such that $P \cap U = f^{-1}(0)$. A correct argument to show that P is not a 1-dimensional submanifold of R^2 is to note that if f is a smooth function on any neighborhood U of $(0, 0)$, or of $(\pm 1 / \sqrt{2}, 0)$, with

$P \cap U = f^{-1}(0)$, then there are two tangent directions in R^2 along which f is zero; hence f cannot be a submersion. The reader may work out the details of this argument in Exercise 2.

3.3   Exercises

1. Determine whether each of the following functions is an immersion and/or a submersion,[1] for the domain given:

    (i) $f(u, v) = (u/v, u, u^2)$; domain $(0, \infty) \times (0, \infty)$.

    (ii) $f(u, v, w) = (uv, vw, wu)$; domain $(0, \infty) \times (0, \infty) \times (0, \infty)$.

    (iii) $f(u, v, w, z) = (uv - wz, u - z)$; domain $(0, \infty) \times (0, \infty) \times (0, \infty) \times (0, \infty)$.

2. Suppose M is made up of the edges of a nondegenerate triangle in R^2. Prove that M is not a 1-dimensional submanifold of R^2.

    Hint: Suppose that f is a smooth function on a neighborhood U of one of the vertices x such that $f^{-1}(0) = U \cap M$. By taking directional derivatives of f in the directions of the two edges meeting at x, show that f cannot be a submersion.

3. Which of the following curves, pictured below, are 1-dimensional submanifolds of R^2? Justify your answers carefully.

    (i) The graph $M = \{ (x, y) : y^2 = x^2 - x^4 \} = \{ (\sin t, \sin 2t) : 0 < t < 2\pi \}$.

    (ii) The circle $\{ (x, y) : x^2 + y^2 = 1 \}$ without the point $(0,1)$.

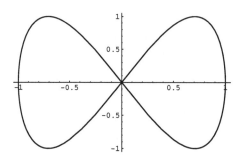

---

[1] A mapping which is both an immersion and a submersion is called a **regular mapping**; obviously domain and range must have the same dimension, and for the map to be regular means that its derivative is a nonsingular matrix at every point.

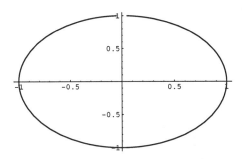

4. Show that the torus T^n, as in (3. 6), is an n-dimensional submanifold of R^{2n}.

3.4 Parametrizations

This notion is a generalization of the notions of parametrized curve and parametrized surface, which the reader may have encountered in multivariable calculus. Suppose M is a subset of R^{n+k}. An **n-dimensional parametrization** of M means a one-to-one immersion $\Psi: W \to U \subseteq R^{n+k}$, where W is an open subset of R^n, U is an open subset of R^{n+k} with $U \cap M \neq \varnothing$, and $\Psi(W) = U \cap M$. If $y \in U \cap M$, we refer to this as an n-dimensional parametrization of M at y, to emphasize that only part of M is actually being parametrized. The image of a 1-dimensional parametrization is called a **parametrized curve**, and the image of a 2-dimensional parametrization is called a **parametrized surface**.

For example $\theta \to (\cos\theta, \sin\theta)$, with domain $(0, 2\pi)$, and range $U = R^2 - (1, 0)$, gives a 1-dimensional parametrization of the unit circle in R^2.

3.4.1 Caution! The Image of a Parametrization Need Not Be a Submanifold[2]

The standard counterexample is the "figure eight" $M = \{(x, y) : y^2 = x^2 - x^4\}$, pictured in Exercise 3. This is not a submanifold of R^2, but it is the image of the parametrization $\Psi(t) = (\sin t, \sin 2t)$, with domain $(0, 2\pi)$. The reader may check that the derivative of Ψ never vanishes.

---

[2] Note that some other authors use a stronger definition of parametrization, so that it is an embedding in the sense of Chapter 5. However, we have chosen the present definition for the sake of consistency with multivariable calculus notions of "parametrized surface," etc.

3.4.2   Some Standard Examples

Note that there is no systematic method of deriving a parametrization; it is simply a knack that one acquires through practice. Here are three standard examples of parametrizations.

3.4.2.1   Parametrization of the 2-Sphere

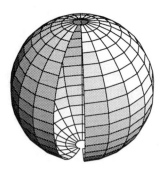

Figure 3. 6   Spherical coordinate parametrization with domain $W = (0.2, 6.08) \times (0.1, 3.04)$

Everyone is familiar with the description of points on a sphere using angles of latitude (i.e., north of the equator) and of longitude (i.e., east of Greenwich). To formalize this into the customary "spherical coordinate" parametrization, take linear transformations of latitude and longitude to obtain new angular variables θ and ϕ, and take

$$\Psi (\theta, \phi) = (\cos\theta \sin\phi, \, \sin\theta \sin\phi, \, \cos\phi), \tag{3. 10}$$

with domain $W = (0, 2\pi) \times (0, \pi)$, and range $U = R^3 - \{ (x, 0, z) : x \ge 0 \}$; that is, U is R^3 with a half–plane deleted. Then Ψ is a parametrization of S^2. It is important to note that the image of Ψ does not cover S^2 (see Figure 3.6); several parametrizations using (3. 10) with different choices of W collectively cover S^2 however.

3.4.2.2   Parametrization of the 2-Hyperboloid

Every point in H^2_{-1} is contained in the image of one or the other of the parametrizations

$$\Psi_i : W_i = (-\infty, \infty) \times (i\pi, (2 + i)\pi) \to R^3, \, i = 0, 1, \tag{3. 11}$$

$$\Psi_i (t, \theta) = (\sinh t, \, \cosh t \cos\theta, \, \cosh t \sin\theta).$$

3.4.2.3   Parametrization of the 2-Torus

Following the pattern of the last two examples, we could use maps of the form $(\theta, \phi) \to (\cos\theta, \sin\theta, \cos\phi, \sin\phi)$, for suitable domain, to parametrize $T^2 \subset R^4$, but this is not very easy to visualize. For any $a > 1$, composing this map with

$$(x_1, x_2, x_3, x_4) \to (y_1, y_2, y_3) = ((a + x_1) x_3, \, (a + x_1) x_4, \, x_2)$$

gives the usual representation (Figure 3. 7) of $T^2 \subseteq R^3$, parametrized by maps of the form

$$\Psi(\theta, \phi) = ((a + \cos\theta)\cos\phi, (a + \cos\theta)\sin\phi, \sin\theta). \qquad (3.12)$$

Figure 3. 7   The 2-torus represented as a subset of R^3

3.4.3 Implicit Function Parametrizations

An n-dimensional **implicit function parametrization** means a parametrization $\Psi: W \subseteq R^n \to U \subseteq R^{n+k}$ of the special form

$$\Psi(x_1, ..., x_n) = (x_1, ..., x_n, h_1(x), ..., h_k(x)), \qquad (3.13)$$

where $x = (x_1, ..., x_n)$, where possibly the entries on the right side are rearranged; here the function h, called the **implicit function** in the parametrization, is necessarily a smooth function from W to an open set $V \subseteq R^k$, and unless otherwise stated we take $U = W \times V$. For example, a 2-dimensional implicit function parametrization at $(0, 0, 1)$ of the sphere

$$S^2 = \{(x, y, z) : x^2 + y^2 + z^2 = 1\}$$

is given by

$$\Psi(x, y) = (x, y, \sqrt{1 - x^2 - y^2})$$

with domain $W = \{(x, y) \in R^2 : x^2 + y^2 < 1\}$ and range $W \times (0, \infty)$. Another such parametrization at $(0, -1, 0)$, with suitable range and domain, is given by

$$\Psi(x, z) = (x, -\sqrt{1 - x^2 - z^2}, z).$$

3.5   Using the Implicit Function Theorem to Parametrize a Submanifold

The Implicit Function Theorem will be used to show that submanifolds of R^{n+k} always have a parametrization at every point; indeed this theorem is precisely the assertion that an implicit function parametrization exists at every point of a submanifold. The starting point of our study will be the Inverse Function Theorem.

3.5.1   Inverse Function Theorem

Suppose U is open in R^m, $F: U \to R^m$ is smooth, $z \in U$, and $DF(z)$ is nonsingular. Then there exists a neighborhood $U' \subseteq U$ of z on which F is a diffeomorphism onto its image; in other words, $F|_{U'}$ is 1-1 with a smooth inverse.

Proof Omitted: Many advanced calculus books prove this theorem. A proof valid for infinite dimensions is given in Lang [1972].

3.5.2   Geometric Formulation of the Implicit Function Theorem

At every point in an n-dimensional submanifold of R^{n+k} there exists an n-dimensional implicit function parametrization.

For example, when $n = 2, k = 1$, the theorem says that around any point in a 2-dimensional submanifold of R^3, the submanifold can be expressed locally as a graph of the form $z = f(x, y)$, or $x = f(y, z)$, or $y = f(x, z)$, for some smooth function f.

3.5.3   Elaborated Version

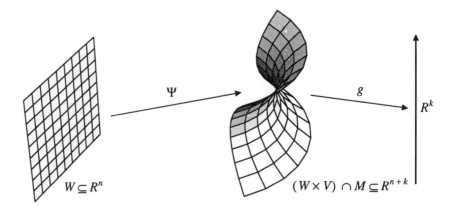

Figure 3. 8   Parametrization of a submanifold

Suppose M is a subset of R^{n+k}, and $g: U \to R^k$ is a submersion on some open subset U of R^{n+k} such that $U \cap M = g^{-1}(0)$. Given $y \in U \cap M$, we can rearrange the labeling of the coordinates in R^{n+k} in such a way that there exist open sets $W \subseteq R^n$ and $V \subseteq R^k$ with $y \in W \times V \subseteq U$, and a smooth function $h: W \to V$ (the "implicit function") such that the map $\Psi: W \to W \times V$ given by

$$\Psi(x) = (x, h(x)) \tag{3.14}$$

is an n-dimensional parametrization of M at y; that is, Ψ is a one-to-one immersion with $\Psi(W) = (W \times V) \cap M$.

Proof (may be omitted): The $k \times (n+k)$ derivative matrix $Dg(y)$ has rank k since g is a submersion. By relabeling the coordinates in R^{n+k} if necessary, we can arrange matters so that the last k columns of $Dg(y)$ are linearly independent. If we express an element of $R^{n+k} \cong R^n \times R^k$ in the new labeling system as (x, z), where $x \in R^n, z \in R^k$, then we may group these last k columns of $Dg(y)$ into a $k \times k$ matrix $D_z g(y)$; the first n columns of $Dg(y)$ are denoted $D_x g(y)$.

Consider the map $F: U \to R^n \times R^k$ given by $F(x, z) = (x, g(x, z))$; differentiating gives

$$DF(x, z) = \begin{bmatrix} I & 0 \\ D_x g & D_z g \end{bmatrix} (x, z),$$

and therefore

$$|DF(y)| = |D_z g(y)| \neq 0.$$

Applying the Inverse Function Theorem 3.5.1, we see that there exists a neighborhood U' of $y = (x_0, z_0)$ with $U' \subseteq U$, such that the restriction of F to U' is a diffeomorphism onto its image $U'' = F(U') \ni (x_0, 0)$. Take open sets $W \subseteq R^n$ and $V \subseteq R^k$ with $W \times \{0\} \subseteq U''$ and $y \in W \times V \subseteq U'$, and define $h: W \to V$ by the equation

$$\Psi(x) = (x, h(x)) = F^{-1}(x, 0). \tag{3.15}$$

We see that Ψ is a one-to-one immersion because F^{-1} is a one-to-one immersion on U''. For every $(x, z) \in (W \times V) \cap M$, we have $g(x, z) = 0$, and therefore $F(x, z) = (x, 0)$; thus $(x, z) = \Psi(x)$ by (3.15).   ¤

3.5.3.1 Corollary
For g and Ψ as above, $\operatorname{Im} d\Psi(x) = \operatorname{Ker} dg(\Psi(x))$ [3] for all $x \in W$.

---

[3] The "kernel" $\operatorname{Ker} dg(y) = \{\xi \in T_y R^{n+k} : dg(y)\xi = 0\}$.

Proof: Apply the Chain Rule to the function $g \bullet \Psi$, which is identically zero on W. We obtain $dg\,(\Psi\,(x)) \bullet d\Psi\,(x) = 0$, which implies $\operatorname{Im} d\Psi\,(x) \subseteq \operatorname{Ker} dg\,(\Psi\,(x))$. However, since Ψ is an immersion and g is a submersion, the vector subspaces $\operatorname{Im} d\Psi\,(x)$ and $\operatorname{Ker} dg\,(\Psi\,(x))$ of $T_{\Psi\,(x)}R^{n+k}$ satisfy

$$\dim\,(\operatorname{Im} d\Psi\,(x)) = n = \dim\,(\operatorname{Ker} dg\,(\Psi\,(x)));$$

therefore they must be identical.   ¤

The notion of the tangent plane to a surface is probably familiar from multivariable calculus. Although one may continue to visualize a "tangent space at a point y" in a submanifold M as if it were actually resting against the submanifold at y, the mathematical definition below of a tangent space implies that it is always a subspace of R^{n+k}, that is, it passes through zero and not necessarily through y. The main idea is that this tangent space at y is something intrinsic to the manifold, and not dependent on the choice of submersion or parametrization used to describe the submanifold around the point y.

3.5.4   The Tangent Space to a Submanifold at a Point

If M is an n-dimensional submanifold of R^{n+k}, and if $f: U \to R^k$ and $g: U' \to R^k$ are two submersions defined on neighborhoods of $y \in M$ such that $U \cap M = f^{-1}\,(0)$ and $U' \cap M = g^{-1}\,(0)$, then the kernel of $df\,(y)$ and the kernel of $dg\,(y)$ are the same n-dimensional subspace of T_yR^{n+k}; in symbols,

$$\operatorname{Ker} df(y) = \{\xi \in T_yR^{n+k}: df(y)\,\xi = 0\} = \operatorname{Ker} dg\,(y) \subseteq T_yR^{n+k}. \qquad \textbf{(3. 16)}$$

This subspace is called the **tangent space** to M at y, denoted T_yM; it may also be expressed as $\operatorname{Im} d\Psi\,(u) = \operatorname{Span}\{D_1\Psi\,(u),\,...,\,D_n\Psi\,(u)\}$, where $u = \Psi^{-1}\,(y)$.

Proof: Take an implicit function parametrization Ψ associated to f as in the proof of 3.5.3. By the reasoning found in the proof of 3.5.3.1, we see that, since $f \bullet \Psi$ and $g \bullet \Psi$ are identically zero on their respective domains, we have

$$\operatorname{Ker} df(y) = \operatorname{Im} d\Psi\,(\Psi^{-1}\,(y)) = \operatorname{Ker} dg\,(y)$$

as desired. The final assertion follows from 3.5.3.1.   ¤

3.5.4.1   Example of Computing a Tangent Space
"Find the tangent space to the torus $T^2 \subset R^4$ at the point $\bar{x} = (0, 1, 0.6, 0.8)$."

Solution: Recall from (3. 6) that $T^2 = g^{-1}\,(0)$ for the submersion

$$g\,(x_1, x_2, x_3, x_4) = (x_1^2 + x_2^2 - 1,\, x_3^2 + x_4^2 - 1).$$

One may easily compute that

$$Dg(\bar{x}) = \begin{bmatrix} 0 & 2 & 0 & 0 \\ 0 & 0 & 1.2 & 1.6 \end{bmatrix}.$$

The tangent space at $\bar{x} = (0, 1, 0.6, 0.8)$ is the kernel of this linear transformation, that is,

$$T_{\bar{x}}M = \{(v_1, v_2, v_3, v_4) : v_2 = 0, 1.2v_3 + 1.6v_4 = 0\}.$$

Alternative Solution: In the parametrization $\Psi: (\theta, \phi) \to (\cos\theta, \sin\theta, \cos\phi, \sin\phi)$, we have

$$D\Psi(\theta, \phi) = (\frac{\partial\Psi}{\partial\theta}, \frac{\partial\Psi}{\partial\phi}) = \begin{bmatrix} -\sin\theta & 0 \\ \cos\theta & 0 \\ 0 & -\sin\phi \\ 0 & \cos\phi \end{bmatrix}.$$

At the specific $(\bar{\theta}, \bar{\phi})$ whose image under Ψ is $\bar{x} = (0, 1, 0.6, 0.8)$, the second description of the tangent space shows that

$$T_{\bar{x}}M = \text{Span }\{\frac{\partial\Psi}{\partial\theta}(\bar{\theta}, \bar{\phi}), \frac{\partial\Psi}{\partial\phi}(\bar{\theta}, \bar{\phi})\} = \text{Span }\left\{\begin{bmatrix} -1 \\ 0 \\ 0 \\ 0 \end{bmatrix}, \begin{bmatrix} 0 \\ 0 \\ -.8 \\ 0.6 \end{bmatrix}\right\},$$

which is the same result as before. ¤

The next assertion is important because it relates n-dimensional submanifolds of R^{n+k} to the abstract manifolds discussed in Chapter 5. It says that, when we "change variables" from one parametrization to another, this change is always smooth.

3.5.5 Switching between Different Parametrizations

Suppose M is an n-dimensional submanifold of R^{n+k}, U_1 and U_2 are neighborhoods of a point y in M, and $\Psi_i: W_i \subseteq R^n \to U_i \subseteq R^{n+k}$ are n-dimensional parametrizations of M at y, for $i = 1, 2$, such that $U_1 \cap U_2 \cap M$ is nonempty. Then both the domain and the range of the map

$$\Psi_2^{-1} \bullet \Psi_1 : W_1 \cap \Psi_1^{-1}(U_2) \to W_2 \cap \Psi_2^{-1}(U_1) \tag{3.17}$$

are open sets in R^n, and the map itself is a diffeomorphism (see Figure 3.9).

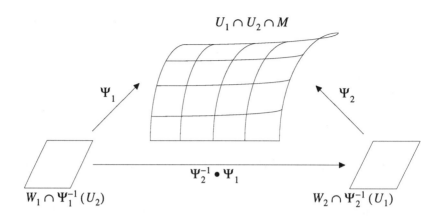

Figure 3. 9   Switching from one parametrization to another

Proof (may be omitted): Since Ψ_1 is continuous and U_2 is open, $\Psi_1^{-1}(U_2)$ is open; therefore so is $W_1 \cap \Psi_1^{-1}(U_2)$, since a finite intersection of open sets is open and W_1 is open. The openness of $W_2 \cap \Psi_2^{-1}(U_1)$ follows similarly. By definition, Ψ_1 and Ψ_2 are 1-1 maps, and therefore the map in (3. 17) is well defined and 1-1. It remains to prove that it is smooth at an arbitrary point $\bar{x}$ in the domain. Let $\bar{y} = \Psi_1(\bar{x}) \in M$. According to the proof of 3.5.2, there exists a local parametrization $\Phi: V \subseteq R^n \to U_0 \cap M \subseteq R^{n+k}$ of M at y of the form (possibly after rearranging variables)

$$\Phi(u) = (u, h(u)),$$

as in (3. 14). Note that the inverse of Φ is the smooth map which projects a vector in R^{n+k} onto its first n components. Therefore if Q denotes $U_0 \cap U_1 \cap U_2 \cap M$, then $\Phi^{-1} \bullet \Psi_2 : \Psi_2^{-1}(Q) \to \Phi^{-1}(Q)$ is a smooth map, and $D(\Phi^{-1} \bullet \Psi_2)(\bar{x})$ is nonsingular because, by the chain rule,

$$D(\Phi^{-1} \bullet \Psi_2)(\bar{x}) = D\Phi^{-1}(\bar{y}) \bullet D\Psi_2(\bar{x}),$$

and both the matrices on the right are of rank n. By the Inverse Function Theorem, the inverse $(\Phi^{-1} \bullet \Psi_2)^{-1} = \Psi_2^{-1} \bullet \Phi$ is a diffeomorphism on a neighborhood of $\Phi(\bar{y})$, and consequently

$$\Psi_2^{-1} \bullet \Psi_1 = (\Psi_2^{-1} \bullet \Phi) \bullet (\Phi^{-1} \bullet \Psi_1)$$

is smooth on a neighborhood of $\bar{x}$ being the composite of two smooth functions.   ¤

3.6   Matrix Groups as Submanifolds

In the sequel, we shall use the notation $R^{n \times m}$ to denote the vector space of real $n \times m$ matrices, which may be identified with R^{nm}. Some of the most important examples of submanifolds of Euclidean spaces are certain infinite groups of matrices known as Lie groups,[4] under the operation of matrix multiplication. Our first example is:

3.6.1 The General Linear Group over R

The **general linear group** $GL_n(R) \subset R^{n \times n}$, and its subgroup $GL_n^+(R)$, consist of non-singular matrices; they are defined by

$$GL_n(R) = \{A \in R^{n \times n} : |A| \neq 0\}, \, GL_n^+(R) = \{A \in R^{n \times n} : |A| > 0\}. \qquad \text{(3. 18)}$$

Since the map $A \to |A|$ is continuous, both of these groups are open subsets of $R^{n \times n}$, and hence are n^2-dimensional submanifolds of $R^{n \times n}$. An important fact about the general linear group is the following:

3.6.2 Smoothness of Matrix Multiplication and Inversion

The maps $(A, B) \to AB$ and $A \to A^{-1}$ are smooth on $GL_n(R) \times GL_n(R)$ and $GL_n(R)$, respectively.

Proof: Matrix multiplication must be smooth, because it consists only of a sequence of multiplications and additions of matrix entries, and polynomials are smooth functions. As for inversion, we note that for any $A \in GL_n(R)$ and any $H \in R^{n \times n}$, and for all sufficiently small ε, $A - \varepsilon H \in GL_n(R)$, and the series expansion below converges

$$(A - \varepsilon H)^{-1} = (I - \varepsilon A^{-1} H)^{-1} A^{-1}$$

$$= (I + \varepsilon A^{-1} H + (\varepsilon A^{-1} H)^2 + \dots) A^{-1};$$

here I denotes the $n \times n$ identity matrix, and convergence of a series of matrices refers to convergence of every entry. This gives an infinite-order Taylor expansion for the map $A \to A^{-1}$, which verifies smoothness.   ¤

3.6.3 Lie Subgroups of $GL_n(R)$

The study of abstract Lie groups is beyond the scope of this book. However, we say that G is a **Lie subgroup of the general linear group** if the following two conditions hold:

---

[4] Named after the Norwegian mathematician Sophus Lie (1842–99).

- G is an algebraic subgroup of $GL_n(R)$, that is, the inverse of every matrix in G and the product of two matrices in G are in G;

- G is a submanifold of $R^{n \times n}$.

Examples include the following.

- The **special linear group** $SL_n(R) = \{A \in R^{n \times n} : |A| = 1\}$; this is an algebraic subgroup because the determinant of the product of two matrices of determinant 1 also has determinant 1; it is a submanifold of dimension $n^2 - 1$ because it is the inverse image of 1 under the map $A \to |A|$, which is a submersion (proof suggested in Exercise 11).

- The **orthogonal group** $O(n) \subset R^{n \times n}$, and its subgroup the **special orthogonal group** $SO(n)$, consist of orthogonal matrices; they are defined by

$$O(n) = \{A \in R^{n \times n} : A^T A = I\}, \tag{3.19}$$

$$SO(n) = \{A \in R^{n \times n} : |A| = 1, A^T A = I\}, \tag{3.20}$$

where A^T denotes A transpose. Of course, $SO(n) = O(n) \cap GL_n^+(R)$. It is clear that the orthogonal and special orthogonal groups are algebraic subgroups of $GL_n(R)$; to show that they are Lie subgroups of $GL_n(R)$, we need:

3.6.4 The (Special) Orthogonal Group Is a Submanifold

$O(n)$ and $SO(n)$ are $n(n-1)/2$-dimensional submanifolds of $R^{n \times n}$.

Proof: We shall carry out the proof for $SO(n)$; Since a similar argument applies to matrices in $O(n)$ with determinant of -1, it follows that $O(n)$ is an $n(n-1)/2$-dimensional submanifold of $R^{n \times n}$ also.

Let $Sym(n)$ denote the symmetric $n \times n$ matrices, that is, $\{B \in R^{n \times n} : B = B^T\}$; this is a vector space of dimension $n(n+1)/2$. Define

$$f : GL_n^+(R) \to Sym(n), f(A) = A^T A - I. \tag{3.21}$$

Recall that $GL_n^+(R)$ is open in $R^{n \times n}$, and note that $SO(n) = f^{-1}(0)$. Hence the result follows if we can show that f is a submersion. Observe that if $\|H\|$ denotes the square root of the sum of squares of entries in the matrix H, and if we write a function $g(x)$ as $o(x)$ to mean that $g(x)/x \to 0$ as $x \to 0$, then

$$f(A+H) - f(A) = (A+H)^T(A+H) - A^T A = H^T A + A^T H + o(\|H\|);$$

$$\therefore Df(A)H = H^T A + A^T H.$$

Since the tangent space to $Sym(n)$ at $B \in Sym(n)$ is just another copy of $Sym(n)$, to show that df is onto it suffices to show that, for every $S \in Sym(n)$ and every $A \in SO(n)$, there exists an H such that $Df(A)H = S$. Choosing $H = AS/2$ gives

$$Df(A)H = (1/2)(SA^TA + A^TAS) = S,$$

which verifies that $Df(A)$ is onto provided $A \in SO(n)$. Moreover it is also of full rank on a neighborhood of A, on which f is therefore a submersion. Since f maps from an open subset of an n^2-dimensional space into an $n(n+1)/2$-dimensional space, it follows that the dimension of $SO(n) = f^{-1}(0)$ is $n(n-1)/2$.   ¤

3.7   Groups of Complex Matrices

Since a complex number may be regarded as a pair of real numbers, the group of nonsingular $n \times n$ matrices with complex entries, or **complex general linear group**, denoted

$$GL_n(C) = \{A \in C^{n \times n} : |A| \neq 0\},$$

can be regarded as an open subset of $R^{n \times n} \oplus R^{n \times n}$, with the associated differentiable structure; here we use the fact that

$$A \in C^{n \times n} \Leftrightarrow A = X + iY, \ X, Y \in R^{n \times n}. \tag{3.22}$$

As in the real case, matrix multiplication and inversion are smooth operations on $GL_n(C) \times GL_n(C)$ and $GL_n(C)$, respectively, and algebraic subgroups of $GL_n(C)$ that are also submanifolds of $R^{n \times n} \oplus R^{n \times n}$ are called Lie subgroups of the complex general linear group. Two of the most important of these are:

- The **complex special linear group**, $SL_n(C) = \{A \in C^{n \times n} : |A| = 1\}$, which is a submanifold of dimension $2n^2 - 2$; for the proof, see Warner [1983].

- The **unitary group**, $U(n) = \{A \in C^{n \times n} : A\overline{A}^T = I\}$, where $\overline{A}$ means the matrix in which every entry of A is replaced by its complex conjugate; this is a submanifold of dimension n^2 (see 3.7.1).

- The **special unitary group**, $SU(n) = U(n) \cap SL_n(C)$, which is a submanifold of dimension $n^2 - 1$; this fact will be proved in the exercises of Chapter 5.

3.7.1 The Unitary Group Is a Submanifold

$U(n)$ is an n^2-dimensional submanifold of $R^{n \times n} \oplus R^{n \times n}$.

Proof: This follows along the same lines as 3.6.4. Let $Sym(n)$ denote the real symmetric $n \times n$ matrices, and let $o(n)$ denote the real antisymmetric $n \times n$ matrices (this is not the same as the "little o" notation in the proof of 3.6.4!); these are vector spaces of dimension $n(n+1)/2$ and $n(n-1)/2$, respectively. Let

$$Sym\,(n) \oplus (\sqrt{-1})\,o\,(n) = \{S + iW \in C^{n \times n} : S \in Sym\,(n),\, W \in o\,(n)\,\}, \qquad \text{(3.23)}$$

which is an n^2-dimensional space of matrices. Define a map

$$f : R^{n \times n} \oplus R^{n \times n} \to Sym\,(n) \oplus (\sqrt{-1})\,o\,(n);$$

$$f(X, Y) = (X - iY)^T (X + iY) - I.$$

(The reader should check that the real part of the image of f is symmetric, and the complex part antisymmetric.) Observe that $f^{-1}(0)$ is identifiable as $U(n)$ under the identification between $(X, Y) \in R^{n \times n} \oplus R^{n \times n}$ and $X + iY \in C^{n \times n}$. A similar calculation to the one in the proof of 3.6.4 shows that

$$Df(X, Y)\,(H, K) = (H - iK)^T (X + iY) + (X - iY)^T (H + iK). \qquad \text{(3.24)}$$

As before, it suffices to show that f is a submersion on the set where

$$(X - iY)^T (X + iY) - I = 0. \qquad \text{(3.25)}$$

Given a specific (X, Y) satisfying (3.25), and an arbitrary $S \in Sym\,(n)$, $W \in o\,(n)$, we choose H and K by the equation:

$$H + iK = \frac{(X + iY)\,(S + iW)}{2}.$$

Using (3.24), (3.25), and the fact that $S^T = S$, $W^T = -W$, the reader may easily verify that for this choice of H and K, $Df(X, Y)\,(H, K) = S + iW$, which verifies that $Df(X, Y)$ is onto as desired. Finally, the dimension of $U(n)$ must be

$$\dim\,(R^{n \times n} \oplus R^{n \times n}) - \dim\,(Sym\,(n) \oplus (\sqrt{-1})\,o\,(n)) = 2n^2 - n^2,$$

giving n^2 as claimed. ¤

3.8   Exercises

5. (i) Verify that H_c^n is an n-dimensional submanifold of R^{n+1} for $c \neq 0$, but is not a submanifold when $c = 0$.

   (ii) Sketch H_1^2, and find a parametrization for H_1^2.

6. Verify that the mapping $\Psi\,(\theta, \phi) = (\,(a + \cos\theta)\cos\phi,\, (a + \cos\theta)\sin\phi,\, \sin\theta)$, used to parametrize the 2-torus in R^3, is indeed an immersion for $a > 1$. Use it to calculate the tangent space to $T^2 \subset R^3$ at the point $(-0.6\,(a + 1),\, -0.8\,(a + 1),\, 0)$.

7. Let $W = \{ (u, \theta) : -1 < u < 1, 0 < \theta < 2\pi \}$.

(i) Verify that $\Phi (u, \theta) = ((1 + u^2) \cos \theta, (1 + u^2) \sin \theta, u)$ is a 2-dimensional parametrized surface obtained by rotating the graph of $x = 1 + z^2$ about the z-axis in R^3.

(ii) Show that $\Psi (u, \theta) = (u^2 \cos \theta, (1 + u^2) \sin \theta, u)$ fails to be an immersion at $(u, \theta) = (0, \pi/2)$.

8. The graph of the smooth function $F : W \subseteq R^n \to V \subseteq R^k$, where W and V are open sets, is the set

$$M = \{ (x, z) \in R^n \times R^k : z = h (x) \} \subset R^{n+k}. \tag{3.26}$$

Show that M is an n-dimensional submanifold of R^{n+k}, and calculate its tangent space at an arbitrary point (x_0, z_0).

Hint: Consider the function $g (x, z) = z - h (x)$.

9. Suppose $\Psi : (a, b) \subset R \to R^2$ is a one-to-one immersion, so $C = \Psi ((a, b))$ is a parametrized curve in R^2 (**Warning!** C is not necessarily a 1-dimensional submanifold of R^2; see 3.4.1). Prove that, for every $t \in (a, b)$, there exists a subinterval (a', b') with $a \leq a' < t < b' \leq b$, such that $C' = \Psi ((a', b'))$ is a 1-dimensional submanifold of R^2.

Hint: Apply the Implicit Function Theorem 3.5.3 to the submersion

$$g (x, y, s) = (\Psi_1 (s) - x, \Psi_2 (s) - y)$$

to construct an implicit function parametrization of C of the form $y = h (x)$ or $x = h (y)$ near $\Psi (t)$; now $g (x, y) = y - h (x)$ or $x - h (y)$ is a submersion with which we can complete the proof.

10. (Continuation) Extend the method of Exercise 9 to show that, given a one-to-one immersion $\Psi : W \subseteq R^n \to R^{n+k}$, and given any $u \in W$, there exists a neighborhood W' of u in W such that $M = \Psi (W')$ is an n-dimensional submanifold of R^{n+k}.

11. (i) Prove that $SL_2 (R)$ is a 3-dimensional submanifold of $GL_2^+ (R)$, and calculate the tangent space to $SL_2 (R)$ in $R^{2 \times 2}$ at the identity.

Hint: Consider $f (x, y, z, w) = xw - yz$.

(ii) By considering $|A + H| - |A|$ when $H = \varepsilon A$, prove that $SL_n (R)$ is a submanifold of $GL_n^+ (R)$ of dimension $n^2 - 1$.

12. Construct a 3-dimensional parametrization for $SO (3)$ at the identity I using the following steps:

(i) Denote the set of skew-symmetric $n \times n$ matrices by

$$o(n) = \{K \in R^{n \times n}: K = -K^T\}. \tag{3.27}$$

Define the **Cayley transform** of $K \in o(n)$ to be $\Psi(K) = (I+K)(I-K)^{-1}$; note that $(I-K)^{-1}$ is certainly well defined for K in a neighborhood of 0, since the mapping $A \to A^{-1}$ is smooth.[5] Prove that if $B = \Psi(K)$, then B is orthogonal, that is, $B^T B = I$.

Hint: Use the fact that $(I+K)(I-K) = (I-K)(I+K)$, and that the inverse of A^T is the transpose of A^{-1}.

(ii) By comparing $\Psi(K+H)$ with $\Psi(K)$, find $D\Psi(K)H$ and show that Ψ is an immersion, at least on a neighborhood of 0.

Hint: This is similar to the calculation in the proof of 3.6.4.

(iii) By finding a formula for K in terms of $B = \Psi(K)$, show that Ψ^{-1} is well defined, at least on $U \cap SO(n)$, where U is a neighborhood of I in $R^{n \times n}$, and therefore if $W = \Psi^{-1}(U \cap SO(n))$, then $\Psi: W \to U \cap SO(n)$ is one-to-one and onto.

(iv) Now verify that the following composite mapping, restricted to a neighborhood of 0 in R^3, yields a 3-dimensional parametrization for $SO(3)$ at the identity:

$$(u, v, w) \in R^3 \to \begin{bmatrix} 0 & u & v \\ -u & 0 & w \\ -v & -w & 0 \end{bmatrix} \to \begin{bmatrix} 1 & u & v \\ -u & 1 & w \\ -v & -w & 1 \end{bmatrix} \begin{bmatrix} 1 & -u & -v \\ u & 1 & -w \\ v & w & 1 \end{bmatrix}^{-1}.$$

13. Show that the tangent space to $SO(3)$ in $R^{3 \times 3}$ at the identity can be identified with $o(3)$ (as defined in (3.27)).

14. Let I_n denote the $n \times n$ identity matrix. The **symplectic group** $Sp(2n)$ is defined by

$$Sp(2n) = \{A \in R^{2n \times 2n}: A^T JA = J\}, J = \begin{bmatrix} 0 & I_n \\ -I_n & 0 \end{bmatrix}.$$

(i) Prove that $Sp(2n)$ is a submanifold of $R^{2n \times 2n}$, and find its dimension.

Hint: Note that the map $A \to A^T JA - J$ takes values in the set $o(2n)$ of $2n \times 2n$ skew-symmetric matrices, that is, the matrices $\{K \in R^{2n \times 2n}: K^T = -K\}$.

(ii) Show that, in the case $n = 1$, $Sp(2)$ is identical with $SL_2(R)$, the set of 2×2 matrices with determinant equal to 1.

15. The **Lorentz group** $O(n, 1)$ is the set of real $(n+1) \times (n+1)$ matrices that preserve the inner product $\langle x|x \rangle = -x_0^2 + x_1^2 + \ldots + x_n^2$; in other words,

---

[5] In fact $I - K$ is nonsingular for all $K \in o(n)$; see Lancaster and Tismenetsky [1985], p. 219.

$$O(n, 1) = \{A \in R^{(n+1) \times (n+1)} : A^T J A = J\}, J = \begin{bmatrix} -1 & 0 \\ 0 & I_n \end{bmatrix}, \qquad \text{(3. 28)}$$

where I_n denotes the $n \times n$ identity matrix. Show that $O(n, 1)$ is an m-dimensional submanifold of $R^{(n+1) \times (n+1)}$ for some m, and find m.

16. Prove that $SL_2(C)$ (i.e., the group of 2×2 matrices with complex entries and with determinant equal to 1) is a 6-dimensional submanifold of $R^{2 \times 2} \oplus R^{2 \times 2}$. Calculate the tangent space to $SL_2(C)$ at the identity.

3.9   Bibliography

A fuller treatment of submanifolds of R^n is given in Berger and Gostiaux [1988]. Of course, all general results about differential manifolds (see Chapter 5) apply in particular to this special case. Conversely, it turns out that any finite-dimensional manifold which can be covered by a countable number of charts can be expressed as a submanifold of R^n for sufficiently large n (Whitney's embedding theorem), and therefore there is only a loss of elegance, not of generality, in studying submanifolds of R^n instead of abstract manifolds. For more on both Implicit Function Theorems and on Lie groups, Warner [1983] is recommended. For a lively account of Lie groups and Lie algebras, with a lot of applications, see Sattinger and Weaver [1986]; for a more advanced treatment, see Helgason [1978].

4 Surface Theory Using Moving Frames

The method of the *repère mobile* (moving frame) – invented by G. Darboux and used extensively by Élie Cartan – is an extremely powerful technique in differential geometry, which we shall not discuss in full generality. Instead we shall focus mainly on surfaces in 3-space, with the intention of building intuition for the later work on connections in vector bundles.

4.1 Moving Orthonormal Frames on Euclidean Space

Recall from Chapter 3 the notion of the special orthogonal group, $SO(n)$. Suppose W is open in R^k, and $\Psi: W \to R^n$ is a one-to-one immersion, in other words a k-dimensional parametrization of $M = \Psi(W)$. A **moving orthonormal frame** on this parametrization (or, more loosely, a moving orthonormal frame on M) simply means a map

$$\Xi: W \to SO(n) \subset R^{n \times n} \tag{4.1}$$

that is smooth, considered as a map into $R^{n \times n}$. A special case of this is where W is an open subset of R^n and Ψ is the identity map, in which case we simply have a moving orthonormal frame on W. Let us give this a geometric interpretation. In terms of a fixed orthonormal basis for R^n, the matrix

$$\Xi(u) = \begin{bmatrix} \xi_1^1(u) & \cdots & \xi_n^1(u) \\ \cdots & \cdots & \cdots \\ \xi_1^n(u) & \cdots & \xi_n^n(u) \end{bmatrix}, u \in W, \tag{4.2}$$

is orthogonal, that is, $\Xi(u)^T \Xi(u) = I$, which is the same as saying that the column vectors

$$\vec{\xi}_i = [\xi_i^1(u), \dots, \xi_i^n(u)]^T, i = 1, 2, \dots, n, \tag{4.3}$$

satisfy $\langle \vec{\xi}_i | \vec{\xi}_j \rangle = \delta_{ij}$; in other words $\{\vec{\xi}_1(u), \dots, \vec{\xi}_n(u)\}$ forms an orthonormal basis for R^n. For the sake of understanding, take the case where W is an open subset of R^n and Ψ is the identity map, and consider these vectors as an orthonormal basis of the tangent space at x by identifying $\vec{\xi}_i$ with the vector field ξ_i, where

$$\xi_i(x) = \xi_i^1(x)\frac{\partial}{\partial x^1} + \dots + \xi_i^n(x)\frac{\partial}{\partial x^n}, \tag{4.4}$$

in terms of an orthonormal coordinate system $\{x^1, \dots, x^n\}$ for R^n. In brief, a moving orthonormal frame should be conceptualized as a smooth assignment of an orthonormal coordinate system for the tangent space at every point of U, as in Figure 4.1.

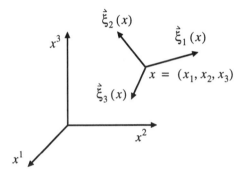

Figure 4.1   A moving orthonormal frame on R^3

Let $\vec{\Psi} = [x^1(u), \dots, x^n(u)]^T$ denote the position vector of the point $x = \Psi(u)$ as a function of $u \in W$. Applying exterior differentiation to each of the functions x^i, and expressing the results as a vector of 1-forms (or "vector-valued 1-form") gives

$$d\vec{\Psi} = [dx^1, \dots, dx^n]^T. \tag{4.5}$$

Define a column vector $\vec{\theta} = [\theta^1, \dots, \theta^n]^T$ of 1-forms by

$$\vec{\theta}(u) = \Xi(u)^T d\vec{\Psi}(u), \tag{4.6}$$

or $\vec{\theta} = \Xi^T d\vec{\Psi}$ for short. Since $\Xi(u)^T \Xi(u) = I$, it follows that $\Xi(u)^T = \Xi(u)^{-1}$, so $\Xi(u)\Xi(u)^T = I$. It follows that $d\vec{\Psi}$ can be expressed in terms of the orthonormal basis $\{\vec{\xi}_1(u), \dots, \vec{\xi}_n(u)\}$ as follows:

$$d\vec{\Psi} = \Xi(u)\Xi(u)^T d\vec{\Psi} = \Xi\vec{\theta}. \tag{4.7}$$

Likewise exterior differentiation of all n component functions of the basis vector $\vec{\xi}_i(u)$ yields a column vector $d\vec{\xi}_i$ of 1-forms; arrange these in the form of a matrix to obtain

$$d\Xi = [d\vec{\xi}_1, ..., d\vec{\xi}_n]. \tag{4.8}$$

Now define an $n \times n$ matrix of 1-forms,

$$\varpi = \begin{bmatrix} \omega_1^1 & \cdots & \omega_n^1 \\ \cdots & \cdots & \cdots \\ \omega_1^n & \cdots & \omega_n^n \end{bmatrix}, \tag{4.9}$$

known as **connection forms**, by the equation

$$\varpi = \Xi^T d\Xi = \begin{bmatrix} \vec{\xi}_1^T \\ \cdots \\ \vec{\xi}_n^T \end{bmatrix} [d\vec{\xi}_1, ..., d\vec{\xi}_n]; \tag{4.10}$$

or, in other words,

$$\omega_j^i = \vec{\xi}_i^T d\vec{\xi}_j = \langle \vec{\xi}_i | d\vec{\xi}_j \rangle. \tag{4.11}$$

We often write the equation defining ϖ in the format:

$$d\Xi = \Xi \varpi. \tag{4.12}$$

Since $\{ \vec{\xi}_1(u), ..., \vec{\xi}_n(u) \}$ is an orthonormal basis, the dot product satisfies $\langle \vec{\xi}_i | \vec{\xi}_j \rangle = \delta_{ij}$, and therefore

$$\omega_j^i + \omega_i^j = \langle \vec{\xi}_i | d\vec{\xi}_j \rangle + \langle d\vec{\xi}_i | \vec{\xi}_j \rangle = d\langle \vec{\xi}_i, \vec{\xi}_j \rangle = 0. \tag{4.13}$$

In other words, ϖ is skew-symmetric: $\varpi + \varpi^T = 0$.

4.2   The Structure Equations

The first and second structure equations are obtained by exterior differentiation[1] of the equations for $d\vec{\Psi}$ and $d\Xi$, respectively. Applying d to (4.7) gives:

---

[1] To reassure the reader about exterior differentiation of wedge products of matrix-valued differential forms, note that if $A = (a_{ij})$ is an $l \times m$ matrix of p-forms, and $B = (b_{jk})$ is an $m \times n$ matrix of q-forms, then

$$d(A \wedge B)_{ik} = d(\sum_j a_{ij} \wedge b_{jk}) = \sum_j da_{ij} \wedge b_{jk} + (-1)^p \sum_j a_{ij} \wedge db_{jk}.$$

$$0 = d(d\vec{\Psi}) = d(\Xi\vec{\theta}) = (d\Xi \wedge \vec{\theta}) + \Xi d\vec{\theta}. \tag{4.14}$$

Using (4.12) and grouping the terms on the right gives

$$\Xi\varpi \wedge \vec{\theta} + \Xi d\vec{\theta} = \Xi(\varpi \wedge \vec{\theta} + d\vec{\theta}) = 0. \tag{4.15}$$

Now premultiply both sides by Ξ^T, and use the identity $\Xi(u)^T\Xi(u) = I$, to obtain the **first structure equation:**

$$d\vec{\theta} = -\varpi \wedge \vec{\theta}. \tag{4.16}$$

Similarly, applying d to (4.12), and applying $d\Xi = \Xi\varpi$ again, gives

$$0 = d(d\Xi) = d\Xi \wedge \varpi + \Xi d\varpi = \Xi\varpi \wedge \varpi + \Xi d\varpi. \tag{4.17}$$

Now premultiplication by Ξ^T as before gives the **second structure equation:**

$$d\varpi = -\varpi \wedge \varpi. \tag{4.18}$$

For ease of reference, let us express the six important equations so far in the form of Table 4.1.

Definitions	Identities	Structure Equations
$d\vec{\Psi} = \Xi\vec{\theta}$	$\Xi^T\Xi = I$	$d\vec{\theta} = -\varpi \wedge \vec{\theta}$
$d\Xi = \Xi\varpi$	$\varpi + \varpi^T = 0$	$d\varpi = -\varpi \wedge \varpi$

Table 4.1   Equations for a moving orthonormal frame in Euclidean space

4.3 Exercises

1. Let $\tau: (a, b) \to R^2$ be a smooth curve parametrized by arc length, which means that $\|\tau'(s)\| = 1$ for all s. Let $\tau(s) = [\tau_1(s), \tau_2(s)]^T$, with exterior derivative

$$d\vec{\tau} = \begin{bmatrix} \tau_1' ds \\ \tau_2' ds \end{bmatrix}. \tag{4.19}$$

(i) Verify that

$$\vec{\xi}_1(s) = \tau'(s), \vec{\xi}_2(s) = \tau''(s)/\|\tau''(s)\| \tag{4.20}$$

defines a moving orthonormal frame on this parametrization (i.e., verify that these unit vectors are orthogonal).

Hint: Differentiate the identity $(\tau_1')^2 + (\tau_2')^2 = 1$ to show the orthogonality.

(ii) Show that $\theta^1 = ds$ and $\theta^2 = 0$.

(iii) Since the matrix of connection forms is skew-symmetric, it clearly takes the form:

$$\omega = \begin{bmatrix} 0 & -\mu \\ \mu & 0 \end{bmatrix} = \begin{bmatrix} 0 & \langle \vec{\xi}_1 | d\vec{\xi}_2 \rangle \\ \langle \vec{\xi}_2 | d\vec{\xi}_1 \rangle & 0 \end{bmatrix}. \tag{4.21}$$

Using the notation $\kappa(s) = \| \tau''(s) \|$, calculate $\langle \vec{\xi}_2 | d\vec{\xi}_1 \rangle$ and show that $\mu = \kappa ds$.

Note: $\kappa(s) = \| \tau''(s) \|$ is the **curvature** of the curve.

2. The setup is the same as in the previous exercise, except that $\tau : (a, b) \to R^3$.

(i) Verify that if $\vec{\xi}_1(s) = \tau'(s), \vec{\xi}_2(s) = \tau''(s)/\| \tau''(s) \|$, and $\vec{\xi}_3 = \vec{\xi}_1 \times \vec{\xi}_2$, then $\Xi = [\vec{\xi}_1, \vec{\xi}_2, \vec{\xi}_3]$ is a moving orthonormal frame[2] on this parametrization.

Hint: Differentiate the identity $(\tau_1')^2 + (\tau_2')^2 + (\tau_3')^2 = 1$ to show the orthogonality.

(ii) In this case the matrix of connection forms takes the form:

$$\omega = \begin{bmatrix} 0 & -\mu^0 & -\mu^1 \\ \mu^0 & 0 & -\mu^2 \\ \mu^1 & \mu^2 & 0 \end{bmatrix} = \begin{bmatrix} 0 & \cdots & \cdots \\ \langle \vec{\xi}_2 | d\vec{\xi}_1 \rangle & 0 & \cdots \\ \langle \vec{\xi}_3 | d\vec{\xi}_1 \rangle & \langle \vec{\xi}_3 | d\vec{\xi}_2 \rangle & 0 \end{bmatrix}. \tag{4.22}$$

Using the notation $\kappa(s) = \| \tau''(s) \|$, prove that

$$\mu^0 = \kappa ds, \mu^1 = 0, \mu^2 = Tds; \tag{4.23}$$

$$T = \frac{(\tau' \times \tau'') \cdot \tau'''}{\tau'' \cdot \tau''}. \tag{4.24}$$

Note: κ and T are called the **curvature** and **torsion**[3] of the curve, respectively.

---

[2] The vectors in this basis are known as the tangent, normal, and binormal vectors, respectively, and the orthonormal frame is called a **Frenet frame**.

[3] See Struik [1961] for clarification of these terms, and more information about the geometry of space curves.

(iii) Show that the **Serret–Frenet formulas** (below) follow from the equation defining the connection forms:

$$d\vec{\xi}_1 = \kappa\vec{\xi}_2 ds, \, d\vec{\xi}_2 = (-\kappa\vec{\xi}_1 + T\vec{\xi}_3)\, ds, \, d\vec{\xi}_3 = -T\vec{\xi}_2 ds. \tag{4.25}$$

3. (Continuation) Calculate the matrix of connection forms explicitly for the Frenet frame associated with the helix

$$\tau(u) = (\frac{\cos u}{\sqrt{2}}, \frac{\sin u}{\sqrt{2}}, \frac{u}{\sqrt{2}}).$$

4.4   An Adapted Moving Orthonormal Frame on a Surface

Let us now specialize to the case where $\Psi: W \subseteq R^2 \to R^3$ is a 2-dimensional parametrization, in other words $M = \Psi(W)$ is a parametrized surface. Let us review the notion of tangent plane to a surface, in the multivariable calculus sense. Let $\{u, v\}$ be a coordinate system on W, and express $x = \Psi(u, v)$ as $(x^1(u, v), x^2(u, v), x^3(u, v))$. The **tangent plane**[4] at x is the plane through the point x spanned by the two vectors:

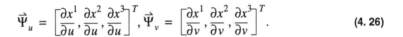

$$\vec{\Psi}_u = \left[\frac{\partial x^1}{\partial u}, \frac{\partial x^2}{\partial u}, \frac{\partial x^3}{\partial u}\right]^T, \vec{\Psi}_v = \left[\frac{\partial x^1}{\partial v}, \frac{\partial x^2}{\partial v}, \frac{\partial x^3}{\partial v}\right]^T. \tag{4.26}$$

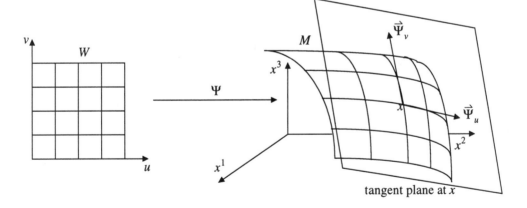

Figure 4. 2   Tangent plane to a parametrized surface

Note that these two vectors are linearly independent because Ψ is an immersion, but they are not necessarily orthogonal. Those who prefer to think in terms of differentials

---

[4] Note that the same tangent plane is obtained for every parametrization, as was proved in Chapter 3.

may note that $\vec{\Psi}_u$ and $\vec{\Psi}_v$ are coefficients of the tangent vectors $d\Psi\,(\partial/\partial u)$ and $d\Psi\,(\partial/\partial v)$ in terms of the basis $\{\partial/\partial x^1, \partial/\partial x^2, \partial/\partial x^3\}$ of the tangent space at x. Figure 4. 2 shows the situation.

4.4.1   Existence of an Adapted Moving Orthonormal Frame on a Parametrized Surface

There exists a moving orthonormal frame $\Xi = [\vec{\xi}_1, \vec{\xi}_2, \vec{\xi}_3]$ *on a parametrized surface M such that* $\vec{\xi}_3$ *is normal to the tangent plane at every point of M, that is, so that*

$$\langle \vec{\xi}_3 | \vec{\Psi}_u \rangle = 0 = \langle \vec{\xi}_3 | \vec{\Psi}_v \rangle. \tag{4.27}$$

Such a frame is called an adapted moving orthonormal frame.

Proof: Let us emphasize again that, since Ψ is an immersion, $\vec{\Psi}_u$ and $\vec{\Psi}_v$ are linearly independent, and therefore neither is ever zero, and that these vectors are smooth functions of $(u, v) \in W$. Hence the prescription

$$\vec{\xi}_1 = \vec{\Psi}_u / \| \vec{\Psi}_u \|,\ \vec{\xi}_3 = \vec{\Psi}_u \times \vec{\Psi}_v / \| \vec{\Psi}_u \times \vec{\Psi}_v \|,\ \vec{\xi}_2 = \vec{\xi}_3 \times \vec{\xi}_1 \tag{4.28}$$

ensures that $\vec{\xi}_3$ is normal to $\vec{\Psi}_u$ and $\vec{\Psi}_v$, that $\{\vec{\xi}_1(u, v), \vec{\xi}_2(u, v), \vec{\xi}_3(u, v)\}$ is a right–handed orthonormal basis of R^3 at every $(u, v) \in W$, and that each basis vector is a smooth function of (u, v).   ¤

4.4.2   Structure Equations for an Adapted Orthonormal Frame

Let us calculate the form of the equations in Table 4.1 for this moving orthonormal frame. Since

$$dx^i = \frac{\partial x^i}{\partial u} du + \frac{\partial x^i}{\partial v} dv,\ i = 1, 2, 3, \tag{4.29}$$

it follows that

$$d\vec{\Psi} = \vec{\Psi}_u du + \vec{\Psi}_v dv; \tag{4.30}$$

$$\vec{\theta} = \begin{bmatrix} \theta^1 \\ \theta^2 \\ \theta^3 \end{bmatrix} = \Xi^T d\vec{\Psi} = \begin{bmatrix} \langle \vec{\xi}_1 | d\vec{\Psi} \rangle \\ \langle \vec{\xi}_2 | d\vec{\Psi} \rangle \\ \langle \vec{\xi}_3 | d\vec{\Psi} \rangle \end{bmatrix} = \begin{bmatrix} \langle \vec{\xi}_1 | \vec{\Psi}_u \rangle du + \langle \vec{\xi}_1 | \vec{\Psi}_v \rangle dv \\ \langle \vec{\xi}_2 | \vec{\Psi}_u \rangle du + \langle \vec{\xi}_2 | \vec{\Psi}_v \rangle dv \\ 0 \end{bmatrix}, \tag{4.31}$$

using (4. 6), (4. 27), and (4. 30); since $\theta^3 = 0$ in the last line, we may write the first of the "Definitions" of Table 4.1 as:

$$d\vec{\Psi} = \theta^1 \vec{\xi}_1 + \theta^2 \vec{\xi}_2. \tag{4.32}$$

This equation is equivalent in classical terminology to calculation of the "First Fundamental Form" (see Exercise 14). Let us lighten the notation for the connection forms (ω^i_j) by writing instead

$$\omega = \begin{bmatrix} 0 & -\eta^0 & -\eta^1 \\ \eta^0 & 0 & -\eta^2 \\ \eta^1 & \eta^2 & 0 \end{bmatrix}. \tag{4.33}$$

Note that the skew-symmetry of ω implies that just three 1-forms suffice to specify ω. These three 1-forms are defined by $d\Xi = \Xi\omega$; in other words:

$$[d\vec{\xi}_1, d\vec{\xi}_2, d\vec{\xi}_3] = [\vec{\xi}_1, \vec{\xi}_2, \vec{\xi}_3] \begin{bmatrix} 0 & -\eta^0 & -\eta^1 \\ \eta^0 & 0 & -\eta^2 \\ \eta^1 & \eta^2 & 0 \end{bmatrix}$$

$$= [\vec{\xi}_2\eta^0 + \vec{\xi}_3\eta^1, -\vec{\xi}_1\eta^0 + \vec{\xi}_3\eta^2, -\vec{\xi}_1\eta^1 - \vec{\xi}_2\eta^2]. \tag{4.34}$$

To calculate the 1-forms η^0, η^1, η^2, take the inner product of the first column with the column vector $\vec{\xi}_2$, and of the first and second columns with $\vec{\xi}_3$, and exploit the orthogonality of the $\vec{\xi}_i$ to obtain:

$$\eta^0 = \langle \vec{\xi}_2 | d\vec{\xi}_1 \rangle, \quad \eta^1 = \langle \vec{\xi}_3 | d\vec{\xi}_1 \rangle, \quad \eta^2 = \langle \vec{\xi}_3 | d\vec{\xi}_2 \rangle. \tag{4.35}$$

As for the structure equations, $d\vec{\theta} = -\omega \wedge \vec{\theta}$ implies

$$\begin{bmatrix} d\theta^1 \\ d\theta^2 \\ 0 \end{bmatrix} = -\begin{bmatrix} 0 & -\eta^0 & -\eta^1 \\ \eta^0 & 0 & -\eta^2 \\ \eta^1 & \eta^2 & 0 \end{bmatrix} \begin{bmatrix} \theta^1 \\ \theta^2 \\ 0 \end{bmatrix} = \begin{bmatrix} \eta^0 \wedge \theta^2 \\ -\eta^0 \wedge \theta^1 \\ -\eta^1 \wedge \theta^1 - \eta^2 \wedge \theta^2 \end{bmatrix}. \tag{4.36}$$

Finally, the second structure equation $d\omega = -\omega \wedge \omega$ gives

$$d\begin{bmatrix} 0 & -\eta^0 & -\eta^1 \\ \eta^0 & 0 & -\eta^2 \\ \eta^1 & \eta^2 & 0 \end{bmatrix} = -\begin{bmatrix} 0 & -\eta^0 & -\eta^1 \\ \eta^0 & 0 & -\eta^2 \\ \eta^1 & \eta^2 & 0 \end{bmatrix} \wedge \begin{bmatrix} 0 & -\eta^0 & -\eta^1 \\ \eta^0 & 0 & -\eta^2 \\ \eta^1 & \eta^2 & 0 \end{bmatrix}. \tag{4.37}$$

The whole of surface theory is in some sense implicit in (4. 31), (4. 34), and the following six identities, which summarize the information in (4. 36) and (4. 37).

First Structure Equation:

$$d\theta^1 = \eta^0 \wedge \theta^2, \, d\theta^2 = -\eta^0 \wedge \theta^1; \tag{4. 38}$$

$$\eta^1 \wedge \theta^1 + \eta^2 \wedge \theta^2 = 0. \tag{4. 39}$$

Gauss's Equation[5]:

$$d\eta^0 = -\eta^1 \wedge \eta^2. \tag{4. 40}$$

Codazzi–Mainardi Equations:

$$d\eta^1 = -\eta^2 \wedge \eta^0, \, d\eta^2 = -\eta^0 \wedge \eta^1. \tag{4. 41}$$

4.4.3   Example: An Adapted Moving Orthonormal Frame on the Sphere

The sphere of radius A, $\{ (x_1, x_2, x_3) : x_1^2 + x_2^2 + x_3^2 = A^2 \}$, admits the parametrization $\Psi (u, v) = (A\cos u \sin v, A\sin u \sin v, A\cos v)$, defined for $(u, v) \in (0, 2\pi) \times (0, \pi)$. One may readily check that

$$\vec{\Psi}_u = \begin{bmatrix} -A\sin u \sin v \\ A\cos u \sin v \\ 0 \end{bmatrix}, \vec{\Psi}_v = \begin{bmatrix} A\cos u \cos v \\ A\sin u \cos v \\ -A\sin v \end{bmatrix}, \vec{\Psi}_u \times \vec{\Psi}_v = \begin{bmatrix} -A^2\cos u \, (\sin v)^2 \\ -A^2\sin u \, (\sin v)^2 \\ -A^2\sin v \cos v \end{bmatrix}. \tag{4. 42}$$

The construction 4.4.1 for an adapted moving orthonormal frame gives

$$\vec{\xi}_1 = \vec{\Psi}_u / \| \vec{\Psi}_u \|, \vec{\xi}_3 = \vec{\Psi}_u \times \vec{\Psi}_v / \| \vec{\Psi}_u \times \vec{\Psi}_v \|, \vec{\xi}_2 = \vec{\xi}_3 \times \vec{\xi}_1. \tag{4. 43}$$

It follows that

$$\vec{\xi}_1 = \begin{bmatrix} -\sin u \\ \cos u \\ 0 \end{bmatrix}, \vec{\xi}_2 = \begin{bmatrix} \cos u \cos v \\ \sin u \cos v \\ -\sin v \end{bmatrix}, \vec{\xi}_3 = \begin{bmatrix} -\cos u \sin v \\ -\sin u \sin v \\ -\cos v \end{bmatrix}. \tag{4. 44}$$

The reader should make a mental check that the three vectors in the previous line indeed form an orthonormal basis at every point. Observe that, in the case of the sphere, the

---

[5] This becomes more like the conventional form of Gauss's equation when the right side is identified with $-K (\theta^1 \wedge \theta^2)$, where K is the Gaussian curvature, as in (4. 67).

normal vector $\vec{\xi}_3 (u, v)$ is parallel to the position vector $\Psi (u, v)$; here it appears as an inward normal.

Exterior differentiation of (4. 44) gives

$$d\vec{\xi}_1 = \begin{bmatrix} -\cos u\, du \\ -\sin u\, du \\ 0 \end{bmatrix}, d\vec{\xi}_2 = \begin{bmatrix} -\sin u \cos v\, du - \cos u \sin v\, dv \\ \cos u \cos v\, du - \sin u \sin v\, dv \\ (-\cos v)\, dv \end{bmatrix}. \tag{4. 45}$$

From (4. 34) we may calculate the matrix of connection forms as follows:

$$\begin{bmatrix} 0 & -\eta^0 & -\eta^1 \\ \eta^0 & 0 & -\eta^2 \\ \eta^1 & \eta^2 & 0 \end{bmatrix} = \begin{bmatrix} 0 & -\langle\vec{\xi}_2|d\vec{\xi}_1\rangle & -\langle\vec{\xi}_3|d\vec{\xi}_1\rangle \\ \cdots & 0 & -\langle\vec{\xi}_3|d\vec{\xi}_2\rangle \\ \cdots & \cdots & 0 \end{bmatrix} = \begin{bmatrix} 0 & \cos v\, du & -\sin v\, du \\ -\cos v\, du & 0 & -dv \\ \sin v\, du & dv & 0 \end{bmatrix}. \tag{4. 46}$$

4.5   The Area Form

The following notation will be convenient. If $\vec{\alpha} = [\alpha^1, \alpha^2, \alpha^3]^T$ and $\vec{\beta}$ are R^3-valued 1-forms, we define an R^3-valued 2-form $\vec{\alpha} \times \vec{\beta}$ by

$$\vec{\alpha} \times \vec{\beta} = \begin{bmatrix} \alpha^2 \wedge \beta^3 - \alpha^3 \wedge \beta^2 \\ \alpha^3 \wedge \beta^1 - \alpha^1 \wedge \beta^3 \\ \alpha^1 \wedge \beta^2 - \alpha^2 \wedge \beta^1 \end{bmatrix}. \tag{4. 47}$$

It follows from this that $(\lambda\vec{\xi}_i) \times (\mu\vec{\xi}_j) = (\lambda \wedge \mu)\, \vec{\xi}_i \times \vec{\xi}_j$ for any 1-forms λ, μ, and that $\times$ distributes over addition. In particular[6]

$$d\vec{\Psi} \times d\vec{\Psi} = (\theta^1\vec{\xi}_1 + \theta^2\vec{\xi}_2) \times (\theta^1\vec{\xi}_1 + \theta^2\vec{\xi}_2)$$

$$= (\theta^1 \wedge \theta^2)\, (\vec{\xi}_1 \times \vec{\xi}_2) + (\theta^2 \wedge \theta^1)\, (\vec{\xi}_2 \times \vec{\xi}_1).$$

It follows that

$$d\vec{\Psi} \times d\vec{\Psi} = 2\, (\theta^1 \wedge \theta^2)\, \vec{\xi}_3. \tag{4. 48}$$

---

[6] Although $v \times v = 0$ for any 3-vector v, it is not necessarily true that $\vec{\alpha} \times \vec{\alpha} = 0$ for vector-valued 1-forms, since, for example, $\alpha^1 \wedge \alpha^2 = -\alpha^2 \wedge \alpha^1$.

This contains important information about the 2-form $\theta^1 \wedge \theta^2$, known as the **area form.**

4.5.1   Properties of the Area Form

The "area form" $\theta^1 \wedge \theta^2$ *is never zero, and is the same for any adapted moving orthonormal frame with the same choice of direction*[7] *for the moving normal vector* $\vec{\xi}_3$*; indeed if* $\vec{\xi}_3$ *is pointing in the direction* $\vec{\Psi}_u \times \vec{\Psi}_v$*, then*

$$\theta^1 \wedge \theta^2 = \left\| \vec{\Psi}_u \times \vec{\Psi}_v \right\| du \wedge dv. \tag{4.49}$$

Proof: According to (4. 30), $d\vec{\Psi} = \vec{\Psi}_u du + \vec{\Psi}_v dv$; therefore (4. 48) gives

$$2 (\theta^1 \wedge \theta^2) \vec{\xi}_3 = (\vec{\Psi}_u du + \vec{\Psi}_v dv) \times (\vec{\Psi}_u du + \vec{\Psi}_v dv),$$

$$(\theta^1 \wedge \theta^2) \vec{\xi}_3 = (du \wedge dv) (\vec{\Psi}_u \times \vec{\Psi}_v). \tag{4.50}$$

Now $\vec{\Psi}_u \times \vec{\Psi}_v \neq 0$ since Ψ is an immersion, and so the 2-form $\theta^1 \wedge \theta^2$ is always nonzero. Equation (4. 49) follows immediately. In equation (4. 48), the terms other than $\theta^1 \wedge \theta^2$ are the same for any choice of adapted moving orthonormal frame with the same choice of normal direction, and hence so is $\theta^1 \wedge \theta^2$.   ¤

4.5.2   The Sphere Example, Continued

For the parametrization of the sphere considered in Example 4.4.3, we have from (4. 42)

$$\left\| \vec{\Psi}_u \right\| = A \sin v, \left\| \vec{\Psi}_u \times \vec{\Psi}_v \right\| = A^2 \sin v, \tag{4.51}$$

$$\theta^1 \wedge \theta^2 = A^2 \sin v \, du \wedge dv. \tag{4.52}$$

4.5.3   Relationship to Surface Integrals

This refers to the integration theory of differential forms, which will be covered thoroughly in Chapter 8. Suppose

$$Y = F_1 \frac{\partial}{\partial x} + F_2 \frac{\partial}{\partial y} + F_3 \frac{\partial}{\partial z} \tag{4.53}$$

is a vector field on an open set $U \subseteq R^3$ with $U \supseteq \Psi(W)$. The flux of Y through the parametrized surface $\Psi(W)$ means the "surface integral" of the 2-form

---

[7] On any parametrized surface, there are two choices for the "outward" normal.

$$\beta = F_1 (dy \wedge dz) + F_2 (dz \wedge dx) + F_3 (dx \wedge dy) \tag{4.54}$$

over $\Psi(W)$, namely the integral

$$\int_{\Psi(W)} \beta = \int_W \Psi^* \beta = \int_W (F_1, F_2, F_3) \cdot (\vec{\Psi}_u \times \vec{\Psi}_v) \, du \, dv. \tag{4.55}$$

Clearly this is the same as

$$\int_W (F_1, F_2, F_3) \cdot \vec{\xi}_3 (\theta^1 \wedge \theta^2). \tag{4.56}$$

4.6   Exercises

4. Calculate the connection forms and the area form for an adapted moving orthonormal frame on the following parametrization of a cylinder, for some $A > 0$:

$$\Psi(u, v) = (A \cos v, A \sin v, u), \quad (u, v) \in (-\infty, \infty) \times (0, 2\pi).$$

5. Consider a surface of revolution of the form

$$\Psi(u, v) = (g(u) \cos v, g(u) \sin v, h(u)), \tag{4.57}$$

on some domain $W \subseteq R^2$, where g and h are smooth functions, and g is strictly positive. Moreover we assume that this is a "canonical parametrization," which means

$$g'(u)^2 + h'(u)^2 = 1. \tag{4.58}$$

(i) Calculate an adapted moving orthonormal frame as in (4.28), and show that (4.31) gives

$$\theta^1 = du, \quad \theta^2 = g \, dv. \tag{4.59}$$

(ii) Prove that the matrix of connection forms is

$$\begin{bmatrix} 0 & -\eta^0 & -\eta^1 \\ \eta^0 & 0 & -\eta^2 \\ \eta^1 & \eta^2 & 0 \end{bmatrix} = \begin{bmatrix} 0 & -g' dv & -(h''g' - h'g'') \, du \\ g' dv & 0 & -h' dv \\ (h''g' - h'g'') \, du & h' dv & 0 \end{bmatrix}. \tag{4.60}$$

6. (Continuation) Now consider any surface of revolution of the form

$$\Psi(t, v) = (r(t) \cos v, r(t) \sin v, q(t)) \tag{4.61}$$

on some domain $W \subseteq R^2$, where r and q are smooth functions, and r is strictly positive. Let $u(t)$ be a smooth reparametrization such that $\dot{u} = \partial u / \partial t > 0$, and if $g(u(t)) = r(t)$ and $h(u(t)) = q(t)$, then $g'(u)^2 + h'(u)^2 = 1$, where $g' = \partial g / \partial u$, etc.

(i) Using the fact that $\dot{g} = \partial g / \partial t = g'\dot{u} = \dot{r}$, show that

$$\dot{u}^2 = \dot{r}^2 + \dot{q}^2. \tag{4.62}$$

(ii) By applying the chain rule, obtain formulas for g', g'', h', h'' in terms of $\dot{r}, \ddot{r}, \dot{q}, \ddot{q}, \dot{u}, \ddot{u}$.

(iii) Show that $\theta^1 = \dot{u} \, dt, \theta^2 = r \, dv$.

(iv) By inserting these expressions into (4.60), show that the matrix of connection forms is

$$\begin{bmatrix} 0 & -\eta^0 & -\eta^1 \\ \eta^0 & 0 & -\eta^2 \\ \eta^1 & \eta^2 & 0 \end{bmatrix} = \begin{bmatrix} 0 & -\dot{r} dv/\dot{u} & -(\ddot{q}\dot{r} - \dot{q}\ddot{r}) \, dt/\dot{u}^2 \\ \dot{r} dv/\dot{u} & 0 & -\dot{q} dv/\dot{u} \\ (\ddot{q}\dot{r} - \dot{q}\ddot{r}) \, dt/\dot{u}^2 & \dot{q} dv/\dot{u} & 0 \end{bmatrix}. \tag{4.63}$$

7. A surface of the form $z = h(x, y)$, with some domain $W \subseteq R^2$, may be expressed using the parametrization $\Psi(u, v) = (u, v, h(u, v))$. Find an adapted moving orthonormal frame as in (4.28), and show that $\dot{\theta}, \varpi$, and the area form are given by the following formulas:

$$\eta^0 = \frac{h_v h_{uu}}{Q} du + \frac{h_v h_{uv}}{Q} dv, \eta^1 = \frac{h_{uu}}{PQ} du + \frac{h_{uv}}{PQ} dv, \tag{4.64}$$

$$\eta^2 = \left(-\frac{h_u h_v h_{uu}}{PQ^2} + \frac{P h_{uv}}{Q^2} \right) du + \left(-\frac{h_u h_v h_{uv}}{PQ^2} + \frac{P h_{vv}}{Q^2} \right) dv; \tag{4.65}$$

$$\theta^1 = P du + \frac{h_u h_v}{P} dv, \theta^2 = \frac{Q}{P} dv, \theta^1 \wedge \theta^2 = Q(du \wedge dv), \tag{4.66}$$

where we use the notation $h_u = \dfrac{\partial h}{\partial u}, h_{uu} = \dfrac{\partial^2 h}{\partial u^2}, P = \sqrt{1 + h_u^2}, Q = \sqrt{1 + h_u^2 + h_v^2}$.

4.7   Curvature of a Surface

Let us proceed toward the definition of curvature. Since the domain W of the parametrization is 2-dimensional, the 2-forms on W at any point (u, v) of W comprise a space with the same dimension as $\Lambda^2 R^2$, which is 1. By 4.5.1, every 2-form on W is simply the area form above multiplied by some smooth function. In particular, there exist smooth functions K and H on W defined by the equations

$$\eta^1 \wedge \eta^2 = K(\theta^1 \wedge \theta^2), \tag{4.67}$$

$$-\theta^2 \wedge \eta^1 + \theta^1 \wedge \eta^2 = 2H(\theta^1 \wedge \theta^2). \tag{4.68}$$

Abusing notation slightly, we may speak of K and H as functions on the surface $\Psi(W)$ by composing them with Ψ^{-1}; in that case, K is called the **Gaussian curvature**, and H is called the **mean curvature**. The relevance of these definitions to intuitive ideas about "curvature" will be explained in the next section, when we express K and H in terms of the principal curvatures.

It appears from the construction we have given that the definitions of K and H depend on a specific choice of adapted moving orthonormal frame; the following calculations will show that this is not so, although the sign of H depends on the choice of normal direction on the surface. In fact Gauss's *Theorema Egregium* says that the values of K and $|H|$ on $\Psi(W)$ are invariant under change of parametrization also. The best way to prove this is to use a more abstract definition of the surface and of curvature, so these functions are defined without reference to any parametrization, as we shall do later. However, a proof "in bad taste" is also given here.

Let us continue to take cross products of vector-valued 1-forms. By (4.34),

$$d\vec{\Psi} \times d\vec{\xi}_3 = (\theta^1 \vec{\xi}_1 + \theta^2 \vec{\xi}_2) \times (-\vec{\xi}_1 \eta^1 - \vec{\xi}_2 \eta^2)$$

$$= -(\theta^1 \wedge \eta^2)(\vec{\xi}_1 \times \vec{\xi}_2) - (\theta^2 \wedge \eta^1)(\vec{\xi}_2 \times \vec{\xi}_1)$$

$$= -(\theta^1 \wedge \eta^2 - \theta^2 \wedge \eta^1)\vec{\xi}_3.$$

Consequently

$$d\vec{\Psi} \times d\vec{\xi}_3 = -2H(\theta^1 \wedge \theta^2)\vec{\xi}_3; \tag{4.69}$$

a similar argument shows that

$$d\vec{\xi}_3 \times d\vec{\xi}_3 = 2(\eta^1 \wedge \eta^2)\vec{\xi}_3 = 2K(\theta^1 \wedge \theta^2)\vec{\xi}_3. \tag{4.70}$$

The last equation already gives some insight into the relationship between K and the intuitive notion of curvature. When a surface is "highly curved," the normal direction $\vec{\xi}_3$ will change very quickly as a function of the parametrization (u,v), which will tend to make the left side of (4.70) have large magnitude; this corresponds to a large value of K on the right side.

4.7.1   Invariance Property

The Gaussian curvature is the same for any two adapted moving orthonormal frames; the mean curvature is the same for any two adapted moving orthonormal frames with the same choice of normal direction.

Proof: Any two adapted moving orthonormal frames, with the same choice of normal direction, share the same vector-valued 1-form $d\vec{\Psi}$ and normal vector $\vec{\xi}_3$. Now (4. 48) shows that the 2-form $\theta^1 \wedge \theta^2$ is the same for both moving orthonormal frames, and hence so are H and K, by equations (4. 69) and (4. 70), respectively. These equations also show that reversing the normal direction $\vec{\xi}_3$ changes the sign of $\theta^1 \wedge \theta^2$, and hence of H, but not of K.   ¤

4.7.2   Gauss's Theorema Egregium

The value of the Gaussian curvature does not depend on the parametrization.

Proof (may be omitted): Suppose $\Phi: V \subseteq R^2 \to R^3$ is another parametrization of the surface M; for simplicity suppose $\Phi(V) = M = \Psi(W)$. Taking $f: V \to W$ to be $\Psi^{-1} \bullet \Phi$, which is smooth by a result in Chapter 3, we have the diagram

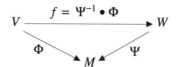

In view of 4.7.1, there is no loss of generality in choosing $\Xi \bullet f$ as the adapted moving orthonormal frame on $\Phi: V \subseteq R^2 \to R^3$. Define $\vec{\vartheta}$ by $d\vec{\Phi} = (\Xi \bullet f)\vec{\vartheta}$; it suffices by (4. 70) to show

$$d(\vec{\xi}_3 \bullet f) \times d(\vec{\xi}_3 \bullet f) = 2(K \bullet f)(\vartheta^1 \wedge \vartheta^2)\vec{\xi}_3 \bullet f. \tag{4. 71}$$

However, applying the pullback of f to the equation $d\vec{\Psi} = \Xi\vec{\theta}$ gives

$$d\vec{\Phi} = f^* d\vec{\Psi} = (\Xi \bullet f)f^*\vec{\theta} = (\Xi \bullet f)\vec{\vartheta}.$$

Thus $\vec{\vartheta} = f^*\vec{\theta}$, and so $f^*(\theta^1 \wedge \theta^2) = \vartheta^1 \wedge \vartheta^2$. Finally we may apply f^* to $d\vec{\xi}_3 \times d\vec{\xi}_3 = 2K(\theta^1 \wedge \theta^2)\vec{\xi}_3$ to obtain (4. 71).   ¤

4.8   Explicit Calculation of Curvatures

Since $\theta^1 \wedge \theta^2$ is never zero, it follows that the 1-forms θ^1 and θ^2 form a basis for the 1-forms on W at every point; in particular there exist smooth functions $a = a(u, v)$, b, c, and $\tilde{c}$ such that

$$\eta^1 = a\theta^1 + c\theta^2, \eta^2 = \tilde{c}\theta^1 + b\theta^2. \tag{4.72}$$

However the third part of the first structure equation says $\eta^1 \wedge \theta^1 + \eta^2 \wedge \theta^2 = 0$, and therefore

$$\eta^1 \wedge \theta^1 = c\theta^2 \wedge \theta^1 = -\eta^2 \wedge \theta^2 = -\tilde{c}\theta^1 \wedge \theta^2, \tag{4.73}$$

giving $c = \tilde{c}$. Thus we may restate (4.72) as:

$$\begin{bmatrix} \eta^1 \\ \eta^2 \end{bmatrix} = \begin{bmatrix} a & c \\ c & b \end{bmatrix} \begin{bmatrix} \theta^1 \\ \theta^2 \end{bmatrix}. \tag{4.74}$$

Knowledge of the terms in this matrix equation is equivalent in classical terminology to calculation of the "Second Fundamental Form"; see Exercise 16. The eigenvalues κ_1, κ_2 of the 2×2 matrix above, which by composition with Ψ^{-1} can be treated as functions on $M = \Psi(W)$, are called the **principal curvatures**.

We shall now derive expressions for the Gaussian and mean curvatures in terms of the principal curvatures. It is immediate from (4.67) and (4.74) that

$$K(\theta^1 \wedge \theta^2) = \eta^1 \wedge \eta^2 = (a\theta^1 + c\theta^2) \wedge (c\theta^1 + b\theta^2) = (ab - c^2)(\theta^1 \wedge \theta^2).$$

Now we see that

$$K = ab - c^2 = \begin{vmatrix} a & c \\ c & b \end{vmatrix} = \kappa_1 \kappa_2, \tag{4.75}$$

using the fact that the determinant of a matrix is the product of the eigenvalues. Similarly, (4.68) gives

$$2H(\theta^1 \wedge \theta^2) = -\theta^2 \wedge \eta^1 + \theta^1 \wedge \eta^2$$

$$= (-\theta^2 \wedge (a\theta^1 + c\theta^2) + \theta^1 \wedge (c\theta^1 + b\theta^2))$$

$$= (a + b)(\theta^1 \wedge \theta^2).$$

Since the trace of a matrix is the sum of the eigenvalues, this shows that

$$H = \frac{a+b}{2} = \frac{\kappa_1 + \kappa_2}{2}. \tag{4.76}$$

To understand the physical meaning of these quantities, consider the "quadric surfaces" $z = 2x^2 + xy/2 - y^2$ and $z = -3x^2 - xy/2 - 5y^2/2$, in the vicinity of $(0, 0)$, shown in Figure 4. 3.

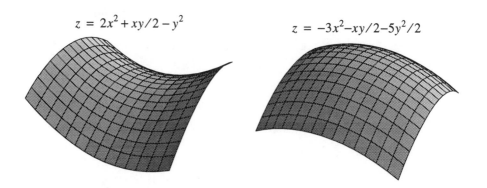

$$z = 2x^2 + xy/2 - y^2 \qquad\qquad z = -3x^2 - xy/2 - 5y^2/2$$

Figure 4. 3   Illustration of negative and positive Gaussian curvatures

As we shall see in Exercise 13, the "saddle" on the left has one positive and one negative principal curvature whatever the direction of the normal vector; hence the Gaussian curvature is negative. For a downward-pointing (resp., upward-pointing) normal vector, the surface on the right has both principal curvatures positive (resp., negative); hence the Gaussian curvature is positive.

4.8.1   Steps for Calculating Curvatures from a Parametrization

Given a parametrized surface $\Psi: W \to R^2$ the first step is to calculate an adapted moving orthonormal frame as in 4.4.1, namely

$$\vec{\xi}_1 = \vec{\Psi}_u / \|\vec{\Psi}_u\|, \vec{\xi}_3 = \vec{\Psi}_u \times \vec{\Psi}_v / \|\vec{\Psi}_u \times \vec{\Psi}_v\|, \vec{\xi}_2 = \vec{\xi}_3 \times \vec{\xi}_1. \tag{4.77}$$

It follows immediately from (4. 31) that the 1-forms θ^1, θ^2 are given by the equations:

$$\begin{bmatrix} \theta^1 \\ \theta^2 \end{bmatrix} = \begin{bmatrix} \langle \vec{\xi}_1 | \vec{\Psi}_u \rangle du + \langle \vec{\xi}_1 | \vec{\Psi}_v \rangle dv \\ \langle \vec{\xi}_2 | \vec{\Psi}_u \rangle du + \langle \vec{\xi}_2 | \vec{\Psi}_v \rangle dv \end{bmatrix} = \begin{bmatrix} \|\vec{\Psi}_u\| du + (\langle \vec{\Psi}_u | \vec{\Psi}_v \rangle / \|\vec{\Psi}_u\|) dv \\ (\|\vec{\Psi}_u \times \vec{\Psi}_v\| / \|\vec{\Psi}_u\|) dv \end{bmatrix}. \tag{4.78}$$

The last entry is obtained from the following formula for vectors $w, z \in R^3$:

$$\frac{z \cdot ((w \times z) \times w)}{\|w \times z\| \|w\|} = \frac{(w \times z) \cdot (w \times z)}{\|w \times z\| \|w\|} = \frac{\|w \times z\|}{\|w\|}. \tag{4.79}$$

As we saw in 4.5.1,

$$\theta^1 \wedge \theta^2 = \left\| \vec{\Psi}_u \times \vec{\Psi}_v \right\| du \wedge dv. \tag{4.80}$$

To calculate the 1-forms η^1, η^2, recall equation (4.34):

$$\eta^1 = \langle \vec{\xi}_3 | d\vec{\xi}_1 \rangle, \eta^2 = \langle \vec{\xi}_3 | d\vec{\xi}_2 \rangle. \tag{4.81}$$

Computation of all kinds of curvatures follows from evaluating the 2-forms on the left side of the following matrix equation (deduced from (4.74)), and reading off the expressions for a, b, c:

$$\begin{bmatrix} \eta^1 \wedge \theta^1 & \eta^1 \wedge \theta^2 \\ \eta^2 \wedge \theta^1 & \eta^2 \wedge \theta^2 \end{bmatrix} = \begin{bmatrix} -c & a \\ -b & c \end{bmatrix} (\theta^1 \wedge \theta^2).$$

For example, as noted above, $K = ab - c^2$ and $H = (a+b)/2$.

4.8.2 Example: The Sphere

This calculation is a continuation of Example 4.4.3. We calculate from (4.78):

$$\begin{bmatrix} \theta^1 \\ \theta^2 \end{bmatrix} = \begin{bmatrix} A \sin v \, du \\ A \, dv \end{bmatrix}; \tag{4.82}$$

$$\theta^1 \wedge \theta^2 = A^2 \sin v \, du \wedge dv. \tag{4.83}$$

From (4.46) comes

$$\begin{bmatrix} \eta^1 \\ \eta^2 \end{bmatrix} = \begin{bmatrix} \sin v \, du \\ dv \end{bmatrix}. \tag{4.84}$$

Now we calculate the following matrix of 2-forms using (4.82), (4.83), and (4.84):

$$\begin{bmatrix} \eta^1 \wedge \theta^1 & \eta^1 \wedge \theta^2 \\ \eta^2 \wedge \theta^1 & \eta^2 \wedge \theta^2 \end{bmatrix} = \begin{bmatrix} 0 & A \sin v \\ -A \sin v & 0 \end{bmatrix} du \wedge dv = \begin{bmatrix} 0 & 1/A \\ -1/A & 0 \end{bmatrix} \theta^1 \wedge \theta^2.$$

It follows that

$$a = 1/A, b = 1/A, c = 0. \tag{4.85}$$

We conclude that both the principal curvatures take the value $1/A$, while the Gaussian curvature is $K = 1/A^2$, and the mean curvature is $H = 1/A$. Note that in this case all these quantities happen to take constant values; in a more general example they will depend on u and v.

4.9   Exercises

8. (Continuation of Exercise 4) Calculate the principal curvatures, the Gaussian curvature, and the mean curvature of a cylinder of radius $A > 0$, using the parametrization

$$\Psi (u, v) = (A\cos v, A\sin v, u), \ (u, v) \in (-\infty, \infty) \times (0, 2\pi).$$

9. On the helicoid, parametrized by $\Psi (u, v) = (u\cos v, u\sin v, v)$, an adapted moving orthonormal frame is given by

$$\vec{\xi}_1 = \begin{bmatrix} \cos v \\ \sin v \\ 0 \end{bmatrix}, \vec{\xi}_2 = \frac{1}{P}\begin{bmatrix} -u\sin v \\ u\cos v \\ 1 \end{bmatrix}, \vec{\xi}_3 = \frac{1}{P}\begin{bmatrix} \sin v \\ -\cos v \\ u \end{bmatrix},$$

where $P = \sqrt{1 + u^2}$. Find the 1-forms θ^1, θ^2 so that $d\vec{\Psi} = \theta^1\vec{\xi}_1 + \theta^2\vec{\xi}_2$ and the matrix ϖ of connection forms, and show that the mean curvature of the helicoid is zero.

10. (Continuation of Exercise 5) For the surface of revolution

$$\Psi (u, v) = (g(u)\cos v, g(u)\sin v, h(u)),$$

where $g'(u)^2 + h'(u)^2 = 1$, and where g and h are smooth functions with g strictly positive, show that the principal curvatures, the Gaussian curvature, and the mean curvature are given by

$$\kappa_1 = h''g' - h'g'', \kappa_2 = \frac{h'}{g}, K = \frac{h'(h''g' - h'g'')}{g}, H = \frac{h''g' - h'g'' + h'/g}{2}.$$

11. (Continuation of Exercise 6) Consider the general surface of revolution

$$\Psi (t, v) = (r(t)\cos v, r(t)\sin v, q(t)), \tag{4.86}$$

on some domain $W \subseteq R^2$, where r and q are smooth functions, and r is strictly positive. As before, let $u(t)$ be a smooth reparametrization such that $\dot{u}^2 = \dot{r}^2 + \dot{q}^2$ and $\dot{u} = \partial u/\partial t > 0$. Show that the principal curvatures are

$$\kappa_1 = \frac{\dot{q}\ddot{r} - \ddot{q}\dot{r}}{(\dot{r}^2 + \dot{q}^2)^{3/2}}, \kappa_2 = \frac{\dot{q}}{r\sqrt{\dot{r}^2 + \dot{q}^2}}. \tag{4.87}$$

Warning: When applied to the sphere, Example 4.8.2, the principal curvatures will turn out to be the negatives of the results we obtained before, because in switching the u and v variables we reversed the direction of the normal vector.

12. Using the results of the last exercise, calculate the principal curvatures, the Gaussian curvature, and the mean curvature of a hyperboloid, with a scale parameter $A > 0$, using the parametrization

$$\Psi(u, v) = (A \sinh u \cos v, A \sinh u \sin v, A \cosh u), \ (u, v) \in (-\infty, \infty) \times (0, 2\pi).$$

13. (i) Suppose A, B, and C are real numbers. Using the results of Exercise 7, show that the principal curvatures of the quadric surface

$$z = \frac{1}{2} \begin{bmatrix} x & y \end{bmatrix} \begin{bmatrix} A & C \\ C & B \end{bmatrix} \begin{bmatrix} x \\ y \end{bmatrix} = Ax^2/2 + Cxy + By^2/2 \tag{4.88}$$

at $(x, y) = (0, 0)$ are the eigenvalues of the matrix $\begin{bmatrix} A & C \\ C & B \end{bmatrix}$.

(ii) Using this result, calculate the Gaussian curvature of the quadric surfaces $z = 2x^2 + xy/2 - y^2$ and $z = -3x^2 - xy/2 - 5y^2/2$ shown in Figure 4.3, at $(x, y) = (0, 0)$.

4.10   The Fundamental Forms: Exercises

The terminology of "fundamental forms" is a restatement of the ideas of this chapter; study of these exercises may help the reader who plans to consult classical geometry books.

14. Given an adapted moving orthonormal frame on a parametrized surface $\Psi : W \subseteq R^2 \to R^3$, the **First Fundamental Form** is a (0,2)-tensor on W, in the sense of Chapter 2, defined by

$$I = \theta^1 \otimes \theta^1 + \theta^2 \otimes \theta^2. \tag{4.89}$$

(i) Using the fact that $\theta^i = \langle \vec{\xi}_i | \vec{\Psi}_u \rangle du + \langle \vec{\xi}_i | \vec{\Psi}_v \rangle dv$ by (4.31), show that

$$I \left(\frac{\partial}{\partial u}, \frac{\partial}{\partial u} \right) = (\theta^1 \cdot \frac{\partial}{\partial u}) \, (\theta^1 \cdot \frac{\partial}{\partial u}) + (\theta^2 \cdot \frac{\partial}{\partial u}) \, (\theta^2 \cdot \frac{\partial}{\partial u}) = \langle \vec{\Psi}_u | \vec{\Psi}_u \rangle.$$

Hint: $\vec{\Psi}_u = \langle \vec{\xi}_1 | \vec{\Psi}_u \rangle \vec{\xi}_1 + \langle \vec{\xi}_2 | \vec{\Psi}_u \rangle \vec{\xi}_2$.

(ii) By evaluating $I(\partial/\partial u, \partial/\partial v)$, etc., show that

$$I\left(A_1\frac{\partial}{\partial u}+A_2\frac{\partial}{\partial v},\ B_1\frac{\partial}{\partial u}+B_2\frac{\partial}{\partial v}\right)\ =\ \begin{bmatrix}A_1 & A_2\end{bmatrix}\begin{bmatrix}\|\vec{\Psi}_u\|^2 & \langle\vec{\Psi}_u|\vec{\Psi}_v\rangle \\ \langle\vec{\Psi}_v|\vec{\Psi}_u\rangle & \|\vec{\Psi}_v\|^2\end{bmatrix}\begin{bmatrix}B_1\\B_2\end{bmatrix}. \tag{4.90}$$

Comment: The First Fundamental Form tells us how distances would be experienced on the surface by someone who was operating in (u, v) coordinates; this is because the square root of $I(\zeta, \zeta)$ is the length in R^3 of the image under the differential of Ψ of the tangent vector ζ at a point in W. The matrix appearing in (4.90) will later be called the Riemannian metric tensor, with respect to the (u, v) coordinate system, of the metric induced by the immersion Ψ. Knowledge of the First Fundamental Form would enable us in principle to calculate "minimal geodesics," that is, paths of shortest length between two points on the surface; see Klingenberg [1982].

15. (Continuation) Calculate the First Fundamental Form of the ellipsoid

$$\frac{x^2}{A^2}+\frac{y^2}{B^2}+\frac{z^2}{C^2}\ =\ 1$$

for the parametrization $\Psi(u, v)\ =\ (A\sin u\cos v,\ B\sin u\sin v,\ C\cos u)$ with domain $(0, \pi)\times(0, 2\pi)$. Show that the Gaussian curvature at (x, y, z) is

$$K\ =\ \frac{1}{A^2 B^2 C^2}\left(\frac{x^2}{A^4}+\frac{y^2}{B^4}+\frac{z^2}{C^4}\right)^{-2}.$$

16. Given an adapted moving orthonormal frame on a parametrized surface $\Psi: W\subseteq R^2\to R^3$, the **Second Fundamental Form** is a $(0,2)$-tensor on W, in the sense of Chapter 2, defined by

$$II\ =\ \eta^1\otimes\theta^1+\eta^2\otimes\theta^2. \tag{4.91}$$

(i) Using the matrix appearing in (4.74), show that

$$II\ =\ a\,(\theta^1\otimes\theta^1)+c\,(\theta^1\otimes\theta^2+\theta^2\otimes\theta^1)+b\,(\theta^2\otimes\theta^2). \tag{4.92}$$

(ii) Using the fact that $\langle\vec{\xi}_i|\vec{\xi}_j\rangle\ =\ \delta_{ij}$, show that $\langle\vec{\xi}_3|d\vec{\xi}_1\cdot\partial/\partial u\rangle\ =\ -\langle(\partial\vec{\xi}_3/\partial u)|\vec{\xi}_1\rangle$.

(iii) Using the formulas $\theta^i\ =\ \langle\vec{\xi}_i|\vec{\Psi}_u\rangle du+\langle\vec{\xi}_i|\vec{\Psi}_v\rangle dv$ and $\eta_i\ =\ \langle\vec{\xi}_3|d\vec{\xi}_i\rangle$ for $i = 1, 2$, as in (4.31) and (4.34), and the fact that $\langle\vec{\xi}_3|\vec{\Psi}_u\rangle\ =\ 0$, show that

$$II\left(\frac{\partial}{\partial u},\frac{\partial}{\partial u}\right)\ =\ \left(\eta^1\cdot\frac{\partial}{\partial u}\right)\left(\theta^1\cdot\frac{\partial}{\partial u}\right)+\left(\eta^2\cdot\frac{\partial}{\partial u}\right)\left(\theta^2\cdot\frac{\partial}{\partial u}\right)\ =\ \langle\vec{\xi}_3|\vec{\Psi}_{uu}\rangle,$$

where $\vec{\Psi}_{uu}\ =\ \partial\vec{\Psi}_u/\partial u$.

(iv) By evaluating $II(\partial/\partial u, \partial/\partial v)$, etc., show that

$$II\left(A_1\frac{\partial}{\partial u}+A_2\frac{\partial}{\partial v}, B_1\frac{\partial}{\partial u}+B_2\frac{\partial}{\partial v}\right) = \begin{bmatrix} A_1 & A_2 \end{bmatrix} \begin{bmatrix} \langle\vec{\xi}_3|\vec{\Psi}_{uu}\rangle & \langle\vec{\xi}_3|\vec{\Psi}_{uv}\rangle \\ \langle\vec{\xi}_3|\vec{\Psi}_{vu}\rangle & \langle\vec{\xi}_3|\vec{\Psi}_{vv}\rangle \end{bmatrix} \begin{bmatrix} B_1 \\ B_2 \end{bmatrix}. \tag{4.93}$$

Comment: The Second Fundamental Form is essentially equivalent to the matrix of functions

$$\begin{bmatrix} a & c \\ c & b \end{bmatrix}$$

appearing in (4. 74), from which all the previous curvature expressions were computed. It is shown by (4. 93) to contain all the information about the components of the second partial derivatives of Ψ in the direction normal to the surface; note incidentally that the sign of the Second Fundamental Form changes if the direction of the normal vector is reversed.

17. (Continuation) Calculate the Second Fundamental Form of the hyperbolic paraboloid

$$z = \frac{x^2}{A^2} - \frac{y^2}{B^2},$$

for the parametrization $\Psi(s,t) = (As, 0, s^2) + t(A, B, 2s)$. Show that the Gaussian curvature is

$$K = -\frac{1}{4A^2B^2}\left(\frac{x^2}{A^2} + \frac{y^2}{B^2} + \frac{1}{4}\right)^{-2}.$$

4.11   History and Bibliography

Gaspard Monge (1746–1818) and C. F. Gauss (1777–1855) may be considered as the founders of the differential geometry of surfaces. The equations bearing his name were discovered by D. Codazzi (1824–75). A comprehensive treatise on surface theory was written by Gaston Darboux (1842–1917). Further historical information may be found in Struik [1961], which is an excellent reference for the classical theory of surfaces; see also Spivak [1979]. The treatment given here is based on Flanders [1989].

5 Differential Manifolds

Until now we have been working on submanifolds of Euclidean spaces, so as to show clearly how exterior calculus is linked to multivariable calculus. However, it is more efficient in the long run to have an "intrinsic" theory and notation for the objects we work with, and to forget about the Euclidean space they may be embedded in. This is the theory of differential manifolds and vector bundles.

5.1 Definition of a Differential Manifold

Suppose M is a set. A pair (U, φ) is called an *n-dimensional chart* on M, or simply a chart, if $U \subseteq M$ and $\varphi: U \to R^n$ is a one-to-one map onto an open set $\varphi(U) \subseteq R^n$. If $p \in U$, we may call (U, φ) a chart for M at p.

Two charts (U, φ) and (V, ψ) are called C^∞-compatible, or simply **compatible**, if either $U \cap V = \varnothing$, or else $U \cap V \neq \varnothing$, $\psi(U \cap V)$ and $\varphi(U \cap V)$ are open in R^n, and $\varphi \bullet \psi^{-1}: \psi(U \cap V) \to \varphi(U \cap V)$ is smooth with a smooth inverse (i.e., is a diffeomorphism).

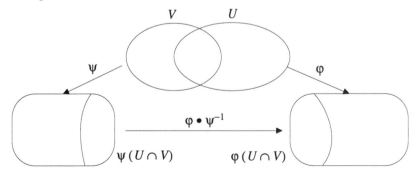

A **smooth atlas** is a family of charts, any two of which are compatible, whose domains cover M. Two atlases are called **equivalent** if their union is an atlas. As the name implies, this is indeed an equivalence relation on atlases (transitivity comes from application of the chain rule from calculus). An equivalence class of atlases on M is called a **smooth differentiable structure** on M. A set M with a smooth differentiable structure is called a (smooth) **differential manifold**. We say that M has dimension n if the dimension of the range of all the chart maps in some (hence any equivalent) atlas is n. The definition above allows M to have separate components with different dimensions, but for simplicity we shall assume henceforward that all our manifolds have a unique dimension.

5.1.1 Example: An Atlas on the Circle

The circle S^1 is the set $\{ (x, y) \in R^2 : x^2 + y^2 - 1 = 0 \}$. Let us show that the following pair of charts (U, φ) and (V, ψ) constitute an atlas for the circle:

$$U = S^1 - (0, 1), \quad \varphi(x, y) = \frac{x}{1-y};$$

$$V = S^1 - (0, -1), \quad \psi(x, y) = \frac{x}{1+y}.$$

First, we show that φ is one-to-one. If $x_1/(1-y_1) = x_2/(1-y_2)$, then squaring both sides and applying the fact that $x_i^2 = 1 - y_i^2$ gives

$$\frac{1-y_1^2}{(1-y_1)^2} = \frac{1-y_2^2}{(1-y_2)^2} \Rightarrow \frac{1+y_1}{1-y_1} = \frac{1+y_2}{1-y_2}.$$

Therefore $y_1 = y_2, x_1 = x_2$, since division by $1 - y_i$ is legitimate on U, where $y_i \neq 1$. Similarly ψ is one-to-one. Observe next that $\varphi(U) = R = \psi(V)$, an open set, and so is

$$\varphi(U \cap V) = (-\infty, 0) \cup (0, \infty) = \psi(U \cap V).$$

Since clearly $S^1 = U \cup V$, it only remains to show that

$$\varphi \bullet \psi^{-1} : \psi(U \cap V) = (-\infty, 0) \cup (0, \infty) \to \varphi(U \cap V) = (-\infty, 0) \cup (0, \infty)$$

is smooth with a smooth inverse (the one-to-one and onto properties are automatic from what we have proved already). To calculate $\varphi \bullet \psi^{-1}$, suppose $w = x/(1-y)$ and $z = x/(1+y)$; then it is easy to check that $wz = 1$, and thus

$$\varphi \bullet \psi^{-1}(z) = 1/z, z \in (-\infty, 0) \cup (0, \infty).$$

This is indeed smooth with a smooth inverse, so (U, φ) and (V, ψ) constitute an atlas.

5.1.2   An Equivalent Atlas on the Circle

Start with the 1-dimensional parametrizations $\Psi_i(\theta) = (\cos\theta, \sin\theta)$, $i = 1, 2$, with domains $W_1 = (-\pi, \pi)$, $W_2 = (0, 2\pi)$, respectively. By performing the same steps as in the previous example, one may show that another atlas for the circle consists of the charts

$$U_1 = S^1 - (-1, 0), \varphi_1 = \Psi_1^{-1};$$

$$U_2 = S^1 - (1, 0), \varphi_2 = \Psi_2^{-1}.$$

To show that $\{(U, \varphi), (V, \psi)\}$ is equivalent to $\{(U_1, \varphi_1), (U_2, \varphi_2)\}$, it now suffices to show that $\varphi \bullet \varphi_1^{-1}, \varphi \bullet \varphi_2^{-1}, \psi \bullet \varphi_1^{-1}, \psi \bullet \varphi_2^{-1}$ are diffeomorphisms wherever defined. This is easy because

$$\varphi \bullet \varphi_i^{-1}(\theta) = \frac{\cos\theta}{1 - \sin\theta}, \psi \bullet \varphi_i^{-1}(\theta) = \frac{\cos\theta}{1 + \sin\theta}.$$

5.1.3   Submanifolds of Euclidean Space

Every n-dimensional submanifold M of R^{n+k}, in the sense of Chapter 3, has a differentiable structure induced by its n-dimensional local parametrizations, and is therefore an n-dimensional differential manifold.

Proof: Given $p \in M$, a result in Chapter 3 shows that there exists a neighborhood U' of p in R^{n+k} and an implicit function parametrization $\Psi: W \subseteq R^n \to U'$; this is by definition one-to-one and onto $U' \cap M$. Taking $(U, \varphi) = (U' \cap M, \Psi^{-1})$ gives a chart for M at p, since W is open by definition. Evidently M can be expressed as the union of domains of such charts. Now the Chapter 3 result on switching between different parametrizations shows that, if $\Psi_i: W_i \subseteq R^n \to U_i' \subseteq R^{n+k}$ are n-dimensional parametrizations of M at y for $i = 1, 2$, then
$W_1 \cap \Psi_1^{-1}(U_2') = \Psi_1^{-1}(U_1' \cap U_2' \cap M)$ and
$W_2 \cap \Psi_2^{-1}(U_1') = \Psi_2^{-1}(U_1' \cap U_2' \cap M)$ are open sets in R^n, and if the last two sets are nonempty then

$$\Psi_2^{-1} \bullet \Psi_1: W_1 \cap \Psi_1^{-1}(U_2') \to W_2 \cap \Psi_2^{-1}(U_1')$$

is a diffeomorphism onto its range. This implies $(U_1' \cap M, \Psi_1^{-1})$ and $(U_2' \cap M, \Psi_2^{-1})$ are compatible charts.   ¤

5.2   Basic Topological Vocabulary

Topology can be loosely described as the study of properties of a space which are invariant under "homeomorphisms," that is, continuous mappings with a continuous inverse. In order to know what the homeomorphisms are, we have to define the class of open sets, since in technical language the homeomorphisms are precisely the mappings which preserve the class of open sets. Although topology is outside the scope of this book, we include here some pieces of vocabulary which will be needed to state later results. Most important, we shall need to know what an open set in a differential manifold M means.

5.2.1 Open Sets in a Differential Manifold

Given a specific atlas $\{ (U_\alpha, \varphi_\alpha), \alpha \in I \}$ for M, we shall say that a subset V of M is **open** in M if $\varphi_\alpha (V \cap U_\alpha)$ is open[1] in R^n for every chart $(U_\alpha, \varphi_\alpha)$. It turns out that the same class of open sets is obtained if $\{ (U_\alpha, \varphi_\alpha), \alpha \in I \}$ is replaced by an equivalent atlas; thus the class of open sets is determined by the differentiable structure on M, not the specific choice of atlas within that structure. Moreover this class of open sets is closed under finite intersections and arbitrary unions, and thus indeed forms a "topology" on M, making M into a "topological space." The proofs of these statements are given in detail in Berger and Gostiaux [1988], pp. 55–61, which offers further information about the topology induced by a differentiable structure.

It is easy to check that every open subset V of a differential manifold M can be given a differentiable structure, using the atlas $\{ (V \cap U_i, \varphi_i|_V), i \in I \}$, where $\{ (U_i, \varphi_i), i \in I \}$ is an atlas for M.

In the case of an n-dimensional submanifold M of R^{n+k}, if $\Psi: W \subseteq R^n \to U' \subseteq R^{n+k}$ is a parametrization of M, our definition implies that $\Psi (W \cap G)$ is open in M for every open set G in R^n. For example, in the case 5.1.1 of the circle, the segment $\{ (x, y) : 0 < x < 1, 0 < y < 1, x^2 + y^2 = 1 \}$, which is the image of the open set $(0, \pi/2)$ under the parametrization $\theta \to (\cos\theta, \sin\theta)$, is open in S^1.

5.2.2 Closed and Compact Sets in a Differential Manifold

As in any topological space, a subset H of M is called **closed** if its complement H^c is open; H is called **compact** if, for every collection $\{ V_\alpha \}$ of open sets in M whose union contains H (known as an "open cover" of H), there is a finite subcollection $\{ V_{\alpha(1)}, ..., V_{\alpha(m)} \}$ whose union contains H. In the case where M is a Euclidean space, a subset of M is compact if and only if it is closed and bounded (i.e., lies entirely within some finite ball centered at the origin); this is the Heine–Borel Theorem.

If M is a submanifold of R^n, then the compact subsets of M are precisely the sets of the form $K \cap M$, for K compact in R^n; this can be deduced from 5.10.1 below. In particular

---

[1] W is open in R^n when, for every $y \in W$, the "open ball" $B (y, \varepsilon) = \{ x : \| x - y \| < \varepsilon \}$ is contained in W for all sufficiently small ε; here $\| \cdot \|$ denotes Euclidean length.

if the submanifold M is itself closed and bounded in R^n, then it is compact. Spheres, tori, and special orthogonal groups are all examples of compact manifolds. To see, for example, that $SO(n)$ is a compact subset of $R^{n \times n}$, note that a nonorthogonal matrix, or an orthogonal one with determinant of -1, lies inside a small open ball in $R^{n \times n}$ which does not intersect $SO(n)$, and so $SO(n)$ is closed; moreover each row in an orthogonal matrix is a unit vector in Euclidean space, and so each entry is constrained to lie within $[-1, 1]$; thus $SO(n)$ is closed and bounded.

In the case of an abstract differential manifold M of dimension n, the compact sets are not so easy to characterize, partly because the notion of boundedness no longer makes sense, since there is no canonical measure of distance.

5.3   Differentiable Mappings between Manifolds

A function from an open subset of Euclidean space into another Euclidean space is called C^r if its derivatives exist up to order r and are continuous; it is called C^∞ if it is C^r for all $r \geq 1$.

5.3.1 Definition of a C^r Mapping

Let $r \geq 1$ be an integer, or else ∞. If M and N are differential manifolds, with dimensions m and n, respectively, $f: M \to N$ is said to be a C^r **mapping** (or C^r **immersion**, or C^r **submersion**) from M to N if, for every $p \in U$, there is a chart (U, φ) for M at p and a chart (V, ψ) for N at $f(p)$ with $f(U) \subseteq V$, such that

$$\psi \bullet f \bullet \varphi^{-1} : \varphi(U) \subseteq R^m \to \psi(V) \subseteq R^n$$

is a C^r map (or a C^r immersion, or C^r submersion, respectively); a C^∞ map is usually called a **smooth** map. Then for any charts (U_1, φ_1) for M at p and (V_1, ψ_1) for N at $f(p)$, restricted if necessary so that $f(U_1) \subseteq V_1$, $\psi_1 \bullet f \bullet \varphi_1^{-1} : \varphi(U_1) \to \psi(V_1)$ is also smooth by the chain rule, because

$$\psi_1 \bullet f \bullet \varphi_1^{-1} = (\psi_1 \bullet \psi^{-1}) \bullet (\psi \bullet f \bullet \varphi^{-1}) \bullet (\varphi \bullet \varphi_1^{-1}).$$

These maps are illustrated below. Naturally the immersion case can only occur when $n = m + k$ for nonnegative k, and the submersion case only when $m = n + k$.

5.3.1.1 Definition of C^r Diffeomorphism

A C^r map f is called a C^r **diffeomorphism** if it is one-to-one and onto, and its inverse f^{-1} is also C^r.

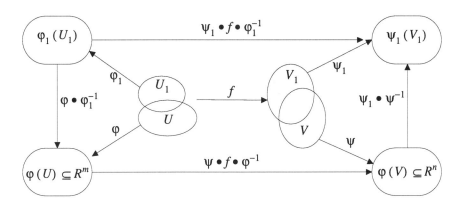

5.3.2 Examples

(i) In the case where M is an open interval J in R, a C^r map $\tau: J \to N$ is called a C^r **curve** in N; this map is an immersion if, at every $s \in J$ and some (hence every) chart (V, ψ) at $\tau(s)$, the derivative $(\psi \bullet \tau)'(s)$ is nonzero. For example, for any $n \times n$ matrix H, there exists $\varepsilon > 0$ such that $\tau(s) = I + sH$, $-\varepsilon < s < \varepsilon$, is a smooth curve in $GL_n^+(R)$, and in fact an immersion if $H \neq 0$.

(ii) Another simple example of a smooth map between two manifolds is the **inclusion map** $\iota: GL_n^+(R) \to GL_{n+1}^+(R)$ which sends the nonsingular $n \times n$ matrix A to the nonsingular $(n+1) \times (n+1)$ matrix

$$\begin{bmatrix} A & 0 \\ 0 & 1 \end{bmatrix}.$$

Since $GL_n^+(R)$ is a subset of a Euclidean space, we can take the trivial chart $(U, \varphi) = (GL_n^+(R), \text{identity})$, and similarly for $GL_{n+1}^+(R)$. The derivative of ι is simply

$$D\iota(A)H = \iota(A+H) - \iota(A) = \begin{bmatrix} H & 0 \\ 0 & 0 \end{bmatrix},$$

which shows that $D\iota(A)$ is one-to-one, and hence ι is an immersion.

(iii) To see an example of a submersion, consider the map f which sends a 3×3 nonsingular matrix A to its first column $A[1, 0, 0]^T$, which is a nonzero vector. Evidently we can treat this as a map $f: GL_3(R) \to N = R^3 - \{(0, 0, 0)\}$ from the general linear group to R^3 minus the origin. The reader is left to calculate $Df(A)H$ as in the previous example, and check that $H \to Df(A)H$ is onto.

5.4   **Exercises**

1. The n-sphere $S^n = \{(x_0, ..., x_n) \in R^{n+1} : x_0^2 + ... + x_n^2 = 1\}$ was discussed in Chapter 3. Generalize Example 5.1.1 to show that the following pair of charts (U, φ) and (V, ψ), known as **stereographic projections**, constitute an atlas for the n-sphere:

$$U = S^n - \{(1, 0, ..., 0)\}, \varphi(x_0, ..., x_n) = \frac{(x_1, ..., x_n)}{1 - x_0};$$

$$V = S^n - \{(-1, 0, ..., 0)\}, \psi(x_0, ..., x_n) = \frac{(x_1, ..., x_n)}{1 + x_0}.$$

2. (Continuation) Let $W \subset R^2$ be the open set $(0, 2\pi) \times (0, \pi)$, and define $\Phi_i : W \to S^2, i = 1, 2$, by

$$\Phi_1(\theta, \phi) = (\cos\theta\sin\phi, \sin\theta\sin\phi, \cos\phi),$$

$$\Phi_2(\theta, \phi) = (-\cos\theta\sin\phi, \cos\phi, \sin\theta\sin\phi).$$

(i) Define charts on S^2 by $(U_i, \psi_i) = (\Phi_i(W), \Phi_i^{-1})$ for $i = 1, 2$. Draw a picture to convince yourself that the domains of these charts cover S^2, and show that they form an atlas.

(ii) For the 2-sphere, describe precisely the steps that have to be performed to show that the atlas $\{(U, \varphi), (V, \psi)\}$ in the previous exercise is equivalent to the atlas $\{(U_1, \psi_1), (U_2, \psi_2)\}$. You need not carry out the whole proof, but at least show that $\varphi \bullet \psi_1^{-1}$ is a diffeomorphism.

3. Suppose $\{(U_\alpha, \varphi_\alpha), \alpha \in I\}$ is an atlas for a set M, and $\{(V_\gamma, \psi_\gamma), \gamma \in J\}$ is an atlas for a set N.

(i) Construct an atlas for $M \times N = \{(p, q) : p \in M, q \in N\}$.

(ii) Show that for the associated differentiable structures on M, N, and $M \times N$, the inclusion map $p \to (p, q)$ from M to $M \times N$, and the projection map $(p, q) \to q$ from $M \times N$ to N are smooth.

4. Suppose M is an n-dimensional submanifold of R^{n+k}. If M is given the differentiable structure described in 5.1.3, show that the identity map from M to R^{n+k} is smooth.

5. Let $W = \{(x, y, z) \in S^2 \subset R^3 : z > 0\}$, in other words the "Northern hemisphere" of S^2.

(i) Show that W is open in S^2 for the differentiable structure described in Exercise 1.

(ii) Let $B(0, 1) = \{(x, y) \in R^2 : x^2 + y^2 < 1\}$. Show that the map

$$f(x, y) = (x, y, \sqrt{1 - x^2 - y^2})$$

from $B(0, 1)$ to S^2 is a smooth diffeomorphism onto W.

6. Prove that if $f: P \to M$ and $g: M \to N$ are C^r maps, then $g \bullet f: P \to N$ is a C^r map.

7. Fix a nonzero vector $v \in R^3$. Show that the map $F: GL_3(R) \to S^2 \subset R^3$ given by $F(A) = (Av) / \|Av\|$ is smooth.

   Hint: Express F as the composite of $A \to Av$ and $w \to w/\|w\|$, and show that both of these maps are smooth.

5.5   Submanifolds

A subset Q of an $(n + k)$-dimensional manifold M is called an **n-dimensional submanifold** of M if, for every $q \in Q$, there exists a chart (U, φ) for M at q such that $\varphi(Q \cap U)$ is an n-dimensional submanifold of R^{n+k}. It follows that Q is itself an n-dimensional manifold, using the atlas obtained by taking chart maps of the form $\Psi^{-1} \bullet \varphi$ with suitable domain in Q, and image in R^n, where Ψ is a parametrization for $\varphi(Q \cap U)$. The integer k is called the **codimension** of the submanifold Q in M. In the case where $M = R^{n+k}$, this definition is seen to be internally consistent, by taking $(U, \varphi) = (R^{n+k}, \text{identity})$.

It could be tedious to use the definition as it stands to check that a subset of M is a submanifold, because one may have to perform calculations in every single chart of some atlas. In the next section, we shall give quicker ways to identify a submanifold. First we develop a useful technical result.

5.5.1   Properties of Maps between Submanifolds

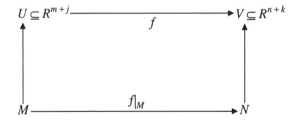

Suppose M is an m-dimensional submanifold of an open subset U of R^{m+j}, N is an n-dimensional submanifold of an open subset V of R^{n+k}, and $f: U \to V$ is a C^r map (or a C^r immersion, or C^r submersion, respectively) such that $f(M) \subseteq N$. Then the

restriction of f to M, denoted $f|_M$, is a C^r map (or a C^r immersion, or C^r submersion, respectively) from M to N.

Proof: Take any $p \in M$, and let $q = f(p) \in N$. According to the Implicit Function Theorem of Chapter 3, there exist smooth parametrizations $\Phi: W \subseteq R^m \to U' \subseteq R^{m+j}$ and $\Psi: Y \subseteq R^n \to V' \subseteq R^{n+k}$ at p and q, respectively, of the form

$$\Phi(x) = (x, z(x)) \in R^m \times R^j, \ \Psi(y) = (y, w(y)) \in R^n \times R^k, \tag{5.1}$$

possibly after rearranging the order of variables. Let us express the map f as

$$f(p) = f(x, z) = (F(x, z), G(x, z)) \in R^n \times R^k. \tag{5.2}$$

The situation is shown in the picture below.

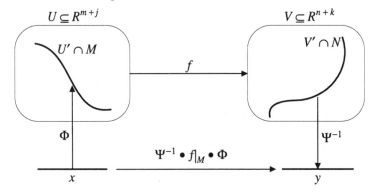

Since $(U' \cap M, \Phi^{-1})$ and $(V' \cap N, \Psi^{-1})$ are charts at p and q respectively, it suffices to show that if $f: U \to V$ is a C^r map (or a C^r immersion, or C^r submersion, respectively), then so is

$$\Psi^{-1} \bullet f|_M \bullet \Phi = F \bullet \Phi: W \to Y,$$

where we use the notation of (5.2).

If f is a C^r map, then so is F and hence so is $F \bullet \Phi$, since Φ is a smooth parametrization. If f is a C^r immersion, then all three components in the derivative

$$D(\Psi^{-1} \bullet f|_M \bullet \Phi)(x) = D\Psi^{-1}(q) \bullet Df(p) \bullet D\Phi(x)$$

are one-to-one linear maps, and so the composite is one-to-one; this proves that $F \bullet \Phi$ is a C^r immersion.

If f is a C^r submersion, write the derivative of f as the $(n + k) \times (m + j)$ matrix

$$Df(p) = \begin{bmatrix} F_x & F_z \\ G_x & G_z \end{bmatrix} \tag{5.3}$$

in the abbreviated notation $F_x = D_x F$, etc., where necessarily $m + j \geq n + k$. Since $Df(p)$ is onto, given $\zeta \in T_y R^n$ there exists a solution $(\xi_1, \xi_2) \in T_x(R^m \times R^j)$ to the linear equation

$$\begin{bmatrix} F_x & F_z \\ G_x & G_z \end{bmatrix} \begin{bmatrix} \xi_1 \\ \xi_2 \end{bmatrix} = \begin{bmatrix} \zeta \\ w_y \zeta \end{bmatrix}. \tag{5.4}$$

However, (5.1), (5.2), and the constraint that $f(M) \subseteq N$ imply by the chain rule that

$$D(F \bullet \Phi)(x) = DF(x, z(x)) = F_x + F_z z_x; \tag{5.5}$$

$$D(G \bullet \Phi)(x) = D(w \bullet F \bullet \Phi)(x) = w_y \bullet (F_x + F_z z_x) = G_x + G_z z_x. \tag{5.6}$$

On comparison of (5.4), (5.5), and (5.6), matrix algebra shows that, once ξ_1 is selected, then setting $\xi_2 = z_x \xi_1$ solves (5.4). In other words, given $\zeta \in T_x R^n$ there exists $\xi_1 \in T_x R^m$ such that

$$(F_x + F_z z_x)\, \xi_1 = \zeta.$$

By (5.5), this is equivalent to saying that $D(F \bullet \Phi)(x)$ is onto, which proves that $\Psi^{-1} \bullet f|_M \bullet \Phi = F \bullet \Phi \colon W \to Y$ is a submersion, as desired. ¤

5.6 Embeddings

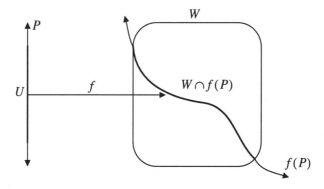

Figure 5. 1   Picture of an embedding

If P and M are differential manifolds, a one-to-one C^∞ immersion $f: P \to M$ is called an **embedding** if, for every open set U in P, there exists an open set W in M such that $f(U) = W \cap f(P)$, as shown in Figure 5. 1.

Note that a map which sends an open interval of R into the curve in R^2 shown in Figure 5. 2 is not an embedding. To see why it is not an embedding, first convince yourself that the image of the map is not a submanifold of R^2, by looking at what happens at the point in the center under a submersion; then apply 5.6.1..

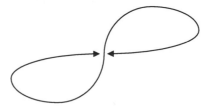

Figure 5. 2   Nonexample of an embedding

Here is the first useful result for obtaining submanifolds without recourse to charts.

5.6.1 Submanifolds Obtained through Embeddings and Submersions

Either of the following conditions implies that Q is a submanifold of M:

- $f: P \to M$ is a C^∞ embedding, and $Q = f(P)$, or
- $g: M \to N$ is a C^∞ submersion, $p \in N$, and $Q = g^{-1}(p)$.

Proof: Suppose $f: P \to M$ is a C^∞ embedding, and $Q = f(P)$. We may suppose that P has dimension n and M has dimension $n + k$. Given any $q \in Q$, let p be the unique point in P such that $q = f(p)$, and take a chart (U, φ) for P at p and a chart (V, ψ) for M at q with $f(U) \subseteq V$, such that

$$\psi \bullet f \bullet \varphi^{-1} : \varphi(U) \subseteq R^n \to \psi(V) \subseteq R^{n+k}$$

is a one-to-one C^r immersion; these charts exist by definition. By a result presented in one of the Exercises of Chapter 3, there exists an open set $W \subseteq \varphi(U)$ which contains $\varphi(p)$ such that $\psi \bullet f \bullet \varphi^{-1}(W)$ is an n-dimensional submanifold of $\psi(V)$. Since f is an embedding, and since $\varphi^{-1}(W)$ is open in P, there is an open set $V' \subseteq V \subseteq M$ such that $f(\varphi^{-1}(W)) = V' \cap Q$. Now (V', ψ) is a chart for M at q such that $\psi(V' \cap Q)$ is an n-dimensional submanifold of R^{n+k}. This proves that Q is a submanifold of M. The proof of the other assertion, which is easier, is left as an exercise.   ¤

It would appear that to verify that a mapping is an embedding is not easy. A useful fact from topology, whose proof (though not difficult) is outside the scope of this book, is:

5.6.2 Compactness Lemma

If $f: P \to M$ is a one-to-one C^r immersion, and P is compact, then f is an embedding.

Another useful topological result is the following.

5.6.3 Relation of Open Sets in a Manifold to Those of a Submanifold

If Q is a submanifold of M, then the open sets of Q are precisely the sets

$$\{U' \cap Q : U' \text{ open in } M\}. \tag{5.7}$$

Proof: This proof is postponed to Section 5.10 because it is somewhat technical. ¤

5.6.4 Inclusion[2] of a Submanifold Is an Embedding

If Q is a submanifold of M, then the inclusion map $\iota : Q \to M$ is an embedding.

Proof: The condition that, for every open set $U \subseteq Q$, there is an open set $U' \subseteq M$ such that $\iota(U) = U' \cap \iota(Q)$, is verified in 5.6.3. Since the inclusion map is already one-to-one, it only remains to prove that it is an immersion. First we claim that, for every $r \in Q$ and every chart (U, φ) for M at r, the inclusion

$$\text{id} : \varphi(U \cap Q) \to \varphi(U) \subseteq R^n$$

is an immersion; this is true by 5.5.1, since $\varphi(U \cap Q)$ is a submanifold of $\varphi(U)$ by assumption, and $\text{id} : \varphi(U \cap Q) \to \varphi(U) \subseteq R^n$ is the restriction of the identity map, which is an immersion on R^n, to a submanifold. Now if we take a k-dimensional chart (V, ψ) for $\varphi(U \cap Q)$ at $\varphi(r)$, then $(U \cap \varphi^{-1}(V), \psi \bullet \varphi)$ is a chart for Q at r, and $\varphi \bullet \iota \bullet (\psi \bullet \varphi)^{-1} = \psi^{-1}$ is an immersion; this verifies the condition for $\iota : Q \to M$ to be an immersion. ¤

The next result is included merely to show that the relation "is a submanifold of" is reasonably well behaved.

5.6.5 Composition of Embeddings

(i) The composition of two embeddings is an embedding.

(ii) M a submanifold of N and N a submanifold of P implies M is a submanifold of P. (Note: The dimensions of M, N, and P could be different.)

---

[2] If S is a subset of a set T, then the inclusion map $\iota : S \to T$ simply means the identity map restricted to S.

Proof: The proof of (i) is left as an exercise. Given (i), we may prove (ii) as follows: By 5.6.4 the inclusion maps $\iota_1 : M \to N$ and $\iota_2 : N \to P$ are embeddings, hence $\iota_2 \bullet \iota_1 : M \to P$ is an embedding, and so M is a submanifold of P by 5.6.1.   ¤

Finally, here is a result that will save us from a lot of calculation when dealing with submanifolds of R^n.

5.6.6   When a Subset of a Submanifold is a Submanifold

If M and N are both submanifolds of R^n, and if $M \subseteq N$, then M is a submanifold of N.

Proof: The identity map $\iota : R^n \to R^n$ is an immersion, and hence its restriction to M is an immersion into N by 5.5.1; it is also one-to-one. To prove that M is a submanifold of N, it suffices by 5.6.1 to prove that $\iota : M \to N$ is an embedding. Given U open in M, we know by 5.6.3 that $U = U' \cap M$ for some U' open in R^n. So $V = U' \cap N$ is open in N by 5.6.3, and $\iota(U) = V \cap M$. Thus $\iota : M \to N$ is an embedding as desired.   ¤

5.7   Constructing Submanifolds without Using Charts

The preceding results are especially useful for dealing with mappings and submanifolds of matrix groups, such as $SO(n)$, for which convenient charts are not available. Here are some examples.

5.7.1   A Sphere as a Submanifold of a Hyperboloid

We may construct a sphere $S^{n-1} \subset R^{n+1}$ as follows:

$$\{x = (0, x_1, \ldots, x_n) : f(x) = x_1^2 + \ldots + x_n^2 - 1 = 0\}$$

Note that this sphere is a submanifold of R^n, and R^n is a submanifold of R^{n+1}, and therefore this sphere is a submanifold of R^{n+1}, by 5.6.5. It is also a subset of the hyperboloid $H_{-1}^n \subset R^{n+1}$, which is the submanifold of R^{n+1} defined by

$$\{x = (x_0, \ldots, x_n) : h(x) = x_1^2 + \ldots + x_n^2 - 1 = x_0^2\}.$$

Think, for example, of the unit circle in the x, y plane inside the figure of a hyperboloid in 3-space in Chapter 3. Since S^{n-1} is a subset of $H_{-1}^n \subset R^{n+1}$, it follows from 5.6.6 that S^{n-1} is a submanifold of H_{-1}^n.

5.7.2 A Submersion of the Special Orthogonal Group

Let $v \in R^n$, $v \neq 0$, and define $f: GL_n^+(R) \to R^n$ by $f(A) = Av$. Since $Df(A)H = Hv$, $Df(A)$ is clearly onto, and so f is a submersion. Hence f restricted to $SO(n)$ is also a submersion by 5.5.1, and therefore by 5.6.1,

$$G = f^{-1}(v) = \{A \in SO(n) : Av = v\}$$

is a submanifold of $SO(n)$. As a special case, take v to be the vector $[0, \ldots, 0, 1]^T$; then $Av = v$ implies that the last column of A is $[0, \ldots, 0, 1]^T$, and by orthogonality of the columns of $A \in SO(n)$, we see that

$$G = \{\begin{bmatrix} B & 0 \\ 0 & 1 \end{bmatrix} : B \in SO(n-1)\} \cong SO(n-1).$$

In this particular case, it would have been easier to prove that G is a submanifold of $SO(n)$ by using 5.6.6, since G is clearly a submanifold of $GL_n^+(R)$, which is open in $R^{n \times n}$.

5.7.3 An Embedding of the Special Orthogonal Group

In 5.3.2 we considered the smooth immersion $\iota: GL_n^+(R) \to GL_{n+k}^+(R)$ which sends the nonsingular $n \times n$ matrix A to the nonsingular $(n+k) \times (n+k)$ matrix

$$\begin{bmatrix} A & 0 \\ 0 & B \end{bmatrix},$$

where B is any matrix in $SO(k)$. We also know from the construction of the Special Orthogonal Group in Chapter 3 that $SO(n)$ is a submanifold of $GL_n^+(R)$, and that ι maps $SO(n)$ into $SO(n+k)$. According to 5.5.1, the restricted map

$$\iota: SO(n) \to SO(n+k)$$

is also an immersion, and it is clearly one-to-one. Since $SO(n)$ is compact, the "Compactness Lemma" 5.6.2 shows that ι is an embedding. Now 5.6.1 shows that the image of $SO(n)$ under ι, which can be identified with $SO(n)$ itself, is a submanifold of $SO(n+k)$.

5.8 Submanifolds-with-Boundary

The material in this section will be needed in Chapter 8 in discussing orientation, and in the statement and proof of Stokes's Theorem. A closed subset P of an n-dimensional manifold M will be called an ***n*-dimensional submanifold-with-boundary** of M if, for every $r \in P$, one of the two following conditions hold:

- there exists an open subset U of M such that $r \in U \subseteq P$;
- r lies in a chart (U, φ) such that $\varphi(P \cap U) = \{ (x_1, ..., x_n) \in \varphi(U) : x_1 \leq 0 \}$.

In the first case r is said to lie in the **interior** P° of P; in the second, the points $r \in P$ such that $\varphi(r) = (0, x_2, ..., x_n)$ are said to lie on the **boundary**[3] ∂P of P.

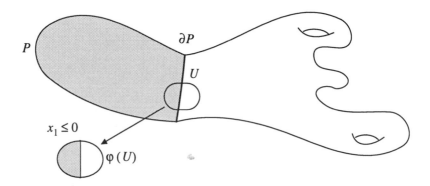

Note the trivial case where $P = M$ and the boundary of P is the empty set. In all other cases, we have the following important property.

5.8.1   The Boundary Is a Submanifold

If P is an n-dimensional submanifold-with-boundary of M whose boundary ∂P is nonempty, then ∂P is an $(n-1)$-dimensional submanifold of M.

Proof: For any point $r \in \partial P$, we may take a chart (U, φ) for M at r such that

$$\varphi(\partial P \cap U) = \{ (x_1, ..., x_n) \in \varphi(U) : x_1 = 0 \} = x^1|_{\varphi(U)}^{-1}(0),$$

where x^1 denotes the projection from R^n onto the first coordinate. Since x^1 is a submersion, this shows that $\varphi(\partial P \cap U)$ is an $(n-1)$-dimensional submanifold of R^n. Thus we have verified the condition for ∂P to be an $(n-1)$-dimensional submanifold of M.   ¤

Following the usual pattern, we would like to find a way to verify that a subset of M is a submanifold-with-boundary without having to take charts. The following result is convenient.

---

[3] Warning! This is not the same as the topological definition of boundary.

5.8.2 Characterization Using Submersions

*P is an n-dimensional submanifold-with-boundary of M if and only if, for every $r \in P$, there exists an open set $U \subseteq M$ containing r so that **either** $r \in U \subseteq P$ **or** there is a submersion $f: U \to R$ with $P \cap U = \{q \in U : f(q) \leq 0\}$; in the second case*

$$\partial P \cap U = \{q \in U : f(q) = 0\}. \tag{5.8}$$

Proof: The "$\Rightarrow$" part of the proof is immediate on taking, for the second case, (U, φ) to be a chart at r of the kind described in the definition, and $f = x^1 \bullet \varphi$ to be the submersion. To prove the "$\Leftarrow$" part, assume we have a submersion $f: U \to R$ with $P \cap U = \{q \in U : f(q) \leq 0\}$ where $U \ni r$; we need to find a chart at r of the special kind mentioned in the definition of submanifold-with-boundary. For this, start with any chart (V, ψ) for M at r, with $V \subseteq U$ without loss of generality, and let $x = \psi(r)$. Observe that $D(f \bullet \psi^{-1})(x) \neq 0$, since $f \bullet \psi^{-1}$ is a submersion on $\psi(V) \subseteq R^n$, and therefore there exists a set consisting of $(n-1)$ of the basis vectors $\{e_1, ..., e_n\}$ (after relabeling, we can suppose the subset to be $\{e_2, ..., e_n\}$) whose span does not include the vector $D(f \bullet \psi^{-1})(x)$. In other words, the following determinant satisfies:

$$\begin{vmatrix} D_1(f \bullet \psi^{-1})(x) & 0 & ... & 0 \\ D_2(f \bullet \psi^{-1})(x) & 1 & ... & 0 \\ ... & ... & ... & ... \\ D_n(f \bullet \psi^{-1})(x) & 0 & ... & 1 \end{vmatrix} \neq 0. \tag{5.9}$$

If we define $\varphi: V \to R^n$ by $\varphi^1 = f$, $\varphi^2 = \psi^2$, ..., $\varphi^n = \psi^n$, then (5.9) says that

$$|D(\varphi \bullet \psi^{-1})(x)| \neq 0.$$

Now we may apply the Inverse Function Theorem (see Chapter 3) to assert that there is a neighborhood $W \subseteq \psi(V)$ of $x \in R^n$ on which $\varphi \bullet \psi^{-1}$ is a diffeomorphism; if we take $U' = \psi^{-1}(W) \subseteq U$, it follows from this that (U', φ) is a chart for M at r such that

$$\varphi(P \cap U') = \varphi(\{q \in U' : f(q) \leq 0\}) = \{(x_1, ..., x_n) \in \varphi(U') : x_1 \leq 0\},$$

which verifies the condition for P to be an n-dimensional submanifold-with-boundary of M. The assertion (5.8) follows immediately. ¤

5.8.3 Examples

5.8.3.1 A Closed Ball in Euclidean Space

Take $M = R^{n+1}$, and $f: R^{n+1} \to R$ to be the map $f(x_0, ..., x_n) = x_0^2 + ... + x_n^2 - 1$. Since f restricted to $U = R^{n+1} - \{0\}$ is a submersion, 5.8.2 shows that the closed ball

$$\overline{B(0,1)} \;=\; \{\,(x_0, \ldots, x_n) : f(x_0, \ldots, x_n) \le 1\}$$

is an $(n+1)$-dimensional submanifold-with-boundary of R^{n+1} whose boundary is the sphere S^n.

5.8.3.2   A General Class of Examples

Suppose U is an open set in R^{n+k}, $g: U \to R^k$ and $f: U \to R$ are submersions, and $a \in R^k$ and $b \in R$ are chosen so that $M = g^{-1}(a)$ and $\{r \in M : f(r) < b\}$ are non-empty. As we saw in Chapter 3, M is an n-dimensional submanifold of R^{n+k}. By 5.5.1, the restriction of f to M is a submersion, and 5.8.2 shows that $P = \{r \in M : f(r) \le b\}$ is an n-dimensional submanifold-with-boundary of M, with boundary

$$\partial P \;=\; \{\,r \in U : g(r) = a, f(r) = b\}. \tag{5.10}$$

For a simple illustration, take $n = k = 2$, $U = R^4 - \{0\}$, and

$$g(x_0, \ldots, x_3) \;=\; (x_0^2 + x_1^2, x_2^2 + x_3^2), f(x_0, \ldots, x_3) \;=\; x_0 - 2x_1 + 3x_2 - 4x_3.$$

If $a = (1,1)$ and $b = -2$, then $M = T^2 \subset R^4$, and ∂P is the intersection of the 2-torus with the hyperplane $x_0 - 2x_1 + 3x_2 - 4x_3 = -2$.

5.9   Exercises

8. Prove the second part of 5.6.1, namely, that if $g: M \to N$ is a C^∞ submersion and $p \in N$, then $Q = g^{-1}(p)$ is an n-dimensional submanifold of M, where M and N are assumed to have dimensions $n+k$ and k, respectively.

   Hint: This follows straight from the definitions.

9. Suppose P is a product of two hyperboloids, and M is the product of two cylinders, as indicated by the following formulas:

$$P \;=\; \{\,(x_0, \ldots, x_5) \in R^6 : x_0^2 - x_1^2 - x_2^2 = 1, x_3^2 - x_4^2 - x_5^2 = -1\};$$

$$M \;=\; \{\,(x_0, \ldots, x_5) \in R^6 : x_1^2 + x_2^2 = 1, x_4^2 + x_5^2 = 1\}.$$

   (i) Show directly that P and M are submanifolds of R^6.

   (ii) Define $P \cap M$, and determine whether it is a submanifold of R^6, P, and M, respectively.

   Hint: Once you have proved the first part by showing that certain maps are submersions, part (ii) can be done by the results proved in this chapter.

10. Show that the rescaled torus

$$T^n = \{(x_1, ..., x_{2n}) \in R^{2n} : x_1^2 + x_2^2 = 1/n, ..., x_{2n-1}^2 + x_{2n}^2 = 1/n\}$$

is a submanifold of $S^{2n-1} = \{(x_1, ..., x_{2n}) \in R^{2n} : x_1^2 + ... + x_{2n}^2 = 1\}$.

11. Show that if $\iota : R^n \to R^{n+k}$ is the inclusion map which takes $(x_1, ..., x_n)$ to $(x_1, ..., x_n, 0, ..., 0)$, then the image of the sphere $S^{n-1} \subset R^n$ under ι is a submanifold of the sphere $S^{n+k-1} \subset R^{n+k}$.

12. Let us accept without proof that the map from $U(n)$ (the Unitary Group discussed in Chapter 3) to C which takes a complex matrix to its determinant is a submersion. Use that fact to show that the Special Unitary Group $SU(n)$ is a submanifold of $U(n)$, and a submanifold of $GL_n(C)$, and find its dimension.

13. Show that if $f : M \to N$ and $g : N \to P$ are embeddings, then $g \bullet f$ is an embedding.
    Hint: There are three things to check; they follow from the definitions.

14. Find an example where M and P are submanifolds of R^2, and $f : P \to M$ is a C^∞ immersion, but $f(P)$ is not a submanifold of M.

15. Let H be a symmetric $n \times n$ matrix of rank n, and J a symmetric $m \times m$ matrix of rank m. Define

$$G_1 = \{A \in GL_{n+m}^+(R) : A^T \begin{bmatrix} 0 & 0 \\ 0 & H \end{bmatrix} A = \begin{bmatrix} 0 & 0 \\ 0 & H \end{bmatrix}\};$$

$$G_2 = \{A \in GL_{n+m}^+(R) : A^T \begin{bmatrix} J & 0 \\ 0 & H \end{bmatrix} A = \begin{bmatrix} J & 0 \\ 0 & H \end{bmatrix}\}.$$

Determine whether G_1 and G_2 are submanifolds of $GL_{n+m}^+(R)$, and whether G_2 is a submanifold of G_1. If so, find the codimension of G_2 in G_1.
Remark: If H and J are both equal to the identity, then G_2 is simply $SO(n+m)$.

16. Fix a nonzero vector $w \in R^n$, and let $P = \{A \in GL_n(R) : w^T A w \le 1\}$. Prove that P is an n^2-dimensional submanifold-with-boundary of $GL_n(R)$, and determine the boundary.

17. Let $M = \{(x_0, ..., x_5) \in R^6 : x_1^2 + x_2^2 = 1, x_4^2 + x_5^2 = 1\}$, and let

$$P = \{(x_0, ..., x_5) \in M : x_0^2 - x_1^2 + x_3^2 - x_4^2 \le 0\}.$$

Determine whether P is a 4-dimensional submanifold-with-boundary of M, and if so determine the boundary.

5.10   Appendix: Open Sets of a Submanifold

The following technical result was needed in the proof of 5.6.3.

5.10.1   Relation of Open Sets in a Manifold to Those of a Submanifold

If Q is a submanifold of M, then the open sets of Q are precisely the sets

$$\{ U' \cap Q: U' \text{ open in } M \}. \tag{5.11}$$

Proof: We shall prove the assertion in three steps.

Step I. Consider first the case where $M = R^{n+k}$, and Q is an n-dimensional submanifold of the form $Q = \Psi(W)$, where $\Psi: W \subseteq R^n \to Q \subset U$ is an n-dimensional implicit function parametrization, that is,

$$\Psi(x_1, ..., x_n) = (x_1, ..., x_n, z_1(x), ..., z_k(x)), \tag{5.12}$$

and W and U are open in R^n and R^{n+k}, respectively. By Definition 5.2.1, V is open in Q if and only if V is of the form $\Psi(W_1)$ for some W_1 open in W; thus

$$V = U_1 \cap Q,$$

where $U_1 \subset R^{n+k}$ is the open set

$$U_1 = \{ (x, y) \in R^n \times R^k : x \in W_1, z_i(x) - 1 < y_i < z_i(x) + 1, 1 \leq i \leq k \};$$

$$x = (x_1, ..., x_n), y = (y_1, ..., y_k).$$

Conversely, if $V = U_1 \cap Q$ and $U_1 \subseteq R^{n+k}$ is open, then V is the image of the projection of U_1 onto the first n components, which is an open set W_1 in W; thus $V = \Psi(W_1)$ is open in Q. Thus the assertion is proved in this special case.

Step II. Next consider the case where Q is any n-dimensional submanifold of $M = R^{n+k}$. By 5.1.3 and the Implicit Function Theorem of Chapter 3, we may take an atlas for Q consisting of charts of the form $\{ (V_\alpha, \Psi_\alpha^{-1}), \alpha \in I \}$, where each

$$\Psi_\alpha: W_\alpha \subseteq R^n \to V_\alpha = U_\alpha \cap Q \subseteq R^{n+k}$$

is a parametrization of the form (5.12) above, maybe with the coordinates rearranged; by definition, Ψ_α is onto, U_α is open in R^{n+k}, and $Q = \bigcup V_\alpha$. By definition 5.2.1,

$$V \text{ open in } Q \Leftrightarrow V \cap V_\alpha \text{ open in } \Psi_\alpha(W_\alpha), \forall \alpha$$

$$\Leftrightarrow V \cap V_\alpha = U_{1,\alpha} \cap Q, \forall \alpha,$$

by Step I, where each $U_{1,\alpha}$ is some open subset of R^{n+k}. This is equivalent to

$$V = \bigcup_\alpha (V \cap V_\alpha) = \bigcup_\alpha (U_{1,\alpha} \cap Q) = \left(\bigcup_\alpha U_{1,\alpha}\right) \cap Q = U_1 \cap Q,$$

where $U_1 \subseteq R^{n+k}$ is open, since it is the union of open sets. Thus the assertion is proved for this case.

Step III. Finally consider the general case. Let $\{(U_\gamma', \varphi_\gamma), \gamma \in J\}$ be an atlas for M. By the definition of submanifold, and of the differentiable structure of Q,

$$V \text{ open in } Q \Leftrightarrow \varphi_\gamma (V \cap U_\gamma') \text{ open in } \varphi_\gamma (Q \cap U_\gamma'), \forall \gamma,$$

$$\varphi_\gamma (V \cap U_\gamma') = U_{1,\gamma} \cap \varphi_\gamma (Q \cap U_\gamma') \subseteq R^{n+k}, \forall \gamma,$$

for some open set $U_{1,\gamma}$ in R^{n+k}, by the result of Step II, since $\varphi_\gamma (Q \cap U_\gamma')$ is a submanifold of R^{n+k}. This is equivalent to

$$V \cap U_\gamma' = \varphi_\gamma^{-1} (U_{1,\gamma}) \cap Q, \forall \gamma,$$

where necessarily $\varphi_\gamma^{-1} (U_{1,\gamma})$ is open in M. This is equivalent to

$$V = U_1' \cap Q, U_1' = \bigcup_\gamma \varphi_\gamma^{-1} (U_{1,\gamma}),$$

and the last set is open in M, so the proof is complete. ¤

5.11 Appendix: Partitions of Unity

This is a technical tool which will be needed in Chapter 8. Suppose $\{U_j, j \in J\}$ is an open cover of a differential manifold M. A (smooth) **partition of unity** subordinate to $\{U_j, j \in J\}$ means an open cover $\{V_\alpha, \alpha \in I\}$ of M together with a collection $\{\upsilon_\alpha : M \to [0,1], \alpha \in I\}$ of smooth functions with the following four properties:

- The support[4] of υ_α is contained in V_α.
- For each V_α there exists a $j(\alpha)$ such that $V_\alpha \subseteq U_{j(\alpha)}$.
- $\{V_\alpha, \alpha \in I\}$ is **locally finite**, which means that every $r \in M$ has a neighborhood U such that $U \cap V_\alpha = \varnothing$ for all but finitely many α.

---

[4] The support of a function f means the smallest closed set containing the set on which $f \neq 0$, that is, the complement of the union of all the open sets on which $f = 0$.

- $\sum_{\alpha \in I} \upsilon_\alpha(r) = 1, \forall r \in M$. (For fixed r, only finitely many summands are nonzero.)

In applications we usually omit mention of the $\{V_\alpha, \alpha \in I\}$, and refer simply to "a partition of unity $\{\upsilon_\alpha, \alpha \in I\}$ subordinate to $\{U_j, j \in J\}$."

5.11.1   Existence of a Partition of Unity

If M is a differential manifold with a countable atlas $\{(U_j, \varphi_j), j \in J\}$, *then there exists a partition of unity subordinate to* $\{U_j\}$.

Proof: The proof is essentially topological, and is therefore outside the scope of these notes; see Berger and Gostiaux [1988] or Spivak [1979], Volume I. A more sophisticated treatment applicable to infinite-dimensional manifolds may be found in Lang [1972]. ¤

5.11.2   Example

The idea of a partition of unity is intuitively simple. Take, for example, the circle $S^1 = \{(x, y) \in R^2 : x^2 + y^2 - 1 = 0\}$ with the atlas $\{(U, \varphi), (V, \psi)\}$, where

$$U = S^1 - (0, 1), \ \varphi(x, y) = \frac{x}{1-y}; \ V = S^1 - (0, -1), \ \psi(x, y) = \frac{x}{1+y}.$$

Take any smooth function $\kappa : R \to [0, 1]$ such that $\kappa(t) = 1$ for $t \in [-1, 1]$, and such that, for some $a > 1$, $\kappa(t) = 0$ for $|t| > a$; a formula for this so-called "bump function" is given in Berger and Gostiaux [1988], p. 13.

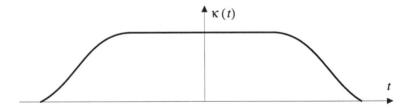

Figure 5. 3   A "bump function"

Define

$$\upsilon_1(x, y) = \kappa(\varphi(x, y)), \ (x, y) \in U; \ \upsilon_1(0, 1) = 0;$$

$$\upsilon_2(x, y) = 1 - \kappa(1/\psi(x, y)), \ (x, y) \in U \cap V; \ \upsilon_2(0, 1) = 1, \upsilon_2(0, -1) = 0.$$

Since $\varphi(x, y) = 1/\psi(x, y)$ on $U \cap V$, it follows that $\upsilon_1 + \upsilon_2 = 1$, and one may check that the support of υ_1 is contained in U, and the support of υ_2 is contained in V.

5.12   History and Bibliography

The notion of a differential manifold emerged in the work of B. Riemann (1826–66), E. Betti (1823–92), H. Poincaré (1854–1912), and others. The modern viewpoint is associated with a 1936 paper by H. Whitney (1907–). A much fuller treatment of these topics may be found in Berger and Gostiaux [1988]. The infinite-dimensional version is in Lang [1972], and in more detail in Abraham, Marsden, and Ratiu [1988].

6   Vector Bundles

The notion of vector bundle gives a powerful and flexible tool for all kinds of calculus on differential manifolds. Important special cases are the tangent bundle of a differential manifold, the cotangent bundle, exterior powers of these bundles, etc. This chapter consists mostly of definitions and constructions.

6.1   Local Vector Bundles

Suppose M is a differential manifold of dimension n, for example, an open subset of R^n, and V is an arbitrary k-dimensional vector space. The product manifold[1] $M \times V$, together with the "projection map" $\pi: M \times V \to M$ such that $\pi(p, v) = p$, is called a **local vector bundle of rank** k over M. We call M the **base manifold**, and $\{p\} \times V$ is called the **fiber over** p, for any $p \in M$; think of a fiber as a copy of the vector space V, sitting on top of the point p, as in the picture on the next page. A convenient general notation is to refer to such a vector bundle as $E = M \times V$, and to refer to the fiber over p as $E_p = \{p\} \times V$.

6.1.1   Sections of Local Vector Bundles

A C^r (resp., C^∞) **section** of a local vector bundle E over M means a C^r mapping[2] $\sigma: M \to M \times R^k$ such that $\sigma(p) \in E_p$ for every p. The idea is simple: σ chooses a point $\sigma(p)$ in the fiber over p for every p, in a way that varies r-times differentiably across the fibers. We denote the set of C^r sections of E by $\Gamma^r(E)$. Every vector bundle has a **zero section** defined by $\sigma(p) = (p, 0) \in \{p\} \times V$.

---

[1] If M is an open subset of R^n, then $M \times R^k$ is an open subset of R^{n+k} with the corresponding differentiable structure; for the general case, see the exercises of Chapter 5.

[2] See Chapter 5 for the meaning of this.

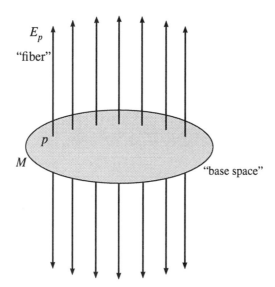

Figure 6. 1   How to visualize a local vector bundle (the fibers shown are 1-dimensional)

6.1.2 Simple Examples

Vector bundle terminology generalizes several constructions we encountered in Chapter 2. Suppose U is an open subset of R^n. First take $V = R^n$, and consider the rank-n vector bundle TU over U, called the **tangent bundle** over U, with fiber T_yR^n, that is, the tangent space at y. The C^∞ sections of this vector bundle are simply the smooth vector fields, or in symbols $\Gamma^\infty(TU) = \mathfrak{I}(U)$.

Next take $V = (R^n)^*$; the **cotangent bundle** T^*U is the rank-n vector bundle over U where the fiber over y is the cotangent space $(T_yR^n)^*$, and whose smooth sections are the 1-forms over U; in symbols, $\Gamma^\infty(T^*U) = \Omega^1 U$.

Name	Rank	Fiber over $y \in U$	Smooth Sections
real line bundle, $U \times R$	1	R	$C^\infty(U)$
tangent bundle, TU	n	T_yR^n	$\mathfrak{I}(U)$, i.e., vector fields
cotangent bundle, T^*U	n	$(T_yR^n)^*$	$\Omega^1 U$, i.e., 1-forms
$\Lambda^q(T^*U)$	$n!/q!\,(n-q)!$	$\Lambda^q((T_yR^n)^*)$ $\cong (\Lambda^q(T_yR^n))^*$	$\Omega^q U$, i.e., q-forms
$T^*U \otimes T^*U$	n^2	$(T_yR^n)^* \otimes (T_yR^n)^*$	(0,2)-tensors

Table 6.1   Examples of local vector bundles over $U \subseteq R^n$, derived from Chapter 2

6.2   Constructions with Local Vector Bundles

6.2.1   Morphisms of Local Vector Bundles

This section may clarify the discussion of differentials and pullbacks in Chapter 2. Suppose $E = M \times V$ is a local rank-k vector bundle over an m-dimensional manifold M, and $E' = N \times W$ is a local rank-q vector bundle over an n-dimensional manifold N. A C^r map $\Phi: E = M \times V \to E' = N \times W$ is called a C^r **(local) vector bundle morphism** if, for some map $\varphi: M \to N$, it takes the form:

$$\Phi(p, v) = (\varphi(p), g(p) v) \in N \times W, \tag{6.1}$$

where $g(p) \in L(V \to W)$ for every y. In other words, Φ is said to be "linear on the fibers," where the linear transformation may vary from fiber to fiber. To clarify the formalism, note that

$$(p, v) \in \{p\} \times V = E_p, \quad \Phi(p, v) \in \{\varphi(p)\} \times W = E'_{\varphi(p)},$$

and therefore

$$(v \to \Phi(p, v)) \in L(E_p \to E'_{\varphi(p)}), \forall p.$$

Multivariable calculus shows that an equivalent condition (in this finite-dimensional case) for a map $\Phi: E \to E'$ to be a C^r local vector bundle morphism is that it takes the form (6.1), and that the maps $\varphi: M \to N$ and $g: M \to L(V \to W)$ are C^r.

We call the map Φ a vector bundle morphism **over the identity** if $M = N$ and the map φ is the identity.

6.2.1.1   Examples: The Tangent Map and Pullback for Mappings of Euclidean Space

From our study of differentials and pullbacks in Chapter 2, some obvious examples spring to mind. Suppose $U \subseteq R^m$ and $U' \subseteq R^n$ are open sets, and $\varphi: U \to U'$ is a smooth onto map. The **tangent map** $T\varphi$ is the morphism from the tangent bundle TU to the tangent bundle TU' which sends a tangent vector $\xi = (y, v)$ at y to

$$T\varphi(y, v) = (\varphi(y), D\varphi(y) v) = d\varphi(y) \xi \in T_{\varphi(y)} R^n. \tag{6.2}$$

Thus for these local tangent bundles, the distinction between the differential and the tangent map is essentially one of formalism: When we view the maps $\{d\varphi(y), y \in U\}$ as a single smooth map between TU and TU', then we obtain $T\varphi$.

The **pullback** of φ is, strictly speaking, a mapping from differential forms on U' (i.e., sections of the exterior powers of the cotangent bundle over U') to differential forms on

U. However we may abuse notation slightly and write φ^* as a map from $T^* U'$ to $T^* U$ given by

$$\varphi^* \left(\varphi(y), a \right) = \left(y, aD\varphi(y) \right) \in \left(T_y R^m \right)^*, \tag{6.3}$$

where we may think of the n-dimensional row vector a as premultiplying the $n \times m$ derivative matrix $D\varphi(y)$ to give an m-dimensional row vector. As in Chapter 2, one may easily write down the corresponding extension of φ^* to any exterior power $\Lambda^q (T^* U')$. Actually the pullback construction is even more general, as we shall see in Exercise 3 in Section 3.5.

6.2.2 Local Vector Bundle Isomorphisms

Consider the special case of a C^r morphism of local vector bundles $\Phi: E \to E'$, where the bundles $E = M \times V$ and $E' = N \times W$ have base manifolds which are diffeomorphic to each other, V and W are isomorphic vector spaces (i.e., of the same dimension k), and where Φ takes the form (6.1) where:

$$\varphi: M \to N \text{ is a } C^r \text{ diffeomorphism, and} \tag{6.4}$$

$$g(p) \in L(V \to W) \text{ is invertible for all } p; \tag{6.5}$$

in other words $g(p)$ is a linear isomorphism between the fiber over p and the fiber over $\varphi(p)$ which varies in a C^r fashion with p. Then we call $\Phi: E \to E'$ a C^r **local vector bundle isomorphism** (over the identity, if φ is the identity map), and we say that E and E' are C^r isomorphic; the usual case is when $r = \infty$, where we often say "E and E' are isomorphic."

Examples would be the tangent map $T\varphi$ and the pullback in Example 6.2.1.1, in the case where $\varphi: U \to U'$ is a diffeomorphism (such as the change of variables map between two coordinate systems on R^n), for in this case the differential of φ is an invertible linear map between respective tangent spaces.

6.2.3 The Homomorphism Bundle

Suppose $E = M \times V$ and $E' = M \times W$ are local vector bundles over the **same** base manifold M. Define the **homomorphism bundle** Hom (E, E') to be the local vector bundle $M \times L(V \to W)$; in other words the fiber over p is

$$\text{Hom}(E, E')_p = L(E_p \to E_p'). \tag{6.6}$$

Clearly the rank of Hom (E, E') is the product of the ranks of E and E', since the dimension of $L(V \to W)$ is dim $V \times$ dim W. Referring to Section 6.2.1, we see that there is a one-to-one correspondence between sections σ of Hom (E, E') and the vector bundle morphisms Φ over the identity from E to E', by the formula

$$\Phi(p, v) = (p, \sigma(p)v) \in M \times W. \tag{6.7}$$

This concept of homomorphism bundle has many applications, of which we now list a few.

6.2.3.1   Dual Bundle

In the case where $E' = M \times R$, the local line bundle over M, the fiber Hom $(E, M \times R)_p$ is simply the dual space E_p^*, and the bundle Hom $(E, M \times R)$ is called the **dual bundle** to E, denoted E^*. For example, the cotangent bundle T^*U described in Section 6.1.2 is the dual of the tangent bundle TU, and a result in Chapter 1 shows that, more generally, the vector bundle $\Lambda^q(T^*U)$ is isomorphic to $(\Lambda^q(TU))^*$.

6.2.3.2   The Differential of a Map as a Section of a Homomorphism Bundle

Suppose $U \subseteq R^m$ and $U' \subseteq R^n$ are open sets, and $\varphi: U \to U'$ is a C^r onto map. Define a new vector bundle

$$\varphi^*(TU') = U \times R^n, \tag{6.8}$$

where the fiber $(\varphi^*(TU'))_y$ is identified with $T_{\varphi(y)}R^n$. This is known as the **pullback** of the tangent bundle TU' under φ. Now the differential $y \to d\varphi(y)$ is a C^{r-1} section of Hom $(TU, \varphi^*(TU'))$, since

$$d\varphi(y) \in L(T_yR^m \to T_{\varphi(y)}R^n) = \text{Hom}(TU, \varphi^*(TU'))_y, \ \forall y \in U, \tag{6.9}$$

and since the map $y \to D\varphi(y)$ has just one degree less of differentiability than φ has.

6.2.3.3   Differential Forms with Values in a Vector Bundle

If E is the tangent bundle TU of an open set $U \subseteq R^m$, then sections of Hom (TU, E') can be called 1-forms on U with values in the vector bundle E'. In the previous example, we see that the differential of φ can be regarded as a 1-form on U with values in the pullback of the tangent bundle of U'. A fuller discussion of bundle-valued forms will be given in Chapter 9.

6.2.3.4   Tensor Product Notation for a Homomorphism Bundle

(May be omitted.) As we discussed in Chapter 2, the tensor product $V^* \otimes W$ is isomorphic to $L(V \to W)$, for vector spaces V and W. Thus if $E = M \times V$ and $E' = M \times W$ are local vector bundles over the same base manifold M, the vector bundle Hom (E, E') is isomorphic to the local vector bundle

$$E^* \otimes E' \equiv M \times (V^* \otimes W), \tag{6.10}$$

a notation which some authors prefer to the "Hom" notation. To illustrate this terminology, one could write the differential in (6.9) in yet another way:

$$d\varphi(x) = \sum_{i,j} D_i \varphi^j(x) \, (dx^i \otimes \frac{\partial}{\partial y^j}), \tag{6.11}$$

where $\{x^i\}$ and $\{y^j\}$ are coordinate systems on U and U', respectively. (In Chapter 9, we shall use a wedge instead of the $\otimes$ notation.)

6.2.4 Other Constructions with Local Vector Bundles

Let $E = M \times V$ and $E' = M \times W$ be local vector bundles over the same base manifold M, with ranks k and k', respectively. Table 6.2 summarizes some of the constructions above, and gives some new ones.

Name	Rank	Fiber over $p \in M$	Smooth Sections
dual bundle $E^* = \text{Hom}(E, M \times R)$	k	E_p^*	–
homomorphism bundle Hom $(E, E') \cong E^* \otimes E'$	kk'	$L(E_p \to E_p')$	vector bundle morphisms over the identity
tensor product bundle $E \otimes E'$	kk'	$E_p \otimes E_p'$	–
exterior power bundle $\Lambda^r(E)$	$k!/r!\,(k-r)!$	$\Lambda^r(E_p)$	–
direct sum bundle $E \oplus E'$	$k + k'$	$E_p \oplus E_p'$	–

Table 6.2   Constructions using local vector bundles over the same base manifold M

6.3   General Vector Bundles

6.3.1 Definition

Suppose M is a differential manifold. A manifold E together with a smooth onto map $\pi: E \to M$ (called the **projection**) is called a C^r **vector bundle of rank** k **over** M (often we refer to E itself as the vector bundle) if the following three conditions hold:

- There exists a k-dimensional vector space V such that, for every $p \in M$, $E_p = \pi^{-1}(p)$ is a real vector space isomorphic to V; E_p is called the **fiber over** p.
- Each point in M is contained in some open set $U \subseteq M$ such that there is a C^r diffeomorphism

$$\Phi_U: \pi^{-1}(U) \to U \times V \tag{6.12}$$

with the property that Φ_U restricted to the fiber E_p maps E_p onto $\{p\} \times V$.

- For any two such open sets U, U' with $U \cap U' \neq \varnothing$, the map

$$\Phi_U \bullet \Phi_{U'}^{-1} \colon (U \cap U') \times V \to (U \cap U') \times V \qquad \text{(6. 13)}$$

is a C^r local vector bundle isomorphism over the identity (cf. Section 6.2.2).

The n-dimensional manifold M is called the **base**, and the manifold E, which has dimension $n + k$, is called the **total space**. The map (6. 12) is called a **local trivialization**, and a collection of local trivializations $\{\Phi_{U_\alpha} : \alpha \in I\}$, where $M \subseteq \bigcup_{\alpha \in I} U_\alpha$, is called a **trivializing cover** for the vector bundle.

Loosely speaking, a vector bundle of rank k is a manifold in which vector spaces of dimension k, called "fibers." are "attached" to all the points of a "base manifold" M, with a differentiable structure such that, over any sufficiently small open set U in M, the collection of fibers attached to points in U is diffeomorphic to the local vector bundle $U \times V$. An important idea to grasp at this stage is that there could be a vector bundle of rank k over M which is not isomorphic to the local vector bundle $M \times V$. To understand this, first take $M = S^1$ and $V = R$; the local vector bundle $S^1 \times R$ is diffeomorphic to the cylinder embedded in R^3 as shown (actually the cylinder is infinite, but only a finite part of it is shown); the fibers make up the sides of the cylinder.

Figure 6. 2   A cylinder – part of a local rank-one vector bundle over the circle

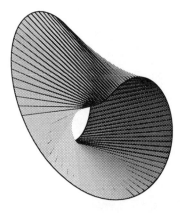

Figure 6. 3   The Möbius strip – part of a nonlocal rank-one vector bundle over the circle

Now consider instead the Möbius strip as a subset of R^3; if one were to extend the sides of the strip to infinity, then here too we have a copy of the real line attached to every point on a circle, and this too is a rank one vector bundle over S^1, because if one looks only at the part of the manifold consisting of fibers attached to some arc U of the circle, this part can be "unrolled" to look like the infinite rectangle $U \times R$. In Section 6.4.2.1 we shall give a formal construction of the so-called "Möbius bundle."

6.3.2 Transition Functions

Suppose Φ_U and $\Phi_{U'}$ are two local trivializations, and $p \in U \cap U'$. The assumptions (6. 12) imply that the map $\Phi_U \bullet \Phi_{U'}^{-1}$, applied to $(p, v) \in (U \cap U') \times V$, is linear in v for fixed p, and thus is of the form shown in the following diagram, where $g_{UU'}(p)$ is a linear map from V to V for each p:

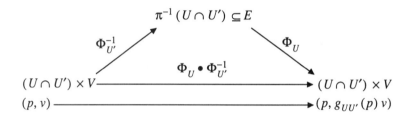

$$\pi^{-1}(U \cap U') \subseteq E$$

$$\Phi_{U'}^{-1} \qquad \Phi_U$$

$$\Phi_U \bullet \Phi_{U'}^{-1}$$

$$(U \cap U') \times V \longrightarrow (U \cap U') \times V$$
$$(p, v) \longrightarrow (p, g_{UU'}(p) v)$$

Moreover since Φ_U and $\Phi_{U'}$ are C^r diffeomorphisms, $\Phi_U \bullet \Phi_{U'}^{-1}$ must be C^r with a C^r inverse, and therefore

$$p \to g_{UU'}(p) \qquad (6.\ 14)$$

is a C^r map from $U \cap U'$ into $GL(V)$, the invertible linear transformations from V to V (a differential manifold isomorphic to $GL_k(R)$, the invertible $k \times k$ matrices); this map is called the **transition function** from the local trivialization Φ_U to the local trivialization $\Phi_{U'}$. It follows straight from the definition that transition functions have the properties:

$$g_{UU}(p) = I; \qquad (6.\ 15)$$

$$g_{UU'}(p) g_{U'U}(p) = I, p \in U \cap U'; \qquad (6.\ 16)$$

$$g_{UU'}(p) g_{U'U''}(p) g_{U''U}(p) = I, p \in U \cap U' \cap U''. \qquad (6.\ 17)$$

6.3.2.1   Example: Transition Functions for the Möbius Band

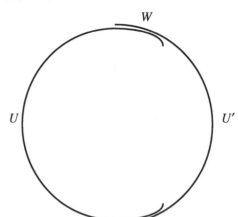

Figure 6. 4   Domains of transition functions for the Möbius bundle

Divide the base manifold, in this case the circle, up into two open arcs U and U' which overlap at both ends, as in Figure 6. 4, and let W and W' denote their regions of overlap. The local trivialization maps identify the parts of the band sitting on top of U and U' with $U \times R$ and $U' \times R$, respectively. The single "twist" observed in the Möbius strip is obtained by letting the transition function, which here takes values in the nonzero reals (i.e., invertible one-by-one matrices!), be -1 on one of the overlap regions, and 1 on the other. To understand this, think of how you would glue together two rectangular pieces of paper, with bases U and U', to make a Möbius band. Specifically:

$$g_{UU'}(p) = g_{U'U}(p) = 1, p \in W, \text{ and } = -1, p \in W';$$

$$g_{UU}(p) = g_{U'U'}(p) = 1. \tag{6. 18}$$

The reader may easily verify that (6. 15), (6. 16), and (6. 17) are satisfied.

6.3.3   Complex Vector Bundles

If V is replaced by a complex vector space throughout this chapter, we obtain a class of vector bundles called the complex vector bundles of rank k over M. Note that E and M are still "real" differential manifolds (not complex manifolds), but the group of invertible linear transformations of V would, for example, become $GL_k(C)$, the manifold of nonsingular $k \times k$ matrices with complex entries, if V were a complex k-dimensional space. For example, a complex vector bundle of rank 1 is called a **complex line bundle**, and the total space has dimension $n + 2$ (since a complex number is represented by two real numbers) if the base has dimension n. Complex vector bundles will figure prominently in Chapter 10.

6.3.4 Vector Bundle Morphisms and Isomorphisms

If $\pi: E \to M$ and $\pi': E' \to M'$ are vector bundles with fibers isomorphic to vector spaces V and V', respectively, and $f: M \to M'$ is a C^r map, then a C^r map $F: E \to E'$ is called a C^r **vector bundle morphism** over f if it maps the fiber E_p linearly into the fiber $E'_{f(p)}$ for each $p \in M$. The following diagram encapsulates the relationship of f and F.

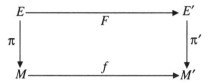

It is easy to see from the definition that the composition of two vector bundle morphisms is a vector bundle morphism. To see how to verify using local trivializations that an arbitrary mapping $F: E \to E'$ is a vector bundle morphism, see Exercise 5 in Section 6.5.

The word "morphism" can be replaced by "isomorphism" if f is a diffeomorphism and F acts as a linear isomorphism on each fiber. Of course this can only occur if M and M' are manifolds of the same dimension, and if dim V = dim V'. If such an isomorphism exists, then the two vector bundles are called C^r **equivalent**; as the name suggests, this sets up an equivalence relation on vector bundles over M.

6.3.5 Subbundles

We say that a C^r vector bundle $\pi': E' \to M$ is a **subbundle** of a C^r vector bundle $\pi: E \to M$ (note that the base manifolds are the same) if $E_p{}'$ is a vector subspace of E_p for every $p \in M$, and if the inclusion map $\iota: E' \to E$ is a C^r vector bundle morphism. An important example will be given in Exercises 11. and 12. on page 140.

6.3.6 Trivial and Nontrivial Bundles

A bundle equivalent to the local vector bundle proj : $M \times V \to M$ (see Section 6.1) is also called a **trivial bundle**; otherwise a bundle is called **nontrivial**. Examples of trivial bundles include:

- a bundle for which a single trivialization suffices as a trivializing cover;
- the tangent bundle to the circle, to be constructed in Exercise 9.in Section 6.7.

Some nontrivial bundles that will be defined later include:

- the Möbius bundle (see Section 6.4.2.1 and Exercise 4 in Section 6.5);
- the tangent bundle to the sphere S^2 (the proof uses a theorem in topology which says "you can't comb a hairy ball," i.e., there is no nonvanishing section).

6.3.7 Sections of Vector Bundles

Following Section 6.1.1, a C^s **section** of a C^r vector bundle $\pi: E \to M$, where $s \le r$, is a C^s mapping $\sigma: M \to E$ such that $\pi \bullet \sigma(p) = p$, $p \in M$; in other words, $\sigma(p) \in E_p$ for all $p \in M$. The set of smooth sections is denoted ΓE. The graph of a section looks like a slice through the vector bundle which cuts each fiber exactly once, as in the following picture.

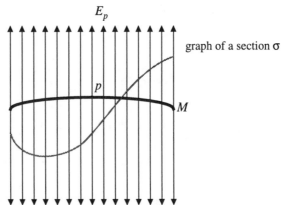

graph of a section σ

6.4   Constructing a Vector Bundle from Transition Functions

Frequently we would like to construct a vector bundle over a particular base manifold M, knowing what vector space V the fibers must be modeled on, and knowing what the transition functions must be, but without knowing in advance what kind of differentiable structure a corresponding vector bundle may have. The following theorem guarantees existence and uniqueness of a differentiable structure of a corresponding vector bundle over M, in the sense described below.

6.4.1 Vector Bundle Construction Theorem

Let V be a k-dimensional vector space (the "fiber"), and let $\{U_\alpha : \alpha \in I\}$ be an open cover of a differential manifold M. Suppose that, for every $\alpha, \gamma \in I$ with $U_\alpha \cap U_\gamma \ne \emptyset$,

$$p \to g_{\alpha\gamma}(p) \equiv g_{U_\alpha U_\gamma}(p) \tag{6. 19}$$

is a C^r map from $U_\alpha \cap U_\gamma$ into $GL(V)$, the invertible linear transformations from V to V such that (6. 15), (6. 16), and (6. 17) hold. Then there exists a C^r vector bundle $\pi: E \to M$ of rank k with the mappings (6. 19) as transition functions, and any other such vector bundle is C^r isomorphic to E.

The proof is in the appendix to this chapter, and need not be mastered yet. The most important example of this construction, namely, construction of the tangent bundle, is in

Section 6.6.2. The following version is sometimes more convenient in the context of constructing new vector bundles from existing ones.

6.4.2 Vector Bundle Construction Theorem – Alternative Version

Let V be a k-dimensional vector space, and for each point p in a manifold M, let E_p be a vector space isomorphic to V. Let $E = \bigcup E_p$ (disjoint union), and $\pi: E \to M$ be the map such that $\pi^{-1}(p) = E_p$. Suppose also that $\{U_\alpha : \alpha \in I\}$ is an open cover of M, and for each α

$$\Phi_\alpha : \pi^{-1}(U_\alpha) \to U_\alpha \times V \qquad (6.20)$$

is a bijection such that, for every $\alpha, \gamma \in I$ with $U_\alpha \cap U_\gamma \neq \varnothing$,

$$\Phi_\alpha \bullet \Phi_\gamma^{-1} : (U_\alpha \cap U_\gamma) \times V \to (U_\alpha \cap U_\gamma) \times V \qquad (6.21)$$

is a C^r local vector bundle morphism over the identity (see Section 6.2.2) such that (6.15), (6.16), and (6.17) hold for the corresponding transition functions. Then there exists a unique differentiable structure on E such that $\pi: E \to M$ is a C^r vector bundle of rank k with the maps (6.20) as a trivializing cover.

The proof is simply Step III in the proof of the previous theorem, in the appendix to this chapter.

6.4.2.1 Example: The Möbius Bundle
Here is the abstract construction of the Möbius bundle, as distinct from the concrete realization in R^3 shown in Figure 6.3. The role of "V" in Theorem 6.4.1 is played by R, the open cover consists of sets U and U' as shown in Figure 6.4, and the transition functions are given by (6.18). The Möbius strip in R^3 with infinite sides is C^∞ equivalent to this bundle, because it has the same transition functions.

6.4.3 Constructions with Vector Bundles

All the constructions with local vector bundles listed in Table 6.2 are valid for any C^r vector bundles.

Proof: For the sake of brevity, we shall only do the construction of Hom (E, E'), where E and E' are C^r vector bundles over the same base manifold M, with fibers isomorphic to vector spaces V and W, respectively. By taking restrictions if necessary, we can assume that the transition functions $\{g_{\alpha\gamma}\}$ for E and $\{g_{\alpha\gamma}{'}\}$ for E' are defined on the same open cover $\{U_\alpha, \alpha \in I\}$ for M.

We wish to apply Theorem 6.4.1, taking $L(V \to W)$ to be the fiber, and taking transition functions:

$$h_{\alpha\gamma}(p) A = g_{\alpha\gamma}'(p) A g_{\gamma\alpha}(p), \quad A \in L(V \to W). \tag{6.22}$$

The right-hand side of (6.22) makes sense as the composition of three linear maps, which are respectively $V \to V$, $V \to W$, and $W \to W$, giving an element of $L(V \to W)$ as desired. We must check the conditions (6.15), (6.16), and (6.17). It is clear that $h_{\alpha\alpha}(p) A = A$ because (6.15) holds for the $\{g_{\alpha\gamma}\}$ and $\{g_{\alpha\gamma}'\}$. Using (6.16) for the $\{g_{\alpha\gamma}\}$ and $\{g_{\alpha\gamma}'\}$,

$$h_{\gamma\alpha}(p) h_{\alpha\gamma}(p) A = g_{\gamma\alpha}'(p) (g_{\alpha\gamma}'(p) A g_{\gamma\alpha}(p)) g_{\alpha\gamma}(p) = A. \tag{6.23}$$

Finally

$$h_{\alpha\beta}(p) h_{\beta\gamma}(p) h_{\gamma\alpha}(p) A = g_{\alpha\beta}'(p) g_{\beta\gamma}'(p) (g_{\gamma\alpha}'(p) A g_{\alpha\gamma}(p)) g_{\gamma\beta}(p) g_{\beta\alpha}(p),$$

which equals A since (6.17) holds for the $\{g_{\alpha\gamma}\}$ and $\{g_{\alpha\gamma}'\}$. Moreover the invertibility of the linear transformations $\{h_{\alpha\gamma}(p)\}$ is shown by (6.23). Thus the conditions of Theorem 6.4.1 are satisfied, giving a construction of the C^r vector bundle Hom (E, E'). ¤

6.4.3.1   Description of the Fibers

There is a natural isomorphism Hom $(E, E')_p \cong L(E_p \to E_p')$, and we shall usually regard these vector spaces as identical. The equivalence class $[p, \alpha, A]$ in Hom $(E, E')_p$, in the sense of (6.50), may be identified with the linear map $[p, \alpha, v] \to [p, \alpha, Av]$ from E_p to E_p'; this map does not depend on the choice of α.

6.5   Exercises

1. Suppose $E = M \times V$ and $E' = M \times W$ are local vector bundles over the **same** base manifold M. Which of the following maps are C^∞ local vector bundle morphisms? Justify your answer.

    (i) $s: E \oplus E \to E$, $(p, (v_1, v_2)) \to (p, v_1 + v_2)$.

    (ii) $t:$ Hom $(E, E') \otimes E \to E'$, $(p, A \otimes v) \to (p, Av)$.

    (iii) Take $W = R$; $u: E \otimes E' \to E$, $(p, v \otimes w) \to (p, v/w)$ when $w \neq 0$, and $(p, 0)$ otherwise.

    (iv) Take $\varphi \in C^\infty(M)$; $s: E \to E$, $(p, v) \to (p, \varphi(p) v)$.

    (v) Take $E = S^2 \times R^3$; $u: E \otimes E \to E$, $(p, v \otimes w) \to (p, (p \cdot v) w)$, considering $S^2 \subset R^3$.

2. Give an example of a C^1 vector bundle morphism that is not a C^∞ vector bundle morphism.

   Hint: Consider part (iv) of Exercise 1 with a suitable φ.

3. Let $\pi: E \to M$ be a C^r vector bundle and let $g: N \to M$ be a C^r map for some manifold N.

   (i) Using Theorem 6.4.1, or otherwise, show that

$$g^* E = \{ (n, \xi) : \pi (\xi) = g (n) \} \subseteq N \times E \tag{6.24}$$

   may be identified with the total space of a vector bundle (the **pullback bundle** under g) over N, with projection $g^* \pi: g^* E \to N$, $g^* \pi (n, \xi) = n$, and where the fiber over n may be identified with $E_{g(n)}$.

   (ii) Show that the map $\Phi: g^* E \to E$, $\Phi (n, \xi) = \xi$, is a C^r vector bundle morphism which is the identity on every fiber.

   (iii) Show that if two vector bundles over M are C^r equivalent, then their pullbacks under $g: N \to M$ are C^r equivalent.

   (iv) If $h: Q \to N$ is also a C^r map, show that the vector bundles

$$(g \bullet h)^* \pi: (g \bullet h)^* E \to Q \text{ and } h^* (g^* \pi): h^* (g^* E) \to Q$$

   are C^r equivalent.

   This problem was adapted from Abraham, Marsden, and Ratiu [1988], p. 193.

4. (i) Prove that if F in the following diagram is a vector bundle morphism, and σ is a section of the bundle $\pi: E \to M$, then $F \bullet \sigma$ is a section of $\pi': E' \to M$.

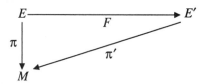

   Moreover if F is a vector bundle isomorphism, then

$$\sigma (r) \neq 0 \forall r \Leftrightarrow F \bullet \sigma (r) \neq 0 \forall r.$$

   (ii) Prove using the transition functions (6.18) that if σ is a section of the Möbius bundle, then $\sigma (r) = 0$ for some $r \in S^1$.

   (iii) Prove using (i) and (ii) that the Möbius bundle is not equivalent to the trivial bundle proj$: S^1 \times R \to S^1$.

5. Let $\pi: E \to M$ and $\pi': E' \to M'$ be C^r vector bundles with fibers isomorphic to vector spaces V and V', respectively, and let $f: M \to M'$ be a C^r map. Also suppose $F: E \to E'$

is an arbitrary mapping, not known to be differentiable, but which takes the fiber E_p into the fiber $E'_{f(p)}$ for each $p \in M$. Prove that F is a C^r vector bundle morphism over f if and only if the following condition holds:

For each $p \in M$, there exist trivializations Φ_U for M and $\Psi_{U'}$ for M' with $U \ni p$ and $U' \ni f(p)$, restricted if necessary so that $f(U) \subseteq U'$, such that

$$\Psi_{U'} \bullet F \bullet \Phi_U^{-1} : U \times V \to U' \times V', \qquad (6.25)$$

represented below, is a C^r local vector bundle morphism over f (see Section 6.2.1).

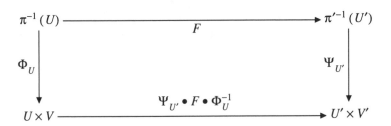

6. Using 6.4.3 as a model, carry out the construction of the following C^r vector bundles, starting from a C^r vector bundle $\pi : E \to M$ with fibers isomorphic to a vector space V:

(i) The dual bundle $\pi_{\text{dual}} : E^* \to M$, where the fiber over p is to be identified with E_p^*.

(ii) The exterior product bundle $\pi_\wedge : \Lambda^2 E \to M$, where the fiber over p is to be identified with $\Lambda^2 E_p$.

6.6   The Tangent Bundle of a Manifold

6.6.1   Tangent Vectors – Intuitive Ideas

When a curve M is a one-dimensional submanifold of R^3, it turns out to be useful to put a differentiable structure on the set of (disjoint) tangent lines to the curve, such that these tangent lines are in some sense "smoothly related" to one another as one moves along the curve. In the same way, one may wish to put a differentiable structure on the set of (disjoint) tangent planes to a parametrized surface M at all points in the surface, and indeed to the union of all the tangent planes to an n-dimensional submanifold M of R^{n+k}. Instead of treating all these cases separately, it is most efficient to define a vector bundle called the "tangent bundle" of an abstract differential manifold. Although the abstract symbolism we are going to use now may seem somewhat removed from the original idea of tangent line or tangent plane, the reader may continue to think of tangent spaces in those terms.

The only way we can identify a "tangent vector" in a manifold is by taking a chart, and picking some tangent vector in the range of the chart. However, we then need some way of keeping track of the same "tangent vector" if we shift to a different chart. This is accomplished by referring to an "equivalence class" of vectors in the ranges of various charts, as we shall now describe.

6.6.2 Formal Construction of the Tangent Bundle

Now we are going to work through the main ideas of the proof of the Vector Bundle Construction Theorem, as presented in Section 6.9.1, for the special case of the tangent bundle. Let $\{(U_\alpha, \varphi_\alpha), \alpha \in I\}$ be an atlas for an n-dimensional smooth differential manifold M. For any chart (U, φ) for M at p, $\varphi(U)$ is an open set in R^n, and therefore the tangent space $T_x R^n$ at $x = \varphi(p)$ is well defined in the sense of Chapter 2. Consider the set of triples

$$T = \{(p, \alpha, \xi) : \alpha \in I, p \in U_\alpha, \xi \in T_{\varphi_\alpha(p)} R^n\}. \tag{6.26}$$

We shall say that two such triples are equivalent, written $(p, \alpha, \xi) \sim (q, \gamma, \zeta)$, if

$$p = q, \zeta = d(\varphi_\gamma \bullet \varphi_\alpha^{-1})(\varphi_\alpha(p))\xi, \tag{6.27}$$

where, as in Chapter 2, $d(\varphi_\gamma \bullet \varphi_\alpha^{-1})$ is the differential of the map

$$\varphi_\gamma \bullet \varphi_\alpha^{-1} : U_\alpha \cap U_\gamma \subseteq R^n \to U_\alpha \cap U_\gamma \subseteq R^n,$$

which is being evaluated at $\varphi_\alpha(p)$, and applied to the tangent vector ξ. To shift into the notation of Theorem 6.4.1, we may write our "transition function" as

$$g_{\gamma\alpha}(p) = d(\varphi_\gamma \bullet \varphi_\alpha^{-1})(\varphi_\alpha(p)) \in GL_n(R), \tag{6.28}$$

which is C^∞ on $U_\alpha \cap U_\gamma \neq \varnothing$. Of the conditions (6.15), (6.16), and (6.17), the first is immediate, while the others follow from the chain rule for differentiation (see Chapter 2), which shows that

$$d(\varphi_\alpha \bullet \varphi_\gamma^{-1})(y) = \{d(\varphi_\gamma \bullet \varphi_\alpha^{-1})(x)\}^{-1},$$

$$d(\varphi_\delta \bullet \varphi_\alpha^{-1})(x) = d(\varphi_\delta \bullet \varphi_\gamma^{-1})(y) \bullet d(\varphi_\gamma \bullet \varphi_\alpha^{-1})(x),$$

where $x = \varphi_\alpha(p)$, $y = \varphi_\gamma(p)$. Thus the conditions of Theorem 6.4.1 are satisfied. In particular, (6.27) defines an equivalence relation on T, as in the proof in Section 6.9.1. The set of equivalence classes is denoted

$$TM = \{[p, \alpha, \xi] : \alpha \in I, p \in U_\alpha, \xi \in T_{\varphi_\alpha(p)} R^n\}, \tag{6.29}$$

and the proof in Section 6.9.1 shows the existence of a C^∞ differentiable structure on TM under which we call $\pi : TM \to M$, $[p, \alpha, \xi] \to p$, the **tangent bundle** of M.

For fixed p, an equivalence class of the form $[p, \alpha, \xi]$ is called a **tangent vector** at p; the set of such vectors is called the **tangent space** at p, denoted $T_p M$.[3] It has the structure of an n-dimensional vector space, induced by the bijection

$$[p, \alpha, \xi] \to \xi, \; T_p M \to T_{\varphi_\alpha(p)} R^n \cong R^n$$

and this structure does not depend on whether we represent a tangent vector as $[p, \alpha, \xi]$ or as $[p, \gamma, \zeta]$, because $d(\varphi_\gamma \bullet \varphi_\alpha^{-1})(\varphi_\alpha(p))$ is a linear isomorphism.

6.6.3   Computing an Atlas for the Tangent Bundle

Note that TM is a $2n$-dimensional manifold, and according to (6. 51), it has a smooth atlas $\{ (\pi^{-1}(U_\alpha), \tilde{\varphi}_\alpha), \alpha \in I \}$, based on the atlas $\{ (U_\alpha, \varphi_\alpha), \alpha \in I \}$ for M, given by

$$\tilde{\varphi}_\alpha([p, \alpha, \xi]) = (\varphi_\alpha(p), \xi) \in R^n \times R^n. \tag{6. 30}$$

Note that, when $U_\alpha \cap U_\gamma \neq \varnothing$, $\tilde{\varphi}_\gamma \bullet \tilde{\varphi}_\alpha^{-1}$ involves the differential of the map $\varphi_\gamma \bullet \varphi_\alpha^{-1}$:

$$\tilde{\varphi}_\gamma \bullet \tilde{\varphi}_\alpha^{-1}(x, \xi) = (y, d(\varphi_\gamma \bullet \varphi_\alpha^{-1})(x)\xi), \tag{6. 31}$$

for $x = \varphi_\alpha(p)$, $y = \varphi_\gamma(p)$, as shown by (6. 27).

6.6.3.1   Example: Tangent Bundle to the Projective Plane
The real **projective plane** $P^2(R)$, here abbreviated to P^2, can be thought of as a manifold consisting of the lines through the origin in R^3, or as the sphere S^2 with opposite points identified. Formally speaking, P^2 is the set of equivalence classes in $R^3 - \{0\}$, under the equivalence relation $\approx$ defined by:

$$x \approx y \text{ if and only if } x \text{ and } y \text{ are collinear}, \tag{6. 32}$$

and with the differentiable structure induced by the atlas $\{ (U_i, \varphi_i), i = 0, 1, 2 \}$ defined as follows. For $i = 0, 1, 2$, let

$$V_i = \{ x = (x_0, x_1, x_2) \in R^3 : x_i \neq 0 \} \tag{6. 33}$$

---

[3] The reader will have the opportunity in Exercise 7 to check that this notation is consistent with earlier definitions.

and define maps $\Phi_i \colon V_i \to R^2$ by

$$\Phi_0(x) = (x_1/x_0, x_2/x_0), \ \Phi_1(x) = (x_0/x_1, x_2/x_1), \ \Phi_2(x) = (x_0/x_2, x_1/x_2).$$

If $[x]$ denotes the equivalence class of x under $\approx$, and if $U_i = \{[x] : x \in V_i\}$, then $\Phi_i(x) = \Phi_i(y) \Leftrightarrow [x] = [y]$, and so we may take $\varphi_i \colon U_i \to R^2$ to be the bijection

$$\varphi_i([x]) = \Phi_i(x). \tag{6.34}$$

The change-of-chart maps take the form

$$\varphi_1 \bullet \varphi_0^{-1}(z_0, z_1) = \left(\frac{1}{z_0}, \frac{z_1}{z_0}\right), \tag{6.35}$$

etc. (see Exercise 8). Thus the tangent bundle TP^2 to the real projective plane is a four-dimensional manifold with an atlas consisting of the three charts $\{(\pi^{-1}(U_i), \tilde{\varphi}_i), i = 0, 1, 2\}$, where, for example,

$$\tilde{\varphi}_0([[x], 0, \xi]) = ((x_1/x_0, x_2/x_0), \xi), \tag{6.36}$$

and where, for $z_0 \neq 0$,

$$\tilde{\varphi}_1 \bullet \tilde{\varphi}_0^{-1}(z_0, z_1, \xi_1, \xi_2) = \left(\frac{1}{z_0}, \frac{z_1}{z_0}, -\frac{\xi_1}{z_0^2}, -\frac{z_1\xi_1}{z_0^2} + \frac{\xi_2}{z_0}\right). \tag{6.37}$$

6.6.3.2 Interpretation of a Tangent Vector on a Surface in Terms of Parametrizations

Consider two different parametrizations $\Psi = (\Psi^1(u, v), \Psi^2(u, v), \Psi^3(u, v))$ and $\Phi = (\Phi^1(r, s), \Phi^2(r, s), \Phi^3(r, s))$ of some surface M in R^3, and suppose that $p = \Psi(\bar{u}, \bar{v}) = \Phi(\bar{r}, \bar{s}) \in M$. Each parametrization gives a chart for M as described in Chapter 5. According to (6.27), to say that

$$\left(p, \Psi^{-1}, a\frac{\partial}{\partial u} + b\frac{\partial}{\partial v}\right) \sim \left(p, \Phi^{-1}, A\frac{\partial}{\partial r} + B\frac{\partial}{\partial s}\right)$$

means that

$$(d(\Phi^{-1} \bullet \Psi))\left(a\frac{\partial}{\partial u} + b\frac{\partial}{\partial v}\right) = A\frac{\partial}{\partial r} + B\frac{\partial}{\partial s},$$

or, in the terminology of Chapter 2, that

$$D(\Phi^{-1} \bullet \Psi)(\bar{u}, \bar{v}) \begin{bmatrix} a \\ b \end{bmatrix} = \begin{bmatrix} A \\ B \end{bmatrix}.$$

The same expression can be written in terms of a Jacobian matrix, that is,

$$\begin{bmatrix} \partial r/\partial u & \partial r/\partial v \\ \partial s/\partial u & \partial s/\partial v \end{bmatrix} \begin{bmatrix} a \\ b \end{bmatrix} = \begin{bmatrix} A \\ B \end{bmatrix},$$

where the Jacobian must be evaluated at $(\bar{u}, \bar{v})$. Loosely speaking, we can say that the (Euclidean) tangent vectors

$$a\frac{\partial}{\partial u} + b\frac{\partial}{\partial v} \text{ at } (\bar{u}, \bar{v}) \text{ and } A\frac{\partial}{\partial r} + B\frac{\partial}{\partial s} \text{ at } (\bar{r}, \bar{s})$$

correspond to the same tangent vector on M, because one transforms into the other under the change of variables formula. The abstract definition of tangent vector is merely intended to formalize this notion.

6.6.4   The Cotangent Bundle

The **cotangent bundle** $T^* M$ is simply the dual bundle to TM, in the sense of Section 6.4.3. A step-by-step construction is suggested in Exercise 15. The fiber over p, denoted $T_p^* M$, is called the **cotangent space** at p, and may be identified with $(T_p M)^*$, as shown in 6.4.3.1. Elements of $T_p^* M$ are called **cotangent vectors** at p.

6.6.5   Example of a Vector Bundle Morphism: The Tangent Map

Given a C^{r+1} map $f: M \to M'$, the **tangent map**

$$Tf: TM \to TM', \tag{6.38}$$

whose restriction to $T_p M$ is denoted $T_p f$, is defined as follows: If $(U_\alpha, \varphi_\alpha)$ and (U_γ', ψ_γ) are charts at $p \in M$ and $f(p) \in M'$, respectively, and $x = \varphi_\alpha(p)$, then

$$Tf([p, \alpha, \xi]) = [f(p), \gamma, d(\psi_\gamma \bullet f \bullet \varphi_\alpha^{-1})(x)\xi], \tag{6.39}$$

which does not depend upon the choice of charts. Beneath the cumbersome notation, the formula (6.39) is saying the same thing as (6.2); $T_p f \in L(T_p M \to T_{f(p)} M')$ is simply a more abstract version of the differential we encountered in Chapter 2, and coincides with it when M is an open subset of R^n. Note that the following diagram commutes (i.e., $f \bullet \pi = \pi' \bullet Tf$):

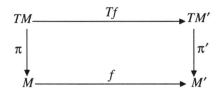

In terms of local trivializations $(\pi^{-1}(U_\alpha), \Phi_\alpha)$ and $(\pi'^{-1}(U_\gamma'), \Psi_\gamma)$ for TM and TM', respectively,

$$\Psi_\gamma \bullet Tf \bullet \Phi_\alpha^{-1}(p, \xi) = (f(p), D(\psi_\gamma \bullet f \bullet \varphi_\alpha^{-1})(x) \xi). \tag{6.40}$$

Thus it follows from Exercise 5 that Tf is a C^r vector bundle morphism.

6.7   Exercises

7. In the case where U is an open subset of R^n, verify that the local vector bundle TU over U, described in Section 6.1.2, coincides with the definition of the tangent bundle in Section 6.6.2, using the identity map as the chart for U.

8. In the projective plane example 6.6.3.1, calculate $\varphi_2 \bullet \varphi_1^{-1}$ and $\tilde{\varphi}_2 \bullet \tilde{\varphi}_1^{-1}$.

9. Consider the circle $S^1 \subset R^2$ with the atlas $\{(U, \varphi), (V, \psi)\}$ as follows; let $\Psi(\theta) = (\cos\theta, \sin\theta) \in S^1$, and let

    $$U = \Psi(0, 2\pi), \varphi = \Psi^{-1}\big|_U;$$

    $$V = \Psi(-\pi, \pi), \psi = \Psi^{-1}\big|_V.$$

    Construct an atlas for the tangent bundle TS^1 to the circle from this atlas, and show that TS^1 is C^∞ equivalent to the trivial bundle $S^1 \times R \to S^1$.

10. In Exercise 1 of Chapter 5, we saw that the pair of charts (U, φ) and (V, ψ) below form an atlas for $S^n = \{(x_0, ..., x_n) \in R^{n+1} : x_0^2 + ... + x_n^2 = 1\}$.

    $$U = S^n - \{(1, 0, ..., 0)\}, \varphi(x_0, ..., x_n) = \frac{(x_1, ..., x_n)}{1 - x_0};$$

    $$V = S^n - \{(-1, 0, ..., 0)\}, \psi(x_0, ..., x_n) = \frac{(x_1, ..., x_n)}{1 + x_0}.$$

    Calculate the associated trivializations Φ_U and Φ_V, and transition functions g_{UV} and g_{VU}, for the tangent bundle TS^n in the way suggested in Section 6.6.2.

11. Suppose that M is an n-dimensional submanifold of R^{n+k}. Prove that $\pi: TM \to M$ is a C^∞ rank-n subbundle (in the sense of Section 6.3.5) of the C^∞ vector bundle $\pi_0: TR^{n+k}\big|_M \to M$ (i.e., the tangent bundle of R^{n+k}, restricted to M).

12. Extend the result of Exercise 11 to show that if M is a submanifold of N, then $\pi: TM \to M$ is a subbundle of $\pi_0: TN|_M \to M$.

13. (Normal Bundle) Suppose M is an n-dimensional submanifold of R^{n+k}. Our goal is to construct a vector bundle of rank k over M, called the **normal bundle**, whose fiber over $p \in M$ can be identified with the "normal space"

$$(T_p M)^\perp = \{p\} \times \{\zeta \in T_p R^{n+k} : \langle \zeta | \xi \rangle = 0, \forall \xi \in T_p M\}, \tag{6.41}$$

where here we are interpreting $T_p M$ as a vector subspace of the Euclidean inner product space $T_p R^{n+k}$. Let

$$T^\perp M = \bigcup_{p \in M} (T_p M)^\perp \tag{6.42}$$

be the disjoint union of these normal spaces, and $\pi: T^\perp M \to M$ be the map such that $\pi^{-1}(p) = (T_p M)^\perp$.

(i) Given $p \in M$, there exists $U' \ni p$ open in R^{n+k} and a submersion $f_U: U' \to R^k$ such that $U = U' \cap M = f_U^{-1}(0)$, by definition of a submanifold. Show that the map

$$\Phi_U: \pi^{-1}(U) \to U \times R^k, \tag{6.43}$$

$$\Phi_U(q, \zeta) = (q, Df_U(q)\,\zeta) \tag{6.44}$$

is a bijection.

(ii) Suppose that $f_V: V' \to R^k$ is a submersion from another open set $V' \subseteq R^{n+k}$ with $V' \ni p$, such that $V = V' \cap M = f_V^{-1}(0)$. Show that there is a well-defined smooth function $g_{UV}: U \cap V \to GL_k(R)$ such that

$$\Phi_U \bullet \Phi_V^{-1}(q, \zeta) = (q, g_{UV}(q)\,\zeta).$$

Hint: The function must satisfy $g_{UV}(q) Df_V(q)\,\zeta = Df_U(q)\,\zeta, \zeta \in (T_q M)^\perp$. For smoothness, extend $Df_V(q)$ to an invertible $(n+k) \times (n+k)$ matrix $A_V(q)$ depending smoothly on q, so that $g_{UV}(q) = Df_U(q) A_V(q)^{-1}$.

(iii) Complete the construction of a vector bundle $\pi: T^\perp M \to M$ with local trivializations and transition functions of the form given in (i) and (ii) above.

14. Suppose M is an n-dimensional submanifold of R^{n+k}, and $\pi: T^\perp M \to M$ is the Normal Bundle discussed in Exercise 13. Prove that, in the case where there exists an open subset $U' \subseteq R^{n+k}$ and a submersion $f: U' \to R^k$ such that $M = f^{-1}(0)$ (spheres, tori, and hyperboloids, for example), then the Normal Bundle is trivial.

15. Carry out a step-by-step construction of the cotangent bundle $T^* M$ (the dual of the tangent bundle) along the lines of Section 6.6.2, starting with the set of triples

$$T^* = \{ (p, \alpha, \lambda) : \alpha \in I, p \in U_\alpha, \xi \in (T_{\varphi_\alpha(p)} R^n)^* \}, \tag{6.45}$$

where two such triples are equivalent, written $(p, \alpha, \lambda) \sim (q, \gamma, \mu)$, if

$$p = q, \lambda = (\varphi_\gamma \bullet \varphi_\alpha^{-1})^* \mu, \tag{6.46}$$

referring to the pullback introduced in Chapter 2.

6.8   History and Bibliography

The theory of fibered spaces, of which vector bundles are a special case, is attributed to H. Hopf (1894–1971), E. Stiefel (1909–78), N. Steenrod (1910–71), and others. For a more "functorial" treatment of vector bundles, see Lang [1972]; for the topological viewpoint, see Husemoller [1975].

6.9   Appendix: Constructing Vector Bundles

6.9.1   Proof of the Vector Bundle Construction Theorem

Let V be a k-dimensional vector space, and let $\{ U_\alpha, \alpha \in I \}$ be an open cover of a differential manifold M. Suppose that, for every $\alpha, \gamma \in I$ with $U_\alpha \cap U_\gamma \neq \varnothing$,

$$p \to g_{\alpha\gamma}(p) \equiv g_{U_\alpha U_\gamma}(p) \tag{6.47}$$

is a C^r map from $U_\alpha \cap U_\gamma$ into $GL(V)$, the invertible linear transformations from V to V such that (6. 15), (6. 16), and (6. 17) hold. Then there exists a C^r vector bundle $\pi : E \to M$ of rank k with the mappings (6. 19) as transition functions, and any other such vector bundle is C^r isomorphic to E.

Step I. Constructing the Fibers. Suppose $\{ U_\alpha, \alpha \in I \}$ is an open cover of the manifold M, and for every $\alpha, \gamma \in I$ such that $U_\alpha \cap U_\gamma \neq \varnothing$, a map $g_{\alpha\gamma} : U_\alpha \cap U_\gamma \to GL_k(R)$ (short for $g_{U_\alpha U_\gamma}$) is a C^r map with the properties stated in the theorem. Define

$$\tilde{E} = \{ (p, \alpha, z) \in M \times I \times V : p \in U_\alpha \}, \tag{6.48}$$

and define a relation $\sim$ on $\tilde{E}$ as follows:

$$(p, \alpha, v) \sim (q, \gamma, w) \Leftrightarrow p = q \text{ and } w = g_{\gamma\alpha}(p) v. \tag{6.49}$$

Reflexivity of this relation follows from the condition $g_{\alpha\alpha}(p) = I$, symmetry from $g_{\alpha\gamma}(p) g_{\gamma\alpha}(p) = I, p \in U_\alpha \cap U_\gamma$, and transitivity from $g_{\alpha\beta}(p) g_{\beta\gamma}(p) g_{\gamma\alpha}(p) = I$,

$p \in U_\alpha \cap U_\beta \cap U_\gamma$; therefore ~ is an equivalence relation. The equivalence class of (p, α, v) is denoted $[p, \alpha, v]$. For p in M, define

$$E_p = \pi^{-1}(p) = \{ [p, \alpha, z] : U_\alpha \ni p, z \in V \}. \tag{6.50}$$

Introduce a vector space structure on E_r by taking

$$c[p, \alpha, z] + [p, \gamma, w] = c[p, \alpha, z] + [p, \alpha, g_{\alpha\gamma}(p) w] = [p, \alpha, cz + g_{\alpha\gamma}(p) w].$$

Clearly this structure does not depend on the specific α chosen, because each $g_{\alpha\gamma}(p)$ is a linear isomorphism.

Step II. Local Trivializations for E. For each α, the map

$$\Phi_\alpha = \Phi_{U_\alpha} : \pi^{-1}(U_\alpha) \to U_\alpha \times V, \Phi_\alpha([p, \alpha, z]) = (p, z)$$

is a bijection onto an open subset of $M \times V$: It is one-to-one because any $[p, \gamma, w]$ can be uniquely expressed as $[p, \alpha, g_{\alpha\gamma}(p) w]$ by nonsingularity of $g_{\alpha\gamma}(p)$. Also

$$\Phi_\alpha \bullet \Phi_\gamma^{-1} : (U_\alpha \cap U_\gamma) \times V \to (U_\alpha \cap U_\gamma) \times V, (p, w) \to (p, g_{\alpha\gamma}(p) w)$$

is a C^r onto map with a C^r inverse $(p, z) \to (p, g_{\gamma\alpha}(p) z)$, and of course $(U_\alpha \cap U_\gamma) \times V$ is open in $M \times V$. In other words, in the diagram

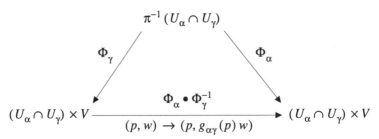

$\Phi_\alpha \bullet \Phi_\gamma^{-1}$ is a diffeomorphism.

Step III. Differentiable Structure for E. It follows that for any atlas $\{ (U'_i, \psi_i), i \in J \}$ for M,

$$\{ (\pi^{-1}(U'_i \cap U_\alpha), (\psi_i \times \text{identity}) \bullet \Phi_\alpha) : i \in J, \alpha \in I, U'_i \cap U_\alpha \neq \varnothing \} \tag{6.51}$$

is an atlas for E, making E into an $(n + k)$-dimensional manifold. With the resulting differentiable structure, each $\Phi_\alpha : \pi^{-1}(U_\alpha) \to U_\alpha \times V$ becomes a local trivialization, because it is a diffeomorphism onto its range with the desired linearity property.

Moreover $\pi: E \to M$ is C^r because it can be expressed as the composition proj $\bullet\, \Phi_\alpha$ in the following diagram.

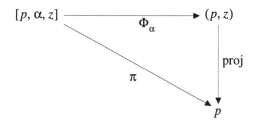

Clearly the transition functions for this vector bundle are exactly the ones that we used to construct it.

Step IV. Uniqueness up to Bundle Equivalence. It remains to prove the following statement:

Two vector bundles with the same transition maps $\{g_{\alpha\gamma}\}$ (which refer to the same open cover $\{U_\alpha\}$ of M) are C^r equivalent.

Proof: Let $\pi: E \to M$ be as above, and let $\pi': E' \to M$ be another vector bundle with the same $\{g_{\alpha\gamma}\}$, but with local trivializations $\{\Psi_\alpha\}$. It is necessary to construct a C^r map F from E to E' such that F is a linear isomorphism from E_p to E'_p for each p. For $p \in U_\alpha$ and $v \in V$, define a map

$$F_\alpha: \pi^{-1}(U_\alpha) \to \pi'^{-1}(U_\alpha), \ [p, \alpha, v] \to \Psi_\alpha^{-1}(p, v). \tag{6.52}$$

Clearly F_α is a linear isomorphism on each fiber. We are going to show that $F_\alpha = F_\gamma$ on $\pi^{-1}(U_\alpha \cap U_\gamma)$. For if $[p, \alpha, v] = [p, \gamma, w]$, then $w = g_{\gamma\alpha}(p)\, v$, and therefore, using the fact that the two vector bundles have the same transition functions,

$$F_\gamma([p, \gamma, w]) = \Psi_\gamma^{-1}(p, w) = \Psi_\gamma^{-1}(p, g_{\gamma\alpha}(p)\, v)$$

$$= \Psi_\alpha^{-1} \bullet (\Psi_\alpha \bullet \Psi_\gamma^{-1})(p, g_{\gamma\alpha}(p)\, v)$$

$$= \Psi_\alpha^{-1}(p, g_{\alpha\gamma}(p)\, g_{\gamma\alpha}(p)\, v)$$

$$= F_\alpha([p, \alpha, v]).$$

Hence it makes sense to define F to be the function $E \to E'$ which equals F_α on $\pi^{-1}(U_\alpha)$. This function is C^r since, for each α, $F_\alpha = \Psi_\alpha^{-1} \bullet \Phi_\alpha$, and a composition of C^r maps is C^r. ¤

7   Frame Fields, Forms, and Metrics

This chapter is intended mainly to supply constructions needed in Chapters 8, 9, and 10. The important subject of Riemannian manifolds will receive only a cursory introduction here; see do Carmo [1992] and Gallot et al. [1990] for further information.

7.1   Frame Fields for Vector Bundles

Suppose $\Phi_U : \pi^{-1}(U) \rightarrow U \times R^k$ is a local trivialization of a rank-k vector bundle $\pi : E \rightarrow M$. We immediately obtain sections

$$s_i(r) = \Phi_U^{-1}(r, e_i), i = 1, 2, ..., k, r \in U, \tag{7.1}$$

of the restricted vector bundle $\pi|_U : \pi^{-1}(U) \rightarrow U$, where $\{e_1, ..., e_k\}$ is the standard basis of R^k. From the definition of a vector bundle, it follows that $\{s_1(r), ..., s_k(r)\}$ forms a basis for the fiber E_r, and we call the map

$$r \rightarrow \{s_1(r), ..., s_k(r)\}$$

a **local frame field** for E over U. Any section $\sigma \in \Gamma E$ has a local expression

$$\sigma(r) = \sigma^1(r) s_1(r) + ... + \sigma^k(r) s_k(r), r \in U, \tag{7.2}$$

where each $\sigma^i \in C^\infty(U)$.

7.1.1   Local Expressions for Vector Fields and 1-Forms

The smooth sections of the tangent bundle $\pi : TM \rightarrow M$ are called **vector fields**, denoted ΓTM or more commonly $\Im(M)$. The smooth sections of the cotangent bundle $\pi : T^*M \rightarrow M$ (no confusion arises from using the same π) are called **first-degree**

differential forms, or simply **1-forms**, denoted $\Gamma T^* M$ or more commonly $\Omega^1 M$. We shall see shortly that these definitions are entirely consistent with those given in Chapter 2 for the case when M is an open subset of a Euclidean space.

7.1.2 Frame Fields for the Tangent Bundle

In the case of the tangent bundle $\pi: TM \to M$, take a chart $(U_\alpha, \varphi_\alpha)$ for M, and let $\{x^1, ..., x^n\}$ be the standard Euclidean coordinate system on $\varphi_\alpha(U_\alpha) \subseteq R^n$. Recall that the associated trivialization Φ_α, as described in Chapter 6, is defined by

$$\Phi_\alpha([r, \alpha, \xi]) = (\varphi_\alpha(r), \xi), \, r \in U, \, \xi \in T_{\varphi_\alpha(r)} R^n. \tag{7.3}$$

If we use the identification between a vector ξ and the directional derivative in direction ξ discussed in Chapter 2, it makes sense to write[1] a frame field for the tangent bundle as:

$$s_i(.) = \Phi_\alpha^{-1}(., e_i) = \frac{\partial}{\partial x^i}, \, i = 1, ..., n. \tag{7.4}$$

Thus a vector field X, which over U_α takes the form $X(r) = [r, \alpha, \xi_\alpha(r)]$, where

$$\xi_\alpha(r) = \xi_\alpha^1(r) e_1 + ... + \xi_\alpha^n(r) e_n \in R^n,$$

can also be expressed over U_α in the familiar form

$$X = \xi_\alpha^1 \frac{\partial}{\partial x^1} + ... + \xi_\alpha^n \frac{\partial}{\partial x^n}, \tag{7.5}$$

where each $\xi_\alpha^i \in C^\infty(U_\alpha)$. This suggests that vector fields can be interpreted as first-order differential operators (i.e., derivations of $C^\infty(M)$), as in the Euclidean case; this is confirmed in 7.1.4.

7.1.3 Frame Fields for the Cotangent Bundle

Given a chart $(U_\alpha, \varphi_\alpha)$ for M, the associated trivializations Ψ_α for the cotangent bundle are given by (see the exercises of Chapter 6 for the equivalence class notations):

$$\Psi_\alpha([r, \alpha, \lambda]) = (\varphi_\alpha(r), \lambda), \, r \in U, \, \lambda \in (T_{\varphi_\alpha(r)} R^n)^*. \tag{7.6}$$

The duality between tangent and cotangent vectors is expressed by:

---

[1] An alternative and nonabusive notation, which is preferable in cases where we want $\partial/\partial x^i$ to refer to a vector field on $\varphi_\alpha(U_\alpha)$, using the "push-forward" defined in Chapter 2, would be

$$(\varphi_\alpha^{-1})_* \frac{\partial}{\partial x^i} \text{ instead of } \frac{\partial}{\partial x^i}.$$

$$[r, \alpha, \lambda] \cdot [r, \alpha, \xi] = \lambda(\xi), \tag{7.7}$$

and this does not depend on α, as discussed in Chapter 6. Let $\{\varepsilon_1, \ldots, \varepsilon_n\}$ be the dual basis of $(R^n)^*$ to $\{e_1, \ldots, e_n\}$. It follows from (7.7) that

$$\{\Psi_\alpha^{-1}(r, \varepsilon_1), \ldots, \Psi_\alpha^{-1}(r, \varepsilon_n)\}$$

is the dual basis to $\{\Phi_\alpha^{-1}(r, e_1), \ldots, \Phi_\alpha^{-1}(r, e_n)\}$, and so a frame field for the cotangent bundle is

$$s_i(.) = \Psi_\alpha^{-1}(., e_i) = dx^i, i = 1, \ldots, n.$$

Hence over U_α, any 1-form ω can be expressed either as $\omega(r) = [r, \alpha, h(r)]$, where $h(r)$ is the row vector with entries $(h_1(r), \ldots, h_n(r))$, or in the familiar form

$$\omega = h_1 dx^1 + \ldots + h_n dx^n,$$

where each $h_i \in C^\infty(U_\alpha)$.

7.1.4 Action of Vector Fields on Functions

For every $X \in \mathfrak{I}(M)$, that is, X is a vector field on M, and for every smooth function f on M, there exists a smooth function Xf on U, where (U, φ) is any chart on M, defined by

$$Xf(r) = (\xi^1 \frac{\partial}{\partial x^1} + \ldots + \xi^n \frac{\partial}{\partial x^n})(f \bullet \varphi^{-1})(\varphi(r)) = (d(f \bullet \varphi^{-1}) \cdot \xi)(r). \tag{7.8}$$

Clearly the right side depends linearly on the value of X at r, and therefore there exists a 1-form, denoted df, on U, specified by

$$(df \cdot X)(r) = Xf(r). \tag{7.9}$$

In fact Xf and df are well defined on the whole of M, because their meanings are unambiguous on the intersection of charts $(U_\alpha, \varphi_\alpha)$ and $(U_\gamma, \varphi_\gamma)$. For suppose that $X(r) = [r, \alpha, \xi_\alpha(r)] = [r, \gamma, \xi_\gamma(r)], r \in U_\alpha \cap U_\gamma$, so $\xi_\gamma = d(\varphi_\gamma \bullet \varphi_\alpha^{-1})\xi_\alpha$; omitting the arguments of certain differentials,

$$df \cdot X = d(f \bullet \varphi_\gamma^{-1}) \cdot \xi_\gamma = d(f \bullet \varphi_\gamma^{-1}) \cdot d(\varphi_\gamma \bullet \varphi_\alpha^{-1})\xi_\alpha$$

$$= (\varphi_\gamma \bullet \varphi_\alpha^{-1})^* (d(f \bullet \varphi_\gamma^{-1})) \cdot \xi_\alpha = d(f \bullet \varphi_\alpha^{-1}) \cdot \xi_\alpha.$$

The mapping $f \rightarrow Xf$ presents X as a differential operator on $C^\infty(M)$; let us formalize this as follows.

7.1.5 Vector Fields as Differential Operators

(i) The set $\mathfrak{I}(M)$ of vector fields on M is in one-to-one correspondence with the set of derivations of $C^\infty(M)$ (i.e., R-linear mappings Z from $C^\infty(M)$ to $C^\infty(M)$ such that $Z(fg) = f \cdot Zg + Zf \cdot g$).

(ii) Every smooth function f on M gives rise to a 1-form df on M characterized by $df \cdot X = Xf$, for $X \in \mathfrak{I}(M)$.

Proof: The first assertion follows from applying the corresponding result in Chapter 2, for vector fields on open subsets of Euclidean space, to (7. 8). The second assertion was proved above. ¤

7.2 Tangent Vectors as Equivalence Classes of Curves

The definition of tangent vectors as equivalence classes of triples has served us so far, but it is clumsy in many calculations. Here is an equivalent characterization which is both concise and intuitively satisfying.

Consider the set of smooth maps γ from open intervals about zero in R to M; such maps are called **smooth curves** in M. We shall say that two such curves are equivalent, written $\gamma_1 \sim \gamma_2$, if $\gamma_1(0) = \gamma_2(0) \in M$ and, for some chart (U, φ) at $r = \gamma_i(0)$,

$$D(\varphi \bullet \gamma_1)(0) = D(\varphi \bullet \gamma_2)(0). \tag{7. 10}$$

Note that this equivalence does not depend on the choice of chart, because for any other chart (U', ψ) at r, and writing $x = \varphi(r)$,

$$D(\psi \bullet \gamma_1)(0) = D(\psi \bullet \varphi^{-1} \bullet \varphi \bullet \gamma_1)(0) = D(\psi \bullet \varphi^{-1})(x) D(\varphi \bullet \gamma_1)(0)$$

$$= D(\psi \bullet \varphi^{-1})(x) D(\varphi \bullet \gamma_2)(0) = D(\psi \bullet \gamma_2)(0).$$

What we are saying here is simply that two curves are equivalent if they pass through the same point, and with the same velocity, at "time" zero. It is left as an exercise for the reader to prove that ~ is an equivalence relation.

7.2.1 Characterization of Tangent Vectors as Equivalence Classes of Curves

Given an atlas $\{(U_\alpha, \varphi_\alpha) : \alpha \in I\}$, equivalence classes $[\gamma]$ of curves γ are in one-to-one correspondence with tangent vectors using the mapping

$$[\gamma] \to [\gamma(0), \alpha, D(\varphi_\alpha \bullet \gamma)(0)], \tag{7. 11}$$

where of course the index α must satisfy $U_\alpha \ni \gamma(0)$.

Proof: Note first that the mapping is unambiguous, because for any indices α, δ

$$[\gamma(0), \delta, D(\varphi_\delta \bullet \gamma)(0)] = [\gamma(0), \alpha, D(\varphi_\alpha \bullet \varphi_\delta^{-1}) D(\varphi_\delta \bullet \gamma)(0)]$$

$$= [\gamma(0), \alpha, D(\varphi_\alpha \bullet \gamma)(0)].$$

Second, the mapping is one-to-one because (7. 10) shows that two classes of curves with the same image are the same class. To show that it is onto, pick any $r \in M$, any α with $U_\alpha \ni r$, and any vector $v \in R^n$, and let $\gamma(t) = \varphi_\alpha^{-1}(\varphi_\alpha(r) + tv) \in M$, for t in an interval about 0 small enough so that $\varphi_\alpha(r) + tv$ always belongs to $\varphi_\alpha(U_\alpha)$. Then (7. 11) implies that $[\gamma] \to [r, \alpha, v]$ is a one-to-one correspondence, as desired.   ¤

7.3   Exterior Calculus on Manifolds

The qth exterior power $\Lambda^q T^* M$ of the cotangent bundle can be constructed using the methods of Chapter 6. As explained in Chapters 1 and 2, we make the identifications

$$(\Lambda^q T^* M)_r \cong \Lambda^q (T_r^* M) \cong (\Lambda^q (T_r M))^*. \tag{7. 12}$$

The **differential** (or **exterior**) **forms of degree** q on M, or q-forms for short, are simply the smooth sections $\Gamma(\Lambda^q T^* M)$ of the qth exterior power of the cotangent bundle; we usually denote them by $\Omega^q(M)$. As before, $\Omega^0(M) = C^\infty(M)$. There are several ways of representing a q-form ω. An elegant but abstract way is to think of a q-form as a multilinear alternating map on q-tuples of vector fields in $\mathfrak{S}(M)$: We can treat ω as a map $\mathfrak{S}(M) \times \dots \times \mathfrak{S}(M) \to C^\infty(M)$, namely,

$$(X_1, \dots, X_q) \to \omega \cdot (X_1 \wedge \dots \wedge X_q), \tag{7. 13}$$

$$(\omega \cdot (X_1 \wedge \dots \wedge X_q))(r) = \omega(r)(X_1(r) \wedge \dots \wedge X_q(r)), r \in M. \tag{7. 14}$$

This makes sense by (7. 12), since $\omega(r)$ is a linear form on the qth exterior product of the tangent space at r.

A local representation of a q-form may be given in terms of a local frame field, as in 7.1.1. Using the notation of local frame fields, we may write

$$\omega(r) = \sum_I b_I(r)(dx^{i(1)} \wedge \dots \wedge dx^{i(q)})(r), r \in U_\alpha, \tag{7. 15}$$

where as usual $I = (i(1) < \dots < i(q))$, and $b_I(.)$ is smooth.

Recall from Chapter 2 that the operations of exterior product and exterior derivative for differential forms on open subsets of Euclidean space "commute with the pullback," in the sense that if $\psi: U \subseteq R^n \to V \subseteq R^n$ is smooth, then

$$\psi^* \, d\omega = d(\psi^* \, \omega), \psi^* \, (\omega \wedge \eta) = (\psi^* \, \omega) \wedge (\psi^* \, \eta). \tag{7.16}$$

In the case where ψ is a diffeomorphism, we note that duality of forms and vector fields, and the Lie derivative of forms, commute with the pullback in the sense that

$$\psi^* \, \omega \cdot (X_1 \wedge \ldots \wedge X_q) = \omega \cdot (\psi_* X_1 \wedge \ldots \wedge \psi_* X_q); \tag{7.17}$$

$$L_X (\psi^* \, \omega) = L_{\psi_* X} \omega \tag{7.18}$$

(the first statement holds by definition; the second can be inferred from the formula for Lie derivative of a form).

7.3.1 Transferring Exterior Calculus to Manifolds

Any operation applied to differential forms on open subsets of Euclidean space that commutes with the pullback in the sense above – for example, exterior product, exterior derivative, and Lie derivative of forms – is well defined for differential forms on manifolds, and conforms to the same calculus; thus for any q-form ω and p-form η,

$$\eta \wedge \omega = -(\omega \wedge \eta), dd\omega = 0; \tag{7.19}$$

$$d(\omega \wedge \eta) = d\omega \wedge \eta + (-1)^{\deg \omega} (\omega \wedge d\eta); \tag{7.20}$$

$$L_X f = Xf, L_X (df) = d(L_X f), f \in C^\infty (M), X \in \mathfrak{I} (M);^2 \tag{7.21}$$

$$L_X (\omega \wedge \eta) = L_X \omega \wedge \eta + \omega \wedge L_X \eta; \tag{7.22}$$

and if $g: P \to M$ is a smooth map, the pullback $g^* : \Omega^q M \to \Omega^q P$, defined by

$$g^* \, \omega \cdot (X_1 \wedge \ldots \wedge X_q) = \omega \cdot (Tg \bullet X_1 \wedge \ldots \wedge Tg \bullet X_q), \tag{7.23}$$

satisfies

$$g^* \, d\omega = d(g^* \, \omega), g^* \, (\omega \wedge \eta) = (g^* \, \omega) \wedge (g^* \, \eta). \tag{7.24}$$

Proof: All that needs to be checked is that if we perform any of these operations on the local representation (7. 15) of one or more forms, then the result does not depend on

---

^{2} See (7. 9) for the meanings of df and Xf.

which local representation we chose. Take the case of exterior differentiation, for example. On $U_\alpha \cap U_\gamma \subseteq M$, a differential form ω may be expressed as

$$r \to [r, \alpha, \lambda(r)] = [r, \gamma, \mu(r)], \lambda = (\varphi_\gamma \bullet \varphi_\alpha^{-1})^* \mu,$$

and so it follows that we may define $d\omega$ on U_α to be $r \to [r, \alpha, d\lambda(r)]$ without fear of ambiguity, because (7. 16) tells us that

$$[r, \alpha, d\lambda(r)] = [r, \alpha, ((\varphi_\gamma \bullet \varphi_\alpha^{-1})^* d\mu)(r)] = [r, \gamma, d\mu(r)], r \in U_\alpha \cap U_\gamma.$$

The other cases follow the same pattern. The formulas (7. 19) to (7. 22) and (7. 24) hold because they hold in every local representation.   ¤

7.3.2 Example

Take $U = R^3 - \{0\}$, which is a three-dimensional submanifold of R^3, and consider the differential form $\omega \in \Omega^2 U$ given by

$$\omega = \frac{x(dy \wedge dz) - y(dx \wedge dz) + z(dx \wedge dy)}{(x^2 + y^2 + z^2)^{3/2}}. \tag{7. 25}$$

The restriction of ω to the sphere S^2 gives a 2-form $\eta \in \Omega^2 S^2$, which can be represented in various ways; for example, since $x^2 + y^2 + z^2 = 1$, it follows that

$$dx = \frac{-y dy - z dz}{x}, \tag{7. 26}$$

and so on the portion of the sphere where

$$x = -\sqrt{1 - y^2 - z^2} \neq 0,$$

we may represent η in the (y, z) coordinate system by

$$\eta = \frac{dy \wedge dz}{x} = -\frac{dy \wedge dz}{\sqrt{1 - y^2 - z^2}}. \tag{7. 27}$$

Similarly, on the part of the sphere where $y \neq 0$, we can write $\eta = -(dx \wedge dz)/y$. One consequence of 7.3.1 is that the exterior derivative $d\eta$ will always be the same (actually it must be zero in this case, because the third exterior power of a 2-dimensional space is always trivial) whether we calculate it in the representation (7. 27) or some other one. For example, to compute $d\eta$ from (7. 27), write

$$d(x\eta) = dx \wedge \eta + x d\eta = d(dy \wedge dz) = 0.$$

Now substitute (7. 26) and (7. 27) into $dx \wedge \eta$ to see that it must be zero, and hence $d\eta$ is zero.

7.4   Exercises

1. Suppose $\Phi_U: \pi^{-1}(U) \to U \times R^k$ and $\Phi_V: \pi^{-1}(V) \to V \times R^k$ are local trivializations of a rank-k vector bundle $\pi: E \to M$, with $U \cap V \neq \emptyset$. Consider the local frame fields $r \to \{s_1(r), ..., s_k(r)\}$ and $r \to \{t_1(r), ..., t_k(r)\}$ over $U \cap V$, where

    $$s_i(r) = \Phi_U^{-1}(r, e_i), \, t_i(r) = \Phi_V^{-1}(r, e_i), \, i = 1, 2, ..., k, \, r \in U \cap V.$$

    Let $g_j^i(r)$ denote the (i, j) entry of the transition function $g_{UV}(r)$. Prove, using the definition of transition function in Chapter 6, that

    $$t_j = g_j^1 s_1 + ... + g_j^k s_k. \tag{7.28}$$

2. Two smooth curves on a manifold M are called equivalent, written $\gamma_1 \sim \gamma_2$, if $\gamma_1(0) = \gamma_2(0) \in M$ and, for some (and hence any) chart (U, φ) at $r = \gamma_i(0)$,

    $$D(\varphi \bullet \gamma_1)(0) = D(\varphi \bullet \gamma_2)(0). \tag{7.29}$$

    Prove that $\sim$ is an equivalence relation.

3. Suppose $f: M \to N$ is a smooth map between two manifolds. Prove that the definition of the tangent map Tf given in Chapter 6 is equivalent to

    $$Tf([\gamma]) = [f \bullet \gamma] \tag{7.30}$$

    for every curve γ on M, using the notation of Exercise 2 and the correspondence given in (7. 11).

4. Let us endow the sphere $S^2 \subset R^3$ with the atlas $\{(U, \varphi), (V, \psi)\}$ as follows:

    $$U = S^2 - \{(0, 0, 1)\}, \varphi(x, y, z) = \frac{(x, y)}{1 - z};$$

    $$V = S^2 - \{(0, 0, -1)\}, \psi(x, y, z) = \frac{(x, y)}{1 + z}.$$

    (i) Consider the curve $\gamma(t) = (1 - t^2/2, 0, t\sqrt{1 - t^2/4}) \in S^2$. Express the equivalence class $[\gamma]$ of this curve as a tangent vector $[r, \alpha, z]$ for $\alpha = 1, 2$ (i.e., with respect to both charts) where $r = (1, 0, 0)$.
    (ii) Define a map $f: S^2 \to S^2$ by taking $\theta(z) = \pi(1 - z^2)$ and

$$f(x, y, z) = (x\cos\theta(z) - y\sin\theta(z), x\sin\theta(z) + y\cos\theta(z), z).$$

Calculate the tangent vector $Tf([\gamma])$ in the form $[f(r), \alpha, z]$, for one or other choice of α.

7.5   Indefinite Riemannian Metrics

From the discussion in Chapter 6 of the tensor product of vector bundles, the reader may deduce that, for any vector bundle $\pi: E \to M$, $E^* \otimes E^* \to M$ is a vector bundle whose fiber over $r \in M$ can be regarded as the set of bilinear maps from $E_r \times E_r$ to R. A smooth section of this bundle, denoted $r \to \langle \cdot | \cdot \rangle_r$, is called an **indefinite metric** if it has the following two properties at every $r \in M$:

(i) **Symmetry:** $\langle \zeta | \xi \rangle_r = \langle \xi | \zeta \rangle_r$, $\forall \xi, \zeta \in E_r$;

(ii) **Nondegeneracy:** $\xi \neq 0 \Rightarrow \langle \xi | \xi \rangle_r \neq 0$.

In other words, an indefinite metric gives an inner product on the fiber over r, which varies smoothly with r. We call $r \to \langle . | . \rangle_r$ a **metric** if it satisfies (i) and the following strengthening of (ii):

(iii) **Positivity:** $\xi \neq 0 \Rightarrow \langle \xi | \xi \rangle_r > 0$.

In the special case where E is the tangent bundle of M, we refer to the map $r \to \langle \cdot | \cdot \rangle_r$ as an **indefinite Riemannian metric**, or as a **Riemannian metric** if (iii) holds. A manifold M with a Riemannian metric (resp., indefinite Riemannian metric) $r \to \langle \cdot | \cdot \rangle_r$ is referred to as "the **Riemannian manifold** (resp., **pseudo-Riemannian manifold**) $(M, \langle \cdot | \cdot \rangle)$." If X and Y are vector fields on M, we customarily abbreviate $\langle X(r) | Y(r) \rangle_r$ to $\langle X | Y \rangle_r$.

7.5.1   Local Expression for an Indefinite Riemannian Metric

Suppose (U, φ) is a chart for an n-dimensional pseudo-Riemannian manifold $(M, \langle \cdot | \cdot \rangle)$, and $\{x^1, ..., x^n\}$ is the standard coordinate system on R^n. The local frame field for TM over U

$$\{ (\varphi^{-1})_* \frac{\partial}{\partial x^1}, ..., (\varphi^{-1})_* \frac{\partial}{\partial x^n} \},$$

which we already abbreviated in (7. 4) to $\{\frac{\partial}{\partial x^1}, ..., \frac{\partial}{\partial x^n}\}$, will be further abbreviated to

$$\{D_1, ..., D_n\}.$$

The local form of the **Riemannian metric tensor** means the map which takes $r \in U$ to the $n \times n$ nondegenerate symmetric matrix $G(r) = (g_{ij}(r))$ given by

$$g_{ij}(r) = \langle D_i | D_j \rangle_r.$$ (7.31)

Of course $G(r)$ is a positive definite matrix in the case where the metric is Riemannian. This gives a convenient formula for the inner product between two vector fields:

$$X = \xi^1 D_1 + \dots + \xi^n D_n, \; Y = \zeta^1 D_1 + \dots + \zeta^n D_n$$

$$\Rightarrow \langle X | Y \rangle_r = \sum_{i,j} g_{ij}(r) \, \xi^i(r) \, \zeta^j(r).$$ (7.32)

A more concise way to write this formula, using some notation from Chapter 2, is

$$\langle \cdot | \cdot \rangle = \sum_{i,j} g_{ij} dx^i \otimes dx^j.$$ (7.33)

It is often most convenient to check the smoothness condition on the metric using (7.31); the proof of the following statement is omitted.

7.5.2 Condition for a Collection of Inner Products to Give an Indefinite Metric

An assignment of an inner product $\langle \cdot | \cdot \rangle_r$ to the tangent space at r for every r in M is an indefinite Riemannian metric if and only if there is an atlas for M in which all of the maps $(g_{ij}(\cdot))$ in (7.31) are smooth.

7.6 Examples of Riemannian Manifolds

7.6.1 The Euclidean Metric

Take M to be an open subset of R^n, and give each tangent space $T_x R^n$ the Euclidean inner product $\langle v | w \rangle_x^E = v_1 w_1 + \dots + v_n w_n$, using the standard coordinate system. In other words, using the chart map $\varphi = $ identity, the local expression of the metric tensor is the identity matrix. Often we want to express the same metric in a different chart. For example, if $U \subset R^2 - \{0\}$ and $\varphi^{-1}(r, \theta) = (r\cos\theta, r\sin\theta)$ (polar coordinate chart), then $dx = \cos\theta dr - r\sin\theta d\theta$ and $dy = \sin\theta dr + r\cos\theta d\theta$, and so (7.33) gives

$$\langle \cdot | \cdot \rangle^E = dx \otimes dx + dy \otimes dy$$

$$= (\cos\theta dr - r\sin\theta d\theta) \otimes (\cos\theta dr - r\sin\theta d\theta) + \text{etc.}$$

$$= dr \otimes dr + r^2 d\theta \otimes d\theta .$$

(Note the absence of $dx \otimes dy$ and $dr \otimes d\theta$ terms for this metric.) Thus in this chart the metric tensor is expressed by the matrix

$$G(r, \theta) = \begin{bmatrix} 1 & 0 \\ 0 & r^2 \end{bmatrix} . \tag{7.34}$$

7.6.2 Induced Metric on Submanifolds of R^{n+k}

If M is an n-dimensional submanifold of R^{n+k}, the inclusion map $\iota : M \to R^{n+k}$ induces a Riemannian metric $\langle \cdot | \cdot \rangle$ on M from the Euclidean metric $\langle \cdot | \cdot \rangle^E$ on R^{n+k}, namely,

$$\langle X | Y \rangle_x = \langle X | Y \rangle_x^E, X, Y \in \Im(M) . \tag{7.35}$$

When $n = 2$ and $k = 1$, the induced metric corresponds intuitively to the notions of length and angle in two dimensions for an insect crawling around on the surface of M.

Typically we study such a metric using a parametrization $\Psi : W \subseteq R^n \to U \subseteq R^{n+k}$ for M. The metric tensor can be expressed in terms of a coordinate system $\{u^1, \dots, u^n\}$ on W as follows: If $\Psi(u) = (\Psi^1(u), \dots, \Psi^{n+k}(u))$, then

$$\langle \cdot | \cdot \rangle = dx^1 \otimes dx^1 + \dots + dx^{n+k} \otimes dx^{n+k}$$

$$= (\frac{\partial \Psi^1}{\partial u^1} du^1 + \dots + \frac{\partial \Psi^1}{\partial u^n} du^n) \otimes (\frac{\partial \Psi^1}{\partial u^1} du^1 + \dots + \frac{\partial \Psi^1}{\partial u^n} du^n) + \dots .$$

It follows that

$$\langle \cdot | \cdot \rangle = \sum_{i,j} \left(\sum_m \frac{\partial \Psi^m}{\partial u^i} \frac{\partial \Psi^m}{\partial u^j} \right) du^i \otimes du^j . \tag{7.36}$$

Thus in the chart induced by the parametrization, the metric tensor is expressed by the matrix

$$G(u) = D\Psi(u)^T D\Psi(u) . \tag{7.37}$$

7.6.2.1 Example: The 2-Sphere in R^3

As a special case of 7.6.2, parametrize the 2-sphere in R^3 using the usual angular coordinates, namely, $\Psi(\phi, \theta) = (\cos\theta \sin\phi, \sin\theta \sin\phi, \cos\phi)$. Elementary calculations using the formula (7.37) show that, in this coordinate system, the metric tensor is expressed by the matrix

$$G(\theta, \phi) = \begin{bmatrix} 1 & 0 \\ 0 & (\sin\phi)^2 \end{bmatrix}. \tag{7.38}$$

7.6.3 Hyperbolic Spaces

A generalization of the Lorentz inner product is the inner product $\langle\cdot|\cdot\rangle^L$ on R^{n+1} given by

$$\langle v|v\rangle^L = -v_0^2 + v_1^2 + \ldots + v_n^2, \, v = (v_0, \ldots, v_n). \tag{7.39}$$

Another way to write the definition is

$$\langle\cdot|\cdot\rangle^L = -dx_0 \otimes dx_0 + dx_1 \otimes dx_1 + \ldots + dx_n \otimes dx_n. \tag{7.40}$$

Note that $(R^{n+1}, \langle\cdot|\cdot\rangle^L)$ is an example of a pseudo-Riemannian manifold. Now define the set H^n to be one of the two components of the hyperboloid H_1^n, which is an n-dimensional submanifold of R^{n+1} defined in Chapter 3, namely,

$$H^n = \{(x_0, \ldots, x_n) \in R^{n+1} : x_0^2 - x_1^2 - \ldots - x_n^2 = 1, x_0 > 0\}. \tag{7.41}$$

It turns out that, although $\langle\cdot|\cdot\rangle^L$ is an indefinite inner product (the signature is $n-1$), it induces a Riemannian metric $\langle\cdot|\cdot\rangle$ on H^n by the formula

$$\langle X|Y\rangle = \langle X|Y\rangle^L, X, Y \in \mathfrak{I}(H^n). \tag{7.42}$$

See Exercise 6 for the proof that $\langle\cdot|\cdot\rangle$ is positive definite. The Riemannian manifold $(H^n, \langle\cdot|\cdot\rangle)$ is called the n-dimensional hyperbolic space. To see what this may be like, consider the case $n = 2$, known as the hyperbolic plane; the part of the set $H^2 \subset R^3$ where $x_0^2 \leq 10$ is shown in Figure 7. 1.

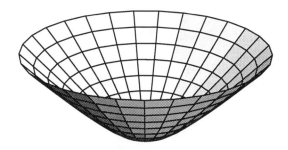

Figure 7. 1   Portion of the hyperbolic plane, without the hyperbolic metric

We can parametrize

$$H^2 = \{ (x, y, z) \in R^3 : z^2 - x^2 - y^2 = 1, z > 0 \} \qquad \text{(7. 43)}$$

by taking polar coordinates, namely, $x = r\cos\theta$, $y = r\sin\theta$, $z = \sqrt{1 + r^2}$. One may easily compute that

$$dx \otimes dx = (\cos\theta)^2 dr \otimes dr - r\cos\theta\sin\theta \, (dr \otimes d\theta + d\theta \otimes dr) + r^2 (\sin\theta)^2 d\theta \otimes d\theta,$$

and similarly for $dy \otimes dy$ and $dz \otimes dz$. Using (7. 40), a routine calculation (taking $z = x_0$) shows that the metric on the hyperbolic plane is

$$\langle \cdot | \cdot \rangle = -dz \otimes dz + dx \otimes dx + dy \otimes dy$$

$$= \frac{dr \otimes dr}{1 + r^2} + r^2 d\theta \otimes d\theta.$$

A useful reparametrization is $r = \sinh s$ (i.e., hyperbolic sine), since a little calculus shows that $ds \otimes ds = dr \otimes dr / (1 + r^2)$; thus in the (s, θ) parametrization,

$$\langle \cdot | \cdot \rangle = ds \otimes ds + (\sinh s)^2 d\theta \otimes d\theta.$$

Thus the form of the metric tensor in the (s, θ) parametrization is

$$G(s, \theta) = \begin{bmatrix} 1 & 0 \\ 0 & (\sinh s)^2 \end{bmatrix}. \qquad \text{(7. 44)}$$

The similarities and the differences among (7. 34), (7. 38), and (7. 44) should be carefully noted.

7.7   Orthonormal Frame Fields

A local frame field $\{ e_1, \ldots, e_n \}$ for TM over U is called an **orthonormal frame field** if $\langle e_i | e_j \rangle_p = \pm\delta_{ij}$ for all i, j and for all $p \in U$. We have already encountered a special case in Chapter 4, where $\{ \xi_1, \xi_2, \xi_3 \}$ was an orthonormal frame field for TR^3 over a parametrized surface M, and $\{ \xi_1, \xi_2 \}$ was an orthonormal frame field for TM.

7.7.1   Existence of Orthonormal Frame Fields

In every chart for a pseudo-Riemannian manifold there exists an orthonormal frame field for the tangent bundle.

Proof: Take a chart (U, φ) around the point p. Consider the inductive hypothesis $H(m)$ that, for every set of vector fields $\{ X_1, \ldots, X_m \}$ on U such that the tangent vectors $\{ X_1(p), \ldots, X_m(p) \} \in T_p M$

are linearly independent for every $p \in U$, there exists another set of vector fields $\{e_1, ..., e_m\}$ on U, with $\langle e_i | e_j \rangle = \pm \delta_{ij}$, that span the same subspace of $T_p M$ for every $p \in U$. Obviously if M is n-dimensional, then $H(n)$ is what we have to prove. Note first that $H(1)$ holds because, given a nonvanishing vector field X_1, we can take

$$e_1 = X_1 / (\sqrt{|\langle X_1 | X_1 \rangle|}). \qquad (7.45)$$

Now assume that $2 \le m \le n - 1$ and that $H(m-1)$ is true. Let $\{X_1, ..., X_m\}$ be vector fields on U such that the tangent vectors $\{X_1(p), ..., X_m(p)\} \in T_p M$ are linearly independent for every $p \in U$. Define e_1 as in (7.45), and define vector fields $\{Y_2, ..., Y_m\}$ on U by

$$Y_j = X_j - \frac{\langle X_j | e_1 \rangle}{\langle e_1 | e_1 \rangle} e_1,$$

which has the consequence that $\langle Y_j | e_1 \rangle = 0$ for $j = 2, 3, ..., m$. Any linear dependence among $\{Y_2(p), ..., Y_m(p)\}$ would imply the same for $\{X_1(p), ..., X_m(p)\}$, which is contrary to the assumptions; hence $\{Y_2, ..., Y_m\}$ satisfy the conditions for the application of $H(m-1)$, which guarantees that there exists a set of vector fields $\{e_2, ..., e_m\}$ on U, with $\langle e_i | e_j \rangle = \pm \delta_{ij}$, such that, for every $p \in U$,

$$\text{Span}\{e_2(p), ..., e_m(p)\} = \text{Span}\{Y_2(p), ..., Y_m(p)\}.$$

It follows in particular that $\langle e_j | e_1 \rangle = 0$ for $j = 2, 3, ..., m$. This construction of the set of vector fields $\{e_1, ..., e_m\}$ verifies $H(m)$, thus completing the induction. ¤

7.7.2 Orthonormal Coframe Fields

Recall that, in our introduction to differential forms on Euclidean space in Chapter 2, the frame field $\{dx^1, ..., dx^n\}$ for the cotangent bundle was defined by taking $\{dx^1(y), ..., dx^n(y)\}$ to be the dual basis of $(T_y R^n)^*$ corresponding to the basis $\{\partial/\partial x^1|_y, ..., \partial/\partial x^n|_y\}$ for the tangent space at $y \in R^n$. Another way to say this is that $\{dx^1, ..., dx^n\}$ is the **dual frame** to $\{\partial/\partial x^1, ..., \partial/\partial x^n\}$.

Likewise, given an orthonormal frame field $\{e_1, ..., e_m\}$ on an open set U in a pseudo-Riemannian manifold $(M, \langle \cdot | \cdot \rangle)$, we immediately have an **orthonormal coframe field** $\{\theta^1, ..., \theta^n\}$, namely, the dual frame to $\{e_1, ..., e_m\}$. This is the set of 1-forms defined by

$$\theta^i \cdot (A^1 e_1 + ... + A^n e_n) = A^i. \qquad (7.46)$$

This is a complete definition because every vector field on U can be expressed in the form $A^1 e_1 + ... + A^n e_n$ for some functions $\{A^i\}$. The smoothness of the sections $\{\theta^1, ..., \theta^n\}$ follows easily from that of $\{e_1, ..., e_m\}$, after taking local coordinates.

7.7.3   Expressing a Frame Field in Terms of an Orthonormal One

Suppose (U, φ) is a chart for an n-dimensional pseudo-Riemannian manifold $(M, \langle \cdot | \cdot \rangle)$, and $\{x^1, \ldots, x^n\}$ is the standard coordinate system on R^n. As before, the local frame field $\{ (\varphi^{-1})_* (\partial / \partial x^1), \ldots, (\varphi^{-1})_* (\partial / \partial x^n) \}$ for TM over U will be abbreviated to $\{D_1, \ldots, D_n\}$, and $\{\varphi^* dx^1, \ldots, \varphi^* dx^n\}$ will be abusively referred to as $\{dx^1, \ldots, dx^n\}$.

Now suppose $\{e_1, \ldots, e_m\}$ is an orthonormal frame field on U. To keep track of the plus ones and minus ones, we introduce a diagonal $n \times n$ matrix

$$\Lambda = \begin{bmatrix} \lambda_1 & 0 & 0 \\ 0 & \ldots & 0 \\ 0 & 0 & \lambda_n \end{bmatrix} = \mathrm{diag}\,(1, \ldots, 1, -1, \ldots, -1), \tag{7.47}$$

where $\lambda_i = \langle e_i | e_i \rangle$. Note that, because of smoothness of the inner product, the same constant matrix Λ applies at all points in the chart. Of course in the Riemannian case, Λ is the identity matrix.

If we try to express the frame field $\{D_1, \ldots, D_n\}$ in terms of the orthonormal frame field, we obtain a function Ξ from U into the nonsingular $n \times n$ matrices, which can be written as $\Xi_{mi}(p) = \xi_i^m(p)$, or else as

$$\Xi(p) = [\vec{\xi}_1(p), \ldots, \vec{\xi}_n(p)], p \in U, \tag{7.48}$$

where the $\{\vec{\xi}_i\}$ are column vectors (cf. Chapter 4), such that

$$D_i = \sum_m \xi_i^m e_m. \tag{7.49}$$

It follows that

$$g_{ij}(p) = \langle D_i | D_j \rangle_p = \sum_{m,q} \xi_i^m(p) \xi_j^q(p) \langle e_m | e_q \rangle_p = \sum_m \xi_i^m(p) \lambda_m \xi_j^m(p);$$

$$G(p) = \Xi(p)^T \Lambda \Xi(p). \tag{7.50}$$

Suppose $(\Xi(p)^{-1})_{ik} = \zeta_k^i$. Postmultiplying (7.49) by this matrix gives

$$e_k = \sum_{m,i} \xi_i^m \zeta_k^i e_m = \sum_i \zeta_k^i D_i. \tag{7.51}$$

Also from (7.46) and (7.49), we see that

$$\theta^m \cdot D_i = \xi_i^m, \therefore \theta^m = \sum_i \xi_i^m dx^i. \tag{7.52}$$

Evidently the construction of orthonormal frame and coframe fields from the original frame field $\{D_1, \ldots, D_n\}$ amounts to finding a "smooth factorization" of the metric tensor G over U of the form (7.50), that is, a smooth map $\Xi: U \to GL_n(R)$ such that $G(p) = \Xi(p)^T \Lambda \Xi(p)$, for then we may define the orthonormal frame using (7.51) and the coframe field using (7.52).

7.7.3.1 Example: Orthonormal Frame Field on a Parametrized Surface

Consider the case where $\Psi: W \subseteq R^2 \to R^3$ is a 2-dimensional parametrization, that is, $M = \Psi(W)$ is a parametrized surface. Let $\{u, v\}$ be a coordinate system on W, and express $\Psi(u, v)$ as $(\Psi^1(u, v), \Psi^2(u, v), \Psi^3(u, v))$. As we saw in 7.6.2, the Euclidean metric on R^3 induces a metric

$$\langle \cdot | \cdot \rangle = \left\| \vec{\Psi}_u \right\|^2 du \otimes du + \langle \vec{\Psi}_u | \vec{\Psi}_v \rangle^E [du \otimes dv + dv \otimes du] + \left\| \vec{\Psi}_v \right\|^2 dv \otimes dv,$$

where $\vec{\Psi}_u = [\partial \Psi^1 / \partial u, \partial \Psi^2 / \partial u, \partial \Psi^3 / \partial u]^T$, etc. Therefore, if $D_u = \Psi_*(\partial/\partial u)$, etc., then

$$\langle D_u | D_u \rangle = \left\| \vec{\Psi}_u \right\|^2, \langle D_u | D_v \rangle = \langle \vec{\Psi}_u | \vec{\Psi}_v \rangle^E, \langle D_v | D_v \rangle = \left\| \vec{\Psi}_v \right\|^2. \tag{7.53}$$

As we saw in Chapter 4, an orthonormal frame field is given by

$$\xi_1 = \left\| \vec{\Psi}_u \right\|^{-1} D_u, \xi_2 = -\frac{\langle \vec{\Psi}_u | \vec{\Psi}_v \rangle^E}{\left\| \vec{\Psi}_u \right\| \left\| \vec{\Psi}_u \times \vec{\Psi}_v \right\|} D_u + \frac{\left\| \vec{\Psi}_u \right\|}{\left\| \vec{\Psi}_u \times \vec{\Psi}_v \right\|} D_v. \tag{7.54}$$

The fact that this is orthonormal will be checked again in Exercise 10.

7.7.3.2 Example: Orthonormal Coframe Field on a Parametrized Surface

In Chapter 4, we used the following expressions for the orthonormal coframe $\{\theta^1, \theta^2\}$ to the orthonormal frame $\{\xi_1, \xi_2\}$ given in (7.54):

$$\theta^1 = \left\| \vec{\Psi}_u \right\| du + \frac{\langle \vec{\Psi}_u | \vec{\Psi}_v \rangle^E}{\left\| \vec{\Psi}_u \right\|} dv, \theta^2 = \frac{\left\| \vec{\Psi}_u \times \vec{\Psi}_v \right\|}{\left\| \vec{\Psi}_u \right\|} dv, \tag{7.55}$$

where du is short for $(\Psi^{-1})^* du$, etc. We leave it to the reader in Exercise 11 to check that $\theta^i \cdot \xi_j = \delta^i_j$.

7.8   An Isomorphism between the Tangent and Cotangent Bundles

7.8.1   Switching between Vector Fields and 1-Forms Using the Metric

Let us now define a vector bundle isomorphism from the tangent bundle to the cotangent bundle by using the map that sends $\zeta \in T_r M$ to the linear form $\xi \to \langle \zeta | \xi \rangle_r$, henceforward abbreviated to $\langle \zeta |$; linearity is obvious, and nondegeneracy of the inner product ensures it is one-to-one and onto. Abusing notation slightly, we extend this to a map from vector fields to 1-forms, where for a vector field X, $\langle X |$ is the 1-form given by

$$\langle X | \cdot Y = \langle X | Y \rangle, \; Y \in \mathfrak{I}(M). \tag{7.56}$$

The inverse to this mapping takes a 1-form ω to the vector field $\omega^\#$, pronounced "omega sharp," where

$$\langle \omega^\# | Y \rangle = \omega \cdot Y, \; Y \in \mathfrak{I}(M). \tag{7.57}$$

By definition, $\langle X |^\# = X$ and $\langle \omega^\# | = \omega$. These operations are already familiar in R^3 with the Euclidean metric; for a vector field $V = v^1 \partial/\partial x + v^2 \partial/\partial y + v^3 \partial/\partial z$, $\langle V | = \varpi_V = v^1 dx + v^2 dy + v^3 dz$, that is, the "work form" referred to in Chapter 2; for a function f on R^3, the vector field $(df)^\# = \operatorname{grad} f$ in the vector calculus sense (hence we use this formula to define $\operatorname{grad} f$ for a function on a pseudo-Riemannian manifold).

7.8.2   Formulas for $\langle X |$, $\omega^\#$, and $\operatorname{grad} f$

(i) $\langle D_i | = \sum_j g_{ij} dx^j$ and $(dx^k)^\# = \sum_j g^{kj} D_j$, where $G(r)^{-1} = (g^{jk}(r))$.

(ii) For any orthonormal frame field $\{e_1, ..., e_m\}$ with orthonormal coframe field $\{\theta^1, ..., \theta^n\}$,

$$\langle e_i | = \theta^i \text{ and } (\theta^i)^\# = e_i. \tag{7.58}$$

(iii) For any smooth function f on M, the vector field $\operatorname{grad} f$ defined by $\operatorname{grad} f = (df)^\#$ has the local expression

$$\operatorname{grad} f = \sum_{j, k} (D_k f \, g^{kj}) D_j. \tag{7.59}$$

Proof: The first and second formulas follow immediately by substituting $X = D_i$, $Y = D_j$, and $\omega = dx^k$ into (7.56) and (7.57), respectively. The formulas in (ii) follow from (7.46), and the formula in (iii) from the second formula in (i).   ¤

7.9   Exercises

5. Calculate the Riemannian metric tensor for the following 2-dimensional submanifolds of R^3, using the induced metric 7.6.2:

(i) A surface of the form $\{ (x, y, z) : z = f(x, y) \}$, using (x, y) parametrization.

(ii) The cylinder $\{ (x, y, z) : x^2 + y^2 = A^2 \}$, using (θ, z) parametrization; here θ is an angular variable in the (x, y) plane.

(iii) The hyperbolic paraboloid

$$z = \frac{x^2}{A^2} - \frac{y^2}{B^2},$$

with the parametrization $\Psi(s, t) = (As, 0, s^2) + t(A, B, 2s)$.

(iv) The ellipsoid

$$\frac{x^2}{A^2} + \frac{y^2}{B^2} + \frac{z^2}{C^2} = 1,$$

for the parametrization $\Psi(u, v) = (A \sin u \cos v, B \sin u \sin v, C \cos u)$.

6. Show that the metric $\langle . | . \rangle^L = - dx_0 \otimes dx_0 + dx_1 \otimes dx_1 + \ldots + dx_n \otimes dx_n$, restricted to

$$H^n = \{ (x_0, \ldots, x_n) \in R^{n+1} : x_0^2 - x_1^2 - \ldots - x_n^2 = 1, x_0 > 0 \},$$

is indeed positive definite, by using the chart $\varphi(x_0, x_1, \ldots, x_n) = (x_1, \ldots, x_n)$, and the fact that

$$dx_0 = \frac{x_1 dx_1 + \ldots + x_n dx_n}{\sqrt{1 + x_1^2 + \ldots + x_n^2}}.$$

7. Calculate the Riemannian metric tensor for the 3-dimensional hyperbolic space H^3 with the metric described in 7.6.3; use a spherical coordinate parametrization (r, ϕ, θ) to begin with, and make the reparametrization $r = \sinh s$ at the end.

8. Derive an orthonormal frame field $\{e_1, e_2\}$ and the corresponding orthonormal coframe field $\{\varepsilon^1, \varepsilon^2\}$ for the hyperbolic plane, starting from the frame field $\{\partial/\partial s, \partial/\partial\theta\}$. Also derive the formula for grad in the (s, θ) coordinate system.

9. Express the Euclidean metric on R^3 in terms of the spherical coordinate system (r, ϕ, θ). Derive the formula for grad in terms of $\{\partial/\partial r, \partial/\partial\phi, \partial/\partial\theta\}$.

10. Verify that the frame field $\{\xi_1, \xi_2\}$ in (7. 54) really is orthonormal under the metric (7. 53). Also calculate the 2×2 matrix-valued function Ξ such that
$$G(p) \; = \; \Xi(p)^T \Xi(p).$$
Hint: Use the identity from Chapter 1 which says $\| v \times w \| = \langle v|v\rangle\langle w|w\rangle - \langle v|w\rangle^2$ for the Euclidean metric on R^3.

11. Show that $\{\theta^1, \theta^2\}$ in (7. 55) is indeed the coframe to the orthonormal frame $\{\xi_1, \xi_2\}$ given in (7. 54), by showing that $\theta^i \cdot \xi_j \; = \; \delta^i_j$.

12. Using (7. 55), calculate an orthonormal coframe field for a surface in R^3 with a parametrization of the type $\Psi(u, v) \; = \; (u, v, h(u, v))$.

13. (i) Given a metric $\langle \cdot | \cdot \rangle$ on a real or complex vector bundle $\pi: E \to M$, with fibers isomorphic to a vector space V, and two orthonormal frame fields $\{s_i\}$ and $\{t_j\}$ related by
$$t_j \; = \; \sum_i g^i_j s_i,$$
show that the matrix g is orthogonal, that is, $g^{-1} = g^T$.

Hint: $\langle t_j | t_k \rangle = \delta_{jk} \Rightarrow \sum_i g^i_j g^i_k = \delta_{jk}$.

(ii) Suppose $\pi: E' \to M$ is another vector bundle, with fibers isomorphic to a vector space V', and with a metric $\langle \cdot | \cdot \rangle^{E'}$. Given sections F and $\tilde{F}$ of the vector bundle Hom (E, E') (i.e., F is a smooth assignment to each $p \in M$ of a linear transformation $F(p) \in L(V \to V')$), let us define
$$\langle F | \tilde{F} \rangle^{\text{Hom}}_p \; = \; \sum_i \langle Fs_i | \tilde{F}s_i \rangle^{E'}_p, \tag{7. 60}$$
where the $\{s_i\}$ form an orthonormal frame field for $\pi: E \to M$. Prove that the left side of (7. 60) does not depend on the choice of frame field, in the sense that if $\{t_j\}$ is another orthonormal frame field, then
$$\sum_j \langle Ft_j | \tilde{F}t_j \rangle_p \; = \; \sum_i \langle Fs_i | \tilde{F}s_i \rangle_p.$$

Hint: $\sum_j g^i_j g^k_j = \delta^{ik}$, by part (i).

(iii) Deduce that (7. 60) defines a metric on Hom (E, E').

(iv) Suppose that, with respect to the orthonormal frame field $\{s_i\}$, $F(p)$ is expressed by the matrix (F^i_j), where $F(p) s_j \; = \; \sum F^i_j s_i$, etc. Prove that

$$\langle F|\tilde{F}\rangle_p^{\text{Hom}} = \sum_i \left(\sum_j F_j^i \tilde{F}_j^i\right) = \text{Tr}\,(F^T \tilde{F})\,(p).\tag{7.61}$$

14. Let (U, φ) be a chart for a Riemannian manifold such that, for some $b > a > 0$, all the eigenvalues of the metric tensor $G(p)$ lie in the set $[a, b]$ for all $p \in U$; note that all these eigenvalues are real because the metric tensor is symmetric. Let c_n be the nth coefficient in the power series expansion of $(1-x)^{1/2}$ around $x = 0$.

(i) Show that the series

$$\Xi(p) = \sqrt{b}\sum_{n=0}^{\infty} c_n\,[I - b^{-1}G(p)]^n\tag{7.62}$$

converges to a symmetric matrix $\Xi(p)$.

Hint: $\left\| [I - b^{-1}G(p)]^n\right\| \leq (1 - a/b)^n$ because, for all $v \in R^n$,

$$0 \leq v^T(I - b^{-1}G(p))\,v = \|v\|^2 - b^{-1}v^T G(p)\,v \leq (1 - a/b)\,\|v\|^2.$$

(ii) Show that $G(p) = \Xi(p)^2 = \Xi(p)^T\Xi(p)$, and that $\Xi: U \to GL_n(R)$ is smooth.
Hint: $G(p)^{1/2} = \sqrt{b}\,(I - [I - b^{-1}G(p)])^{1/2}$. Also $\Xi(p)$ is an analytic[3] function of $G(p)$, which is smooth in p.

(iii) Use these results to obtain an alternative proof of the existence of an orthonormal frame field at any point of a Riemannian manifold.

7.10 History and Bibliography

Basic ideas of Riemannian geometry are due to B. Riemann (1826–66), E. Cartan (1869–1951), and others. The books of do Carmo [1992], Klingenberg [1982], and Gallot, Hulin, and Lafontaine [1990] are highly recommended.

---

[3] A function is called analytic if it is given by a convergent power series at every point. The set of analytic functions is contained in the set of smooth functions.

8   Integration on Oriented Manifolds

In this chapter we come to one of the main uses of differential forms, which is to provide a multidimensional, coordinate-free theory of integration on manifolds. In the process, we shall prove a version of Stokes's Theorem, which is a general form of the fundamental theorem of calculus, and uncover the geometric meaning of the mysterious "exterior derivative."

8.1   Volume Forms and Orientation

A **volume form** on an n-dimensional differential manifold M simply means an n-form σ on M such that $\sigma(r) \neq 0$ at every $r \in M$, or in other words a "nowhere vanishing" n-form. For example, the "area form" calculated from an adapted moving frame in Chapter 4 is a volume form on a surface, and $f(dx \wedge dy \wedge dz)$ is a volume form on R^3 for every smooth function f on R^3 such that f never equals zero.

In vector algebra, the bases $\{e_1, e_2, e_3\}$ and $\{e_2, e_3, e_1\}$ are said to have the "same orientation" because they are both "right-handed," whereas $\{e_2, e_1, e_3\}$ has the opposite orientation because it is "left-handed." This corresponds in exterior calculus to the fact that $dx \wedge dy \wedge dz = dy \wedge dz \wedge dx = -dy \wedge dx \wedge dz$. In higher dimensions, the best way to describe the notion of orientation is in terms of the equality or change of sign in going from one volume form to another, as we shall now delineate.

Two volume forms σ and ρ on M are said to be **equivalent** if $\sigma = f\rho$ for some $f \in C^\infty(M)$ with $f(r) > 0$ at every $r \in M$.

8.1.1   Example of Equivalence of Volume Forms

The volume forms $dx \wedge dy \wedge dz$ and $dr \wedge d\theta \wedge d\phi$ are not equivalent on the manifold

$$U = R^3 - \{ (x, 0, z) : x \geq 0 \},$$

where $x = r\cos\theta\sin\phi$, $y = r\sin\theta\sin\phi$, and $z = r\cos\phi$ (i.e., standard spherical coordinates, with $(r, \theta, \phi) \in (0, \infty) \times (0, 2\pi) \times (0, \pi)$), because as we noted in Chapter 2,

$$dx \wedge dy \wedge dz = -r^2\sin\phi \, (dr \wedge d\theta \wedge d\phi)$$

and $-r^2\sin\phi < 0$. On the other hand, $dx \wedge dy \wedge dz$ and $dr \wedge d\phi \wedge d\theta$ are equivalent.

8.1.2 Orientable Manifolds

An **orientation** for a manifold M means an equivalence class of volume forms. If any volume form exists on M, then M is said to be **orientable**, and the choice of an equivalence class of volume forms makes M **oriented**. A **nonorientable** manifold means one on which no volume form exists. In the language of vector bundles, an n-dimensional manifold M is orientable if and only if the nth exterior power bundle $\Lambda^n (T^* M)$ is trivial, because a line bundle is trivial if and only if it has a nowhere vanishing section (which in this case means a volume form); see the exercises for Chapter 6.

Note that a manifold may be orientable even if the tangent bundle is nontrivial; for example, the 2-form η in the example at the end of Chapter 6 is a volume form on the sphere S^2, so S^2 is orientable; however, S^2 has no nowhere vanishing vector field (the "hairy ball" theorem in topology), and hence TS^2 is not trivial.

A chart (U, φ) for an oriented n-dimensional manifold M is said to be **positively oriented** if $\varphi^* (dx^1 \wedge \ldots \wedge dx^n)$ is equivalent on U to the chosen class of volume forms; as we shall see in Section 8.2, an oriented manifold has an atlas consisting of positively oriented charts.

We would like to have criteria to determine whether a given manifold is orientable. The best kind of criterion would be one that can be verified without taking charts, such as the following:

8.1.3 Orientability of a Level Set of a Submersion

Every n-dimensional submanifold of R^{n+k} of the form $M = f^{-1}(0)$, where $f: U \subseteq R^{n+k} \to R^k$ is a submersion on an open set $U \supset M$, is orientable.

This result shows, for example, that spheres, tori, and hyperboloids of arbitrary dimension, and Lie subgroups of the (real or complex) general linear group, are orientable.

Proof: The Case $k = 1$. Take the Euclidean inner product on R^{n+1}, and apply the associated Hodge star operator to the 1-form $df \in \Omega^1 U$ to obtain an n-form $*df \in \Omega^n U$. We claim that the restriction of $*df$ to M (technically the pullback of $*df$ under the inclusion map from M into R^{n+1}, but the proliferation of stars could be confusing) is a volume form on M.

Fix $r \in M$ and let $\Psi: W \subseteq R^n \to U' \cap M \subset R^{n+1}$ be an n-dimensional parametrization of M at r, in the sense of Chapter 3, with $\Psi(0) = r$. We know from an exercise in that chapter that, in the tangent space to R^{n+1} at r, $\operatorname{Im} d\Psi = \operatorname{Ker} df$. Another way of saying this is that, in terms of coordinates $(x^0, \ldots, x^n)$ for R^{n+1}, the $(n+1) \times (n+1)$ matrix of derivatives

$$
\begin{bmatrix}
D_0 f(r) & D_1 \Psi^0(0) & \cdots & D_n \Psi^0(0) \\
\cdots & \cdots & \cdots & \cdots \\
D_n f(r) & D_1 \Psi^n(0) & \cdots & D_n \Psi^n(0)
\end{bmatrix}
$$

has the property that the first column (i.e., the gradient of f at r) is orthogonal to the tangent space at r (see Chapter 3), and hence to the space spanned by the last n columns. However the last n columns constitute the derivative $D\Psi(0)$, which has a column rank of n, since Ψ is an immersion. It follows from this that the entire matrix is nonsingular, and therefore its determinant is nonzero, or in other words

$$
\sum_{j=0}^n (-1)^j D_j f(r) \left| D\hat{\Psi}^j(0) \right| \neq 0,
\tag{8.1}
$$

where $D\hat{\Psi}^j(0)$ means the $n \times n$ matrix obtained from $D\Psi(0)$ by deleting the jth row, for $0 \leq j \leq n$. It is a simple exercise using the methods of Chapter 1 to verify that

$$
*df(r) = \sum_{j=0}^n (-1)^j D_j f(r) \, (dx^0 \wedge \ldots \wedge d\hat{x}^j \wedge \ldots \wedge dx^n),
\tag{8.2}
$$

where $dx^0 \wedge \ldots \wedge d\hat{x}^i \wedge \ldots \wedge dx^n$ is short for $dx^0 \wedge \ldots \wedge dx^{i-1} \wedge dx^{i+1} \wedge \ldots \wedge dx^n$. On reviewing the section in Chapter 1 on exterior powers of a linear transformation, the reader may verify that (8.1) says precisely that

$$
*df(r) \, (\Lambda^n \Psi_*) \, (\frac{\partial}{\partial u^1} \wedge \ldots \wedge \frac{\partial}{\partial u^n}) \neq 0,
$$

where $\{u^1, \ldots, u^n\}$ is the coordinate system on W. This proves that $*df$ is a nowhere vanishing n-form, as desired.

The Case of Arbitrary k. Express $f(x)$ as $(f^1(x), ..., f^k(x))$. Take the Euclidean inner product on R^{n+k}, and apply the associated Hodge star operator to the k-form $df^1 \wedge ... \wedge df^k \in \Omega^k U$ to obtain an n-form $*(df^1 \wedge ... \wedge df^k) \in \Omega^n U$. The restriction of this differential form to M is a volume form because, by reasoning similar to that above,

$$* (df^1 \wedge ... \wedge df^k) (r) (\Lambda^n \Psi_*) (\frac{\partial}{\partial u^1} \wedge ... \wedge \frac{\partial}{\partial u^n})$$

$$= \left| \left[\begin{array}{cccccc} D_1 f^1(r) & ... & D_1 f^k(r) & D_1 \Psi^1(0) & ... & D_n \Psi^1(0) \\ ... & ... & ... & ... & ... & ... \\ D_{n+k} f^1(r) & ... & D_{n+k} f^k(r) & D_1 \Psi^{n+k}(0) & ... & D_n \Psi^{n+k}(0) \end{array} \right] \right|,$$

which can never be zero. ¤

The previous result is not the most natural way to obtain the orientability of matrix groups; another way is as follows.

8.1.4 Orientability of Matrix Groups

All the Lie subgroups of $GL_n(R)$ and of $GL_n(C)$ are orientable.

Proof: Suppose G is a Lie subgroup of $GL_n(R)$ or $GL_n(C)$ of dimension m. Take a chart (V, ψ) for G at the identity I, and define an m-form σ on V by $\sigma = \psi^*(dx^1 \wedge ... \wedge dx^m)$. Now define an m-form ρ on the whole of G by taking $\rho(A^{-1}) = (L_A^* \sigma)(A^{-1})$, where $L_A: G \to G$ is the diffeomorphism induced by left multiplication by the element $A \in G$, so $L_A(A^{-1}) = I$. Since σ is nonzero at I, and since the differential $(L_A)_*$ is of full rank, it follows that ρ is never zero, and hence is a volume form. ¤

8.2   Criterion for Orientability in Terms of an Atlas

8.2.1 Positivity of Determinants of Derivatives of the Change-of-Charts Maps

Suppose M is an n-dimensional manifold with a countable atlas $\{(U_j, \psi_j), j \in J\}$. M is orientable if and only if there is an equivalent atlas $\{(U_j, \varphi_j), j \in J\}$ such that, whenever $r \in U_i \cap U_j \neq \emptyset$,

$$\Lambda^n(d(\varphi_i \bullet \varphi_j^{-1})) = \left| D(\varphi_i \bullet \varphi_j^{-1})(\varphi_j(r)) \right| > 0; \tag{8.3}$$

that is, the derivative of every change-of-charts map has a positive determinant.

Proof: Assume that M has a countable atlas satisfying (8. 3). By the theorem on partitions of unity (Chapter 5), we may take a partition of unity $\{\upsilon_\alpha : M \to [0, 1], \alpha \in I\}$ subordinate to the atlas $\{(U_j, \varphi_j), j \in J\}$.

Let σ be the canonical volume form $dx^1 \wedge \ldots \wedge dx^n$ on R^n. We assert that

$$\rho(r) = \sum_{\alpha \in I} \upsilon_\alpha(r) (\varphi_{j(\alpha)}{}^* \sigma)(r)$$

is a volume form on M. The fact that it is an n-form is assured by the second and third properties of a partition of unity; given $r \in M$ there is a neighborhood U of r on which ρ takes the form of a finite sum of n-forms multiplied by smooth functions, and hence ρ is a smooth section of the bundle of n-forms. It remains to show that $\rho(r) \neq 0$ for all r.

Fix $r \in M$, and select $\gamma \in I$ such that $\upsilon_\gamma(r) > 0$. For any $\alpha \neq \gamma$, we have

$$\upsilon_\alpha(r) (\varphi_{j(\alpha)}{}^* \sigma)(r) = \upsilon_\alpha(r) ((\varphi_{j(\alpha)} \bullet \varphi_{j(\gamma)}^{-1} \bullet \varphi_{j(\gamma)})^* \sigma)(r)$$

$$= \upsilon_\alpha(r) (\varphi_{j(\gamma)}{}^* ((\varphi_{j(\alpha)} \bullet \varphi_{j(\gamma)}^{-1})^* \sigma))(r).$$

Hence if $X_k = (\varphi_{j(\gamma)}^{-1})_* (\partial/\partial x^k)$ (vector field on $U_{j(\gamma)}$), then

$$(\varphi_{j(\gamma)}{}^* \sigma) \cdot (X_1 \wedge \ldots \wedge X_n) = 1;$$

$$\therefore \rho \cdot (X_1 \wedge \ldots \wedge X_n)(r)$$

$$= \upsilon_\gamma(r) + \sum_{\alpha \neq \gamma} \upsilon_\alpha(r) ((\varphi_{j(\alpha)} \bullet \varphi_{j(\gamma)}^{-1})^* \sigma) \cdot (\frac{\partial}{\partial x^1} \wedge \ldots \wedge \frac{\partial}{\partial x^n})$$

$$= \upsilon_\gamma(r) + \sum_{\alpha \neq \gamma} \upsilon_\alpha(r) \Lambda^n (d(\varphi_{j(\alpha)} \bullet \varphi_{j(\gamma)}^{-1})) > 0,$$

where, in the last line, we used the formula for the nth exterior power of a linear transformation given in Chapter 1, and condition (8. 3).

As for the converse, assume that M has a volume form ρ. Let $\varsigma: R^n \to R^n$ be the map $\varsigma(x_1, \ldots, x_n) = (-x_1, x_2, \ldots, x_n)$, and let $\{(U_j, \psi_j), j \in J\}$ be any countable atlas for M. If $\psi_j{}^* \sigma$ and ρ are equivalent as volume forms on U_j, then let $\varphi_j = \psi_j$; if not, let $\varphi_j = \varsigma \bullet \psi_j$. As a result, $\{(U_j, \varphi_j), j \in J\}$ is an equivalent atlas satisfying (8. 3).   ¤

8.2.2 Applications

8.2.2.1 Möbius Strip

It is intuitively clear that the Möbius strip $\ddot{M}$, the two-dimensional submanifold of R^3 depicted in Chapter 6, has an atlas consisting of two charts which fail to satisfy (8. 3). It is possible to show that no atlas for $\ddot{M}$ satisfies (8. 3); alternatively topological arguments can be used to show that $\ddot{M}$ is indeed nonorientable.

8.2.2.2 Products of Orientable Manifolds Are Orientable

If M and N are manifolds with atlases $\{\, (U_i, \varphi_i)\,, i \in I\}$ and $\{\, (V_j, \psi_j)\,, j \in J\}$, respectively, both satisfying (8. 3), then the obvious "product atlas" for the product manifold $M \times N$ also satisfies (8. 3); the details are left as an exercise.

For example, the torus T^n is orientable because it is the product $S^1 \times \ldots \times S^1$ of circles, each of which is orientable by 8.1.3.

8.2.2.3 Tangent Bundles Are Orientable

For any manifold M (even a nonorientable one!), the tangent bundle TM is an orientable manifold; see Exercise 8, and the hints attached.

8.3 Orientation of Boundaries

8.3.1 In an Oriented Manifold, a Boundary Has a Canonical Orientation

Suppose P is an n-dimensional submanifold-with-boundary (see Chapter 5) of an n-dimensional oriented manifold M, where $n \geq 2$. If the boundary ∂P is nonempty, then it is an orientable $(n-1)$-dimensional submanifold of M with a natural orientation determined by that of M.

Proof: We already proved in Chapter 5 that ∂P is an $(n-1)$-dimensional submanifold of M; it only remains to prove that it is orientable, etc. By definition of a submanifold-with-boundary, we can take an atlas for M such that charts (U, φ) that intersect ∂P are of the special form $\varphi(P \cap U) = \{\, (x_1, \ldots, x_n) \in \varphi(U) : x_1 \leq 0\}$, and $\varphi(\partial P \cap U) = \{\, (x_1, \ldots, x_n) \in \varphi(U) : x_1 = 0\}$. Take a partition of unity $\{\upsilon_\alpha : M \to [0, 1]\,, \alpha \in I\}$ subordinate to this atlas. Recall from the exercise on subbundles in Chapter 6 that a section of the vector bundle

$$\pi : TM|_{\partial P} \to \partial P \qquad (8. 4)$$

means a smooth assignment of a tangent vector in $T_r M$ to every $r \in \partial P$: Note in particular that this tangent vector need not be tangent to the submanifold ∂P, whose tangent spaces have one less dimension than those of M. We define a smooth section ς of this vector bundle as follows:

$$\varsigma(r) = \sum_{\alpha \in I} \upsilon_\alpha(r) \left((\varphi_{j(\alpha)}^{-1}) * \frac{\partial}{\partial x^1} \right)(r).$$

To understand the idea behind the formula, look at the following picture; we are taking the "$\partial/\partial x^1$" vector fields in all the charts along the boundary, and combining them smoothly together using the partition of unity.

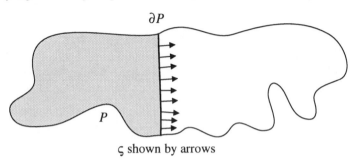

ς shown by arrows

We assert that this section ς is never zero; the proof is similar to part of the proof of 8.2.1. We may write

$$(\varphi_{j(\alpha)}) * \varsigma = \upsilon_\alpha(.) \frac{\partial}{\partial x^1} + \sum_{\gamma \neq \alpha} \upsilon_\gamma(.) \left((\varphi_{j(\alpha)} \bullet \varphi_{j(\gamma)}^{-1}) * \frac{\partial}{\partial x^1} \right). \tag{8.5}$$

The special features of the atlas ensure that $\varphi_{j(\gamma)}(U_{j(\gamma)} \cap \partial P) \subseteq \{0\} \times R^{n-1}$ for every γ; it follows that the map

$$\psi = \varphi_{j(\alpha)} \bullet \varphi_{j(\gamma)}^{-1}$$

sends every point of the form $(0, x_2, ..., x_n)$ to a point of the form $(0, y_2, ..., y_n)$, and therefore $\psi^1(0, x_2, ..., x_n) = 0$. This implies that its derivative matrix looks like this:

$$D\psi(0, x_2, ..., x_n) = \begin{bmatrix} D_1\psi^1 & 0 & ... & 0 \\ D_1\psi^2 & & & \\ & ... & & A \\ D_1\psi^n & & & \end{bmatrix},$$

for some $(n-1) \times (n-1)$ matrix A. A change-of-chart map cannot have a singular derivative, so $0 \neq |D\psi| = (D_1\psi^1)|A|$. It follows that $D_1\psi^1(0, x_2, ..., x_n) \neq 0$. Moreover $\psi^1(x_1, x_2, ..., x_n) < 0$ for $x_1 < 0$, and so

$$D_1\psi^1(0, x_2, ..., x_n) > 0.$$

A glance at the section in Chapter 2 about the local expression for the differential of a map will verify that

$$(\varphi_{j(\alpha)} \bullet \varphi_{j(\gamma)}^{-1})_* \frac{\partial}{\partial x^1} = (D_1 \psi^1) \frac{\partial}{\partial x^1} + \ldots + (D_1 \psi^n) \frac{\partial}{\partial x^n}.$$

Now we see from (8. 5) that $(\varphi_{j(\alpha)})_* \varsigma$ is a strictly positive multiple of $\partial/\partial x^1$, plus some terms which are linearly independent of $\partial/\partial x^1$; here of course we use the fact that

$$\sum_\gamma \upsilon_\gamma(\cdot) = 1.$$

Hence $(\varphi_{j(\alpha)})_* \varsigma \neq 0$, for every α.

The canonical orientation of ∂P is constructed as follows. If ρ is a volume form on M which belongs to the orientation of M, the formula

$$\lambda \cdot (X_1 \wedge \ldots \wedge X_{n-1}) = \rho \cdot (\varsigma \wedge X_1 \wedge \ldots \wedge X_{n-1}), X_1, \ldots, X_{n-1} \in \Im(\partial P), \quad \text{(8. 6)}$$

(here $\Im(\partial P)$ denotes the set of vector fields on ∂P) defines a volume form on ∂P, because ς is nonvanishing. The equivalence class of this volume form is the canonical orientation of ∂P.   ¤

8.3.2 Examples

8.3.2.1 Boundary of a Half-Space in Euclidean Space
Suppose $M = R^n$ with the orientation given by $dx^1 \wedge \ldots \wedge dx^n$, and take $P = \{(x_1, \ldots, x_n) \in R^n : x_1 \leq 0\}$. Then $\varsigma = \partial/\partial x^1$, and taking $X_i = \partial/\partial x^{i+1}$ in (8. 6) shows that the orientation of the boundary $\partial P = \{(x_1, \ldots, x_n) \in R^n : x_1 = 0\}$ is given by the volume form $dx^2 \wedge \ldots \wedge dx^n$.

8.3.2.2 Boundary of a Disk in the Plane
Let $M = R^2 - \{0\}$ with the orientation given by $dr \wedge d\theta$, where (r, θ) is the usual system of polar coordinates; of course we will need two charts, with different θ-domains, to cover M. Let $P = \{(r, \theta) \in M : r \leq 1\}$, which is a 2-dimensional submanifold-with-boundary because $(r, \theta) \rightarrow r$ is a submersion; see the criterion for a submanifold-with-boundary in Chapter 5. Here we may write $\varsigma = \partial/\partial r$, and (8. 6) shows that the orientation of the boundary $\partial P = S^1 = \{(r, \theta) \in M : r = 1\}$ is given by the 1-form λ where

$$\lambda \cdot \frac{\partial}{\partial \theta} = dr \wedge d\theta \cdot (\frac{\partial}{\partial r} \wedge \frac{\partial}{\partial \theta}) = 1;$$

that is, $\lambda = d\theta$. In geometric terms, the orientation of M can be considered as a choice of normal direction to the plane, and the orientation of the boundary circle as a choice of a direction around the circle.

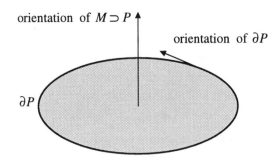

Figure 8. 1   Orientation of the boundary

8.4   **Exercises**

1.   Show that the 2-form $x\,(dy \wedge dz) - y\,(dx \wedge dz) + z\,(dx \wedge dy) \in \Omega^2 R^3$, restricted to the hyperboloid $H^2_{-1} = \{\,(x, y, z) : x^2 - y^2 - z^2 + 1 = 0\}$, gives a volume form on the hyperboloid.
     Hint: Use the (x, r, θ) parametrization, and charts such as $\varphi\,(x, r, \theta) = (x, \theta)$.

2.   (Continuation) Let $P = \{\,(x, y, z) \in H^2_{-1} : x \le 0\}$. Show that P is a 2-dimensional submanifold-with-boundary of H^2_{-1}, and find the orientation on the boundary ∂P induced by the orientation of the hyperboloid represented by the volume form of Exercise 1.

3.   The 2-torus $T^2 \subset R^4$ can be regarded as $f^{-1}\,(0)$, where

$$f(x_1, x_2, x_3, x_4) = (x_1^2 + x_2^2 - 1, x_3^2 + x_4^2 - 1),$$

     which is a submersion when restricted to a suitable neighborhood of T^2. Calculate explicitly the volume form $*\,(df^1 \wedge df^2)$, as in 8.1.3, by expressing its pullback under the parametrization $\Psi\,(\theta, \phi) = (\cos\theta, \sin\theta, \cos\phi, \sin\phi)$ in terms of the 2-form $d\theta \wedge d\phi$.

4.   (Continuation) Let $P = \{\,(x_1, x_2, x_3, x_4) \in T^2 : x_1 - x_2 \ge 0\}$. Show that P is a 2-dimensional submanifold-with-boundary of the 2-torus, and find the orientation on the boundary ∂P induced by the orientation of T^2 represented by the volume form $d\theta \wedge d\phi$.

5. The special linear group $SL_2(R) \subset R^{2 \times 2}$ can be regarded as $f^{-1}(0)$, where

$$f\left(\begin{bmatrix} x & y \\ z & w \end{bmatrix}\right) = xw - yz - 1,$$

which is a submersion when restricted to $GL_2(R)$.

(i) Calculate explicitly the volume form $*df$ in terms of the (x, y, z) parametrization and show $*(df)(I) = -2(dx \wedge dy \wedge dz)$.

(ii) Show that the "left-invariant" volume form $\rho(A^{-1}) = (L_A^*(*df))(A^{-1})$ is $-((x^2 + yz)/x)(dx \wedge dy \wedge dz)$ in terms of the (x, y, z) parametrization, where

$$L_A\left(\begin{bmatrix} s & t \\ u & v \end{bmatrix}\right) = \begin{bmatrix} x & y \\ z & w \end{bmatrix}\begin{bmatrix} s & t \\ u & v \end{bmatrix}, \text{ if } A = \begin{bmatrix} x & y \\ z & w \end{bmatrix}.$$

6. (i) Show that $P = \{A \in GL_2(R) : |A| \leq 1\}$ is a 4-dimensional submanifold-with-boundary of the general linear group $GL_2(R)$, with boundary $SL_2(R)$ as described in Exercise 5.

(ii) If $GL_2(R)$ is given the volume form $dx \wedge dy \wedge dz \wedge dw$, for the coordinates given in Exercise 5, find the orientation induced on $SL_2(R)$ as the boundary of P.

Hint: As a first step, change from $dx \wedge dy \wedge dz \wedge dw$ to an equivalent form where the first entry in the wedge product is df.

(iii) Determine whether the orientation of part (ii) is the same as the one represented by the volume form $*df$ of Exercise 5.

7. If M and N are manifolds with atlases $\{(U_i, \varphi_i), i \in I\}$ and $\{(V_j, \psi_j), j \in J\}$, respectively, both satisfying (8.3), show that the product manifold $M \times N$ is orientable, by considering the atlas $\{(U_i \times V_j, \varphi_i \times \psi_j), (i, j) \in I \times J\}$.

8. Let $\{(U_\alpha, \varphi_\alpha), \alpha \in I\}$ be an atlas for an n-dimensional differential manifold M. Recall that the tangent bundle $\pi: TM \to M$, as constructed in Chapter 6, has an atlas given by $\{(\pi^{-1}(U_\alpha), \psi_\alpha), \alpha \in I\}$, where

$$\psi_\alpha([r, \alpha, v]) = (\varphi_\alpha(r), v) \in R^n \times R^n. \tag{8.7}$$

Also recall that $[r, \alpha, v] = [r, \gamma, w] \Rightarrow w = D(\varphi_\gamma \bullet \varphi_\alpha^{-1})(\varphi_\alpha(r))v$, and so, for every $\alpha, \gamma \in I$ such that $U_\alpha \cap U_\gamma \neq \varnothing$, the change-of-chart map is given by

$$\psi_\gamma \bullet \psi_\alpha^{-1}(x, v) = (\varphi_\gamma \bullet \varphi_\alpha^{-1}(x), D(\varphi_\gamma \bullet \varphi_\alpha^{-1})(x)v). \tag{8.8}$$

Show that the manifold TM is always orientable by showing that the derivatives of the maps (8.8) always satisfy (8.3).

Hint: If $A, B, C \in R^{n \times n}$, then $\left\|\begin{bmatrix} A & 0 \\ B & C \end{bmatrix}\right\| = |A||C|$.

8.5   Integration of an *n*-Form over a Single Chart

Let (U, φ) be a positively oriented chart on an oriented manifold M of dimension n. Suppose that K is a subset of U such that $\varphi(K)$ is a bounded subset of R^n, and is "measurable" in the sense that it qualifies as a domain of integration under the reader's preferred theory of integration;[1] the boundedness condition is included so that we do not have to be concerned about infinite integrals. Suppose also that ρ is an n-form on M. In that case $(\varphi^{-1})^* \rho$ makes sense as an n-form on $\varphi(U)$, and it may be written as $(\varphi^{-1})^* \rho = h(dx^1 \wedge \ldots \wedge dx^n)$, for some smooth function h on $\varphi(U)$. Let us define the **integral of the *n*-form** ρ over K as:

$$\int_K \rho = \int_{\varphi(K)} (\varphi^{-1})^* \rho = \int_{\varphi(K)} h(x_1, \ldots, x_n)\, dx_1 \ldots dx_n. \tag{8.9}$$

For example, if $\varphi(K)$ is a rectangle such as $(s_1, t_1) \times \ldots \times (s_n, t_n)$, the last integral becomes the multiple integral

$$\int_{s_1}^{t_1} \ldots \int_{s_n}^{t_n} h(x_1, \ldots, x_n)\, dx_1 \ldots dx_n.$$

Note how we have dropped the wedge symbols on the right side of (8.9) in shifting from differential forms to multiple integrals. The key idea which makes this possible is that the change of variables formula for multiple integrals is entirely analogous to the way that differential forms transform under pullback, as we shall now demonstrate.

8.5.1   The Integral of an *n*-Form in One Chart Is Intrinsic

If $\rho \in \Omega^n M$, if (U, φ) and (V, ψ) are positively oriented charts, and if $K \subseteq U \cap V$ is a set such that $\varphi(K)$ and $\psi(K)$ are bounded and measurable, then

$$\int_{\varphi(K)} (\varphi^{-1})^* \rho = \int_{\psi(K)} (\psi^{-1})^* \rho. \tag{8.10}$$

In other words, the definition of $\int_K \rho$ does not depend on the chart.

Proof: Suppose $W = U \cap V$, and the coordinate system $\{x^1, \ldots, x^n\}$ on $\varphi(W)$ is mapped to the coordinate system $\{y^1, \ldots, y^n\}$ on $\psi(W)$ under the diffeomorphism $F = \varphi \bullet \psi^{-1} : \psi(W) \subseteq R^n \to \varphi(W) \subseteq R^n$. Let

---

[1] Readers familiar with Riemann integration can think of $\varphi(K)$ as a finite union of open or closed rectangles. For readers who know measure theory, $\varphi(K)$ can be any bounded, measurable set. You do not need to know measure theory to read this section.

$$(\varphi^{-1})^* \rho = h(dx^1 \wedge \ldots \wedge dx^n),$$

for some function $h = h(x^1, \ldots, x^n)$ on $\varphi(W)$. The key idea is that, by the calculations on exterior powers of a linear transformation in Chapter 1 and on pullbacks in Chapter 2,

$$(\psi^{-1})^* \rho = (\varphi \bullet \psi^{-1})^* (\varphi^{-1})^* \rho = F^*(h(dx^1 \wedge \ldots \wedge dx^n))$$

$$= (F^*h)(\Lambda^n DF)(dy^1 \wedge \ldots \wedge dy^n)$$

$$= h \bullet F|DF|(dy^1 \wedge \ldots \wedge dy^n).$$

Using the change of variable formula for multiple integrals,

$$\int_{\varphi(K)} (\varphi^{-1})^* \rho = \int_{\varphi(K)} h\, dx_1 \ldots dx_n,$$

$$= \int_{\psi(K)} h \bullet F|DF|\, dy_1 \ldots dy_n,$$

$$= \int_{\psi(K)} (\psi^{-1})^* \rho,$$

as desired.   ¤

The integral constructed above is already familiar in the guise of the line integral and surface integral in vector calculus, which we shall now review.

8.5.2   Line Integrals of Work Forms

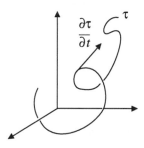

Figure 8. 2   Parametrized Ccurve

Suppose I is an open interval of the real line, and $\tau: I \to U \subseteq R^3$ is a parametrization of an oriented 1-dimensional submanifold of R^3, such that the induced chart $(\tau(I), \tau^{-1})$

is positively oriented. In vector calculus, τ would be called a parametrized curve, and can be expressed in (x, y, z) coordinates as $\tau(t) = (\tau_1(t), \tau_2(t), \tau_3(t))$. As we saw in Chapter 2, to any vector field

$$X = F^1 \frac{\partial}{\partial x} + F^2 \frac{\partial}{\partial y} + F^3 \frac{\partial}{\partial y} \tag{8.11}$$

on U there is associated a "work form" $\varpi_X \in \Omega^1 U$, namely,

$$\varpi_X = F^1 dx + F^2 dy + F^3 dz, \tag{8.12}$$

and by the calculations of Chapter 2,

$$\tau^* \varpi_X = [F^1, F^2, F^3] \begin{bmatrix} \tau_1' \\ \tau_2' \\ \tau_3' \end{bmatrix} dt.$$

According to (8.9), the "line integral" of ϖ_X along the path $\tau(I)$ is:

$$\int_{\tau(I)} \varpi_X = \int_I \tau^* \varpi_X = \int_I (F^1 \tau_1' + F^2 \tau_2' + F^3 \tau_3') \, dt, \tag{8.13}$$

$$= \int_I (\vec{F} \cdot \dot{\vec{\tau}}') \, dt, \tag{8.14}$$

which coincides with the vector calculus definition of the line integral of the vector field X along the curve. The significance of the term "work form" is that its integral measures the work done, in a precise physical sense, by the vector field in moving a particle along the curve. The advantage of the differential-form expression is that it is coordinate-free (as 8.5.1 proves), whereas the vector calculus version appears to depend on Cartesian coordinates and the Euclidean inner product. Note that choosing a parametrization of the opposite orientation, that is, running the opposite way along the curve, would change the sign of the integral.

8.5.3 Surface Integral of a Flux Form

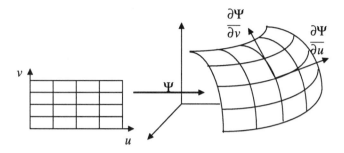

Figure 8. 3   Parametrized surface

Suppose $\Phi\colon W \subseteq R^2 \to U \subseteq R^3$ is a parametrization of an oriented 2-dimensional submanifold of R^3, such that the induced chart $(\Phi(W), \Phi^{-1})$ is positively oriented. In vector calculus, $\Phi(W)$ would be called a parametrized surface. As we saw in Chapter 2, to the vector field X on U in (8. 11) there is associated a "flux form" $\phi_X \in \Omega^2 U$, namely,

$$\phi_X = F^1 (dy \wedge dz) + F^2 (dz \wedge dx) + F^3 (dx \wedge dy). \qquad (8.\,15)$$

If we use the notation $\Phi(u, v) = (\Phi^1, \Phi^2, \Phi^3)$ and $\Phi^1_u = \partial\Phi^1/\partial u$, etc., then we see as in Chapter 2 that

$$\Phi^* dx = \Phi^1_u du + \Phi^1_v dv, \ \Phi^* dy = \Phi^2_u du + \Phi^2_v dv, \ \Phi^* dz = \Phi^3_u du + \Phi^3_v dv; \qquad (8.\,16)$$

$$\Phi^* \phi_X = \left| \begin{bmatrix} F^1 & F^2 & F^3 \\ \Phi^1_u & \Phi^2_u & \Phi^3_u \\ \Phi^1_v & \Phi^2_v & \Phi^3_v \end{bmatrix} \right| du \wedge dv = \vec{F} \cdot (\vec{\Phi}_u \times \vec{\Phi}_v) \, du \wedge dv, \qquad (8.\,17)$$

where $\vec{F} = (F^1, F^2, F^3)^T$, and $\vec{\Phi}_u = (\Phi^1_u, \Phi^2_u, \Phi^3_u)^T$, etc. By (8. 9), the "surface integral" of the flux form ϕ_X over the surface $\Phi(W)$ is

$$\int_{\Phi(W)} \phi_X = \int_W \Phi^* \phi_X = \int_W \vec{F} \cdot (\vec{\Phi}_u \times \vec{\Phi}_v) \, du dv, \qquad (8.\,18)$$

which coincides with the vector calculus notion of integrating the normal component of the vector field X over the surface $\Phi(W)$; note that $\vec{\Phi}_u$ and $\vec{\Phi}_v$ are linearly independent (since Φ is an immersion), and are "tangent" to the surface (see Chapter 4), so their cross product is a nonzero vector normal to the surface. The choice of an orientation for this 2-dimensional submanifold is equivalent to a choice of normal direction; a parametrization with the opposite orientation would reverse the sign of the integral, because it would be measuring the flux in the opposite direction through the surface.

Thus we now discover the geometric meaning of a flux form in R^3; its integral across a surface measures the flux of the associated vector field through the surface, with a positive or negative sign determined by the orientation of the surface.

8.5.4   Computational Examples

8.5.4.1   Example of a Line Integral

Suppose $\tau(t) = (0, \sqrt{t}, 1/t), t \in (0, 1)$; then

$$\int_{\tau(0,1)} y\,dz = \int_{(0,1)} \sqrt{t}\,(-1/t^2)\,dt = -\infty$$

8.5.4.2   Example of a Surface Integral

If $\Phi(r, \theta) = (r\cos\theta, r\sin\theta, r^2/2)$, for $0 < r < 1, 0 < \theta < \pi/2$, then (8. 17) implies that

$$\Phi^*(dz \wedge dx) = \left\| \begin{bmatrix} 0 & 1 & 0 \\ \cos\theta & \sin\theta & r \\ -r\sin\theta & r\cos\theta & 0 \end{bmatrix} \right\| dr \wedge d\theta = -r^2\sin\theta\,(dr \wedge d\theta);$$

and so, if $W = (0, 1) \times (0, \pi/2)$, then

$$\int_{\Phi(W)} dz \wedge dx = -\int_0^1 r^2 dr \int_0^{\pi/2} \sin\theta d\theta = -1/3.$$

8.6   Global Integration of *n*-Forms

Suppose that $\{(U'_j, \varphi'_j), j \in J\}$ is a positively oriented atlas on an oriented manifold M of dimension n, and ρ is an n-form on M. By splitting the domains $\{U'_j\}$ up into smaller domains if necessary, while using the same maps $\{\varphi'_j\}$ on suitably restricted domains, we can obtain an equivalent positively oriented atlas $\{(U_j, \varphi_j), j \in J\}$ such that $\varphi_j(U_j)$ is a bounded subset of R^n for every j.

Suppose that K is a subset of M such that

- K is contained in some compact subset K_0 of M, so there is some finite set of charts, whose domains can be relabeled $\{U_1, \ldots, U_m\}$, such that $K \subseteq \bigcup_{1 \le i \le m} U_i$; and

- $\varphi_i(K \cap U_i)$ is a measurable[2] subset of R^n for every $i \in \{1, 2, ..., m\}$.

Notice that $\varphi_i(K \cap U_i)$ is a bounded subset of R^n by construction, and therefore an expression such as

$$\sum_{i=1}^{m} \left(\int_{K \cap U_i} \rho \right)$$

makes sense. However it would not be an appropriate formula for the integral of the n-form ρ over K, because, for example, if a region H is in the intersection of exactly r of these charts, then we will have integrated over H not once but r times. The easiest way around this problem for computational purposes is to appeal to the "inclusion–exclusion" formula from combinatorics, which we will presently justify. Define the integral[3] of the n-form ρ over K as

$$\int_K \rho = \sum_{i=1}^{m} A_i - \sum_{i<j} A_{ij} + \sum_{i<j<k} A_{ijk} - ... + (-1)^{m+1} A_{12...m}, \tag{8.19}$$

$$A_{i(1)...i(r)} = \int_{K \cap U_{i(1)} \cap ... \cap U_{i(r)}} \rho. \tag{8.20}$$

Note that the integral (8. 20) is well defined by (8. 9), and the sum in (8. 19) makes sense because there are only a finite number of summands. To see why (8. 19) is the only possible formula consistent with the one when K is contained in a single chart, consider the case of a region H contained in the intersection of exactly r of $\{U_1, ..., U_m\}$; the integral over H is included r times in the first sum of (8. 19), and

---

[2] See footnote 1 for the interpretation of "measurable."

[3] Those who know measure theory and topology will find a more complete treatment of integration in Berger and Gostiaux [1988]. In particular, here is how to extend the integral to noncompact sets. A volume form σ which belongs to the equivalence class of the orientation induces a Radon measure μ on M as follows. Given a continuous function f on M with compact support K, define

$$\mu(f) = \int_K f\sigma,$$

where the right side is defined by (8. 19). By the Riesz Representation Theorem, μ is a Radon measure on M. For any other n-form ρ, there exists a smooth function g on M such that $\rho = g\sigma$. If g is a μ-integrable function, we may define, for any measurable set $H \subseteq M$,

$$\int_H \rho = \int_H g d\mu,$$

and it can be proved that the left side is the same whatever volume form σ in the equivalence class of the orientation was selected at the beginning.

$$\binom{r}{k} = \frac{r!}{k!\,(r-k)!}$$

times in the kth sum, for $k = 1, 2, \ldots, r$. Taking the signs into account, we see that the number of times it is included in the integral of ρ over K is

$$\binom{r}{1} - \binom{r}{2} + \ldots + (-1)^{k+1}\binom{r}{k} + \ldots + (-1)^{r+1} = 1 - (1-1)^r = 1,$$

using the binomial theorem. In other words, the integral over H has been counted exactly once, as desired, and so (8. 9) and (8. 19) are consistent in the case where K is contained in one of $\{U_1, \ldots, U_m\}$. It is also true that for disjoint K and K' satisfying the conditions above,

$$\int_{K \cup K'} \rho = \int_K \rho + \int_{K'} \rho, \tag{8.21}$$

the details of which are left as Exercise 15. Now we need to check that the integral of the n-form ρ over K is not dependent on the choice of atlas. Let $\{(V_\gamma, \psi_\gamma), \gamma \in J'\}$ be another positively oriented atlas with $\psi_\gamma(V_\gamma)$ bounded for every γ, and suppose $K \subseteq V_1 \cup \ldots \cup V_q$. We obtain a similar formula for the integral, namely,

$$\int_K \rho = \sum_{s=1}^{q} (-1)^{s+1} \sum_{i'(1) < \ldots < i'(s)} B_{i'(1)\ldots i'(s)}, \tag{8.22}$$

$$B_{i'(1)\ldots i'(s)} = \int_{K \cap V_{i'(1)} \cap \ldots \cap V_{i'(s)}} \rho. \tag{8.23}$$

8.6.1   The Integral of an n-Form Is Intrinsic

The value of the integrals (8. 19) and (8. 22) is the same.

Proof: The integral

$$C_{i(1)\ldots i(r),\,i'(1)\ldots i'(s)} = \int_{K \cap U_{i(1)} \cap \ldots \cap U_{i(r)} \cap V_{i'(1)} \cap \ldots \cap V_{i'(s)}} \rho$$

does not depend on which of the maps $\{\varphi_{i(1)}, \ldots, \varphi_{i(r)}, \psi_{i'(1)}, \ldots, \psi_{i'(r)}\}$ is used to parametrize the domain of integration, by 8.5.1. It follows from the definition that

$$A_{i(1)\ldots i(r)} = \sum_{s=1}^{q} (-1)^{s+1} \sum_{i'(1) < \ldots < i'(s)} C_{i(1)\ldots i(r),\,i'(1)\ldots i'(s)},$$

and similarly for $B_{i'(1)\ldots i'(s)}$. Hence the integral in (8. 19) is

$$\sum_{r=1}^{m} (-1)^{r+1} \sum_{i(1)<\ldots<i(r)} A_{i(1)\ldots i(r)}$$

$$= \sum_{r=1}^{m} (-1)^{r+1} \sum_{i(1)<\ldots<i(r)} \sum_{s=1}^{q} (-1)^{s+1} \sum_{i'(1)<\ldots<i'(s)} C_{i(1)\ldots i(r),\, i'(1)\ldots i'(s)}$$

$$= \sum_{s=1}^{q} (-1)^{s+1} \sum_{i'(1)<\ldots<i'(s)} B_{i'(1)\ldots i'(s)},$$

by rearranging the order of summation, which gives (8. 22), as desired.   ¤

8.7   The Canonical Volume Form for a Metric

Suppose $(M, \langle\cdot|\cdot\rangle)$ is an oriented pseudo-Riemannian manifold (see Chapter 7). In the domain of each chart, we may construct an orthonormal frame field (see Chapter 7), and the corresponding orthonormal coframe field, denoted $\{\varepsilon^1, \ldots, \varepsilon^n\}$. This is a field of 1-forms with the property that

$$\varepsilon^m = \sum_i \xi^m_i dx^i, \tag{8. 24}$$

where the matrix $\Xi = (\Xi_{mi}) = (\xi^m_i)$ of functions satisfies $G(p) = \Xi(p)^T \Lambda \Xi(p)$; here $G = (g_{ij})$ is the local expression of the metric tensor, and Λ is a diagonal matrix whose diagonal entries are all ± 1, in which the number of minus ones is determined by the signature of the inner product.

An orthonormal coframe field will be called **positively oriented** if $\sigma = \varepsilon^1 \wedge \ldots \wedge \varepsilon^n$ is in the equivalence class of the orientation of M; by switching the sign of ε^1 if necessary, any orthonormal coframe field may be made positively oriented.

8.7.1 The Canonical Volume Form Is Intrinsic

If $\{\varepsilon^1, \ldots, \varepsilon^n\}$ and $\{\tilde{\varepsilon}^1, \ldots, \tilde{\varepsilon}^n\}$ are two positively oriented orthonormal coframe fields on an open set U in an oriented pseudo-Riemannian manifold $(M, \langle\cdot|\cdot\rangle)$, then $\varepsilon^1 \wedge \ldots \wedge \varepsilon^n = \tilde{\varepsilon}^1 \wedge \ldots \wedge \tilde{\varepsilon}^n$; therefore the formula

$$\sigma|_U = \varepsilon^1 \wedge \ldots \wedge \varepsilon^n \tag{8. 25}$$

defines unambiguously a volume form σ *on the whole of M, called the* **canonical volume form** *on* $(M, \langle \cdot | \cdot \rangle)$. *The local expression for* $\varepsilon^1 \wedge \ldots \wedge \varepsilon^n$, *in any chart* (U, φ) *such that* $\varphi^* (dx^1 \wedge \ldots \wedge dx^n)$ *is equivalent to the orientation of M, is*

$$\varepsilon^1 \wedge \ldots \wedge \varepsilon^n = \sqrt{|G| |\Lambda|} \, \varphi^* (dx^1 \wedge \ldots \wedge dx^n). \tag{8.26}$$

Proof: As usual, we write $dx^1 \wedge \ldots \wedge dx^n$ instead of $\varphi^* (dx^1 \wedge \ldots \wedge dx^n)$, etc. From (8.24) and our discussion of determinants in Chapter 1, we see that

$$\varepsilon^1 \wedge \ldots \wedge \varepsilon^n = \sum_{i(1)} \xi^1_{i(1)} dx^{i(1)} \wedge \ldots \wedge \sum_{i(n)} \xi^n_{i(n)} dx^{i(n)} = |\Xi| dx^1 \wedge \ldots \wedge dx^n$$

where $G(p) = \Xi(p)^T \Lambda \Xi(p)$. Taking determinants, we see that $|G| = |\Xi|^2 |\Lambda|$, which can be rewritten as $|\Xi|^2 = |G| |\Lambda|$ because $|\Lambda| = \pm 1$. This gives (8.26). Since the same equation holds for $\tilde{\varepsilon}^1 \wedge \ldots \wedge \tilde{\varepsilon}^n$, the conclusion follows.   ¤

8.7.2   Examples

8.7.2.1   Euclidean 3-Space
$\{dx, dy, dz\}$ is an orthonormal coframe field with respect to the Euclidean inner product. The canonical volume form $dx \wedge dy \wedge dz$ can be expressed in spherical polar coordinates (see Example 8.1.1) as $r^2 \sin \phi \, (dr \wedge d\theta \wedge d\phi)$.

8.7.2.2   4-Space with the Lorentz Inner Product
For the Lorentz inner product, the metric tensor with respect to coordinates $\{x, y, z, t\}$ is diagonal with diagonal entries $(1, 1, 1, -c^2)$, and here $\Lambda = \text{diag}(1, 1, 1, -1)$. $\{dx, dy, dz, cdt\}$ is an orthonormal coframe field, and $dx \wedge dy \wedge dz \wedge cdt$ is the canonical volume form.

8.7.2.3   The 2-Sphere
As in Chapter 7, the Euclidean inner product on R^3 induces a Riemannian metric on the 2-sphere. Using the "latitude-longitude" parametrization (see the example in Chapter 7), $|G| = (\sin \phi)^2$, and the canonical volume form is $\sin \phi \, (d\phi \wedge d\theta)$ (with sign depending on the orientation). Note that this is the same as the area form discussed in Chapter 4.

8.7.2.4   The Hyperbolic Plane
For the hyperbolic plane with (s, θ) parametrization (see Chapter 7), the canonical volume form can be written, for example, as

$$\frac{r}{\sqrt{1 + r^2}} (dr \wedge d\theta),$$

or as $\sinh s \, (ds \wedge d\theta)$.

8.7.3 Volume of a Pseudo-Riemannian Manifold

As one might expect, the **Riemannian volume** of an oriented pseudo-Riemannian manifold $(M, \langle \cdot | \cdot \rangle)$ is simply the integral of the canonical volume form over M, denoted $\mathrm{vol}\, M$. In the case of a 1- or 2-dimensional submanifold of R^3, this is simply the Euclidean length or surface area, respectively. For example, let M be the portion of the hyperbolic plane for which $0 < s < 1$ in the (s, θ) parametrization (see Chapter 7), which might be called the punctured unit disc in hyperbolic 2-space. Its Riemannian volume (i.e., area) is (omitting the proper pullback notation in the first integral)

$$\int_M \sinh s \,(ds \wedge d\theta) \;=\; \int_0^{2\pi} d\theta \int_0^1 \sinh s\, ds \;=\; 2\pi\,(\cosh 1 - 1) \;=\; 1.0861613...\pi.$$

It is interesting to note that M has an area greater than that of the unit disc in Euclidean R^2; indeed the disc of radius r has an area increasing exponentially with r, whereas in Euclidean R^2 the area increases as the square of r.

8.8 Stokes's Theorem

Here is the main theorem about integration of differential forms on an oriented submanifold-with-boundary. It plays a fundamental role in areas of applied mathematics such as fluid dynamics and electromagnetic theory; see the final chapter of Abraham, Marsden, and Ratiu [1988] and of Flanders [1989] for examples.

8.8.1 Stokes's Theorem for a Compact Submanifold-with-Boundary

Let M be an oriented n-dimensional manifold and P a compact n-dimensional submanifold-with-boundary of M, whose boundary ∂P is oriented as described in 8.3.1. For every $\omega \in \Omega^{n-1} M$,

$$\int_P d\omega \;=\; \int_{\partial P} \iota^* \omega, \tag{8.27}$$

where $\iota^* \omega$ denotes the pullback of ω under the inclusion map $\iota\colon \partial P \to M$, that is, the restriction of ω to ∂P. In particular, if $\partial P = \varnothing$ (e.g., if M is compact and $P = M$), then

$$\int_P d\omega \;=\; 0, \ \forall \omega \in \Omega^{n-1} M. \tag{8.28}$$

The proof will be found in Section 8.12. To understand Stokes's Theorem, it helps to recall three special cases. For the first, recall that 0-forms are simply smooth functions on M; hence the $n = 1$ case of the theorem does not require proof, since it is contained in:

8.8.1.1 $n = 1$: Fundamental Theorem of Calculus

For every smooth function g on an open interval of the real line containing a closed interval $P = [a, b]$ ($\therefore \partial P = \{a, b\}$),

$$\int_a^b dg = \int_a^b g'(x)\, dx = g(b) - g(a).$$

The reader should glance at the review of line integrals and surface integrals in 8.5.2 and 8.5.3 and verify that 8.8.1 includes the following vector calculus results:

8.8.1.2 $n = 2$: Green's Theorem

For every compact oriented 2-dimensional submanifold-with-boundary P of the plane, and every vector field X on a neighborhood of P,[4]

$$\int_P d\varpi_X = \int_P \phi_{\operatorname{curl} X} = \int_{\partial P} \varpi_X. \tag{8.29}$$

8.8.1.3 $n = 3$: Divergence Theorem

For every compact oriented 3-dimensional submanifold-with-boundary P of R^3, and every vector field X on a neighborhood of P,

$$\int_P d\phi_X = \int_P \operatorname{div} X\, (dx \wedge dy \wedge dz) = \int_{\partial P} \phi_X. \tag{8.30}$$

For the generalization of the Divergence Theorem to the case of an arbitrary oriented pseudo-Riemannian manifold $(M, \langle \cdot | \cdot \rangle)$, with canonical volume form σ, see Exercise 19.

8.9   The Exterior Derivative Stands Revealed

This section attempts to describe, using Stokes's Theorem, what might be called the geometric meaning[5] of exterior differentiation. Let ω be an $(n - 1)$-form on a neighborhood of 0 in R^{n+k}. Given a set of linearly independent vectors $\{v_1, \ldots, v_n\}$ in R^{n+k}, we would like to know the geometric interpretation of $d\omega(0)\,(v_1 \wedge \ldots \wedge v_n)$. First of all, it suffices to restrict ω to the tangent space at 0 to the copy of R^n spanned by

---

[4] See (8.12) and (8.15) for the "work form" and "flux form" notation.
[5] These ideas come from unpublished notes of John Hubbard (Cornell).

the vectors $\{v_1, ..., v_n\}$; therefore we shall now regard ω as an $(n-1)$-form on a neighborhood of 0 in R^n.

Next consider the parallelepiped P with edges prescribed by the vectors $\{v_1, ..., v_n\}$. See the picture below for the case $n = 3$. For $\varepsilon > 0$, let εP stand for the parallelepiped with edges prescribed by the vectors $\{\varepsilon v_1, ..., \varepsilon v_n\}$, and let $\partial(\varepsilon P)$ denote the "boundary" of εP; of course, εP is not a submanifold-with-boundary of R^n because it has sharp edges, but there is a modified version of the theory (see Spivak [1979], Abraham, Marsden, and Ratiu [1988], etc.) which applies to the piecewise smooth case; certainly $\partial(\varepsilon P)$ is a finite union of oriented $(n-1)$-dimensional submanifolds of R^n.

A fact of geometry is that the Riemannian volume of P is given by the formula

$$v_1 \wedge ... \wedge v_n = (\text{vol } P) \, e_1 \wedge ... \wedge e_n. \tag{8.31}$$

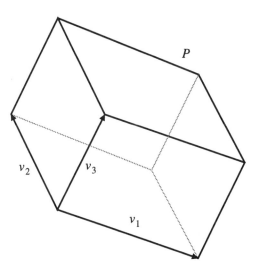

Figure 8. 4   Parallelepiped in 3-space

Necessarily the constant n-form $d\omega(0)$ obtained by evaluating $d\omega$ at 0 is expressible as $d\omega(0) = a(dx^1 \wedge ... \wedge dx^n)(0)$ for some real number a, so

$$d\omega(0)(v_1 \wedge ... \wedge v_n) = a(dx^1 \wedge ... \wedge dx^n)((\text{vol } P) \, e_1 \wedge ... \wedge e_n)$$

$$= a(\text{vol } P)$$

$$= a\int_P dx^1 \wedge ... \wedge dx^n$$

$$= \int_P d\omega\,(0).$$

Since $d\omega$ is smooth, and $\operatorname{vol}\varepsilon P = \varepsilon^n\,(\operatorname{vol}P)$,

$$d\omega\,(0)\,(v_1 \wedge \ldots \wedge v_n) = \lim_{\varepsilon \to 0} \varepsilon^{-n} \int_{\varepsilon P} d\omega\,(0) = \lim_{\varepsilon \to 0} \varepsilon^{-n} \int_{\varepsilon P} d\omega.$$

The middle integral refers to the constant form $d\omega\,(0)$, and the last one to the nonconstant form $d\omega$. Applying Stokes's Theorem (or rather a modified version allowing piecewise smooth boundaries – see Spivak [1979], Abraham, Marsden, and Ratiu [1988], etc.), we come to the following realization.

8.9.1 Geometric Meaning of the Exterior Derivative

$$d\omega\,(0)\,(v_1 \wedge \ldots \wedge v_n) = \lim_{\varepsilon \to 0} \varepsilon^{-n} \int_{\partial\,(\varepsilon P)} \omega, \quad \forall \omega \in \Omega^{n-1} U. \tag{8.32}$$

Look at the right side carefully: We are integrating over a boundary whose volume is of order ε^{n-1}, and then dividing by ε^n. This is correct, because the contributions to the integral from opposite faces of $\partial\,(\varepsilon P)$ almost cancel, leaving a term of order ε^n. Thus the right side of formula (8.32) measures the infinitesimal variation of ω in each of the directions that make up the edges of the parallelepiped P.

8.9.2 Example: $n = 2$

Suppose ω is a 1-form on R^2, which we may write as $\omega = A\,dx + B\,dy$. For simplicity, take $v_1 = e_1, v_2 = e_2$. If we take the usual orientation on R^2, the right side of formula (8.32) entails a line integral counterclockwise around the little square shown below:

Indeed the right side of formula (8.32) becomes

$$\lim_{\varepsilon \to 0} \varepsilon^{-2} \left\{ \int_0^\varepsilon A\,(x, 0)\,dx + \int_0^\varepsilon B\,(\varepsilon, y)\,dy + \int_\varepsilon^0 A\,(x, \varepsilon)\,dx + \int_\varepsilon^0 B\,(0, y)\,dy \right\}$$

$$= \lim_{\varepsilon \to 0} \varepsilon^{-2} \left\{ \int_0^\varepsilon \left(B\left(\varepsilon, y\right) - B\left(0, y\right) \right) dy - \int_0^\varepsilon \left(A\left(x, \varepsilon\right) - A\left(x, 0\right) \right) dx \right\}$$

$$= \frac{\partial B}{\partial x}\left(0, 0\right) - \frac{\partial A}{\partial y}\left(0, 0\right),$$

after about two lines of calculation, which the reader of Chapter 2 will recognize as $d\left(A\,dx + B\,dy\right)\left(0, 0\right)\left(e_1 \wedge e_2\right)$.

8.10   Exercises

In the following exercises, a submanifold of Euclidean space is assumed to have the induced Riemannian metric (see Chapter 7), unless otherwise stated.

9. Using the usual parametrization $\Psi\left(\phi, \theta\right) = \left(\cos\theta\sin\phi, \sin\theta\sin\phi, \cos\phi\right)$ of the 2-sphere S^2, take M to be the subset of S^2 with $0 < \phi < 1$ (a sort of punctured unit disc). Find the Riemannian volume of M using the canonical volume form $\sin\phi\,\left(d\phi \wedge d\theta\right)$.

10. Find the induced Riemannian metric on the ellipse $x^2/a^2 + y^2/b^2 = 1$, its canonical volume form, and its Riemannian volume (i.e., length).

11. Calculate the induced Riemannian metric of the hyperboloid $x^2 + y^2 - z^2 = 1$ in R^3, and the canonical volume form, using the parametrization

$$\Phi\left(t, \theta\right) = \left(\cosh t\cos\theta, \cosh t\sin\theta, \sinh t\right) = \left(x, y, z\right).$$

Calculate the Riemannian volume of the portion M of the hyperboloid on which $0 < z < 1$.

Remark: This metric is **not** the same as that of the hyperbolic plane!

12. Following the model of Example 8.9.2, verify formula (8. 32) for the case where ω is a 2-form on an open subset of R^3.

13. Let B^{n+1} denote the submanifold-with-boundary in R^{n+1} consisting of points whose distance from the origin is ≤ 1; thus $\partial B^{n+1} = S^n$. By applying Stokes's Theorem to the n-form $x^0\left(dx^1 \wedge \dots \wedge dx^n\right)$, prove that

$$\text{vol}\, S^n = \left(n + 1\right)\text{vol}\, B^{n+1}. \tag{8. 33}$$

14. Suppose M and N are oriented submanifolds of R^m and R^n, respectively. Prove that $M \times N$, which we know from Exercise 7 to be an oriented submanifold of R^{m+n}, satisfies $\text{vol}\left(M \times N\right) = \text{vol}\, M \times \text{vol}\, N$.

Hint: The canonical volume form on $M \times N$ can be taken to be the wedge product of the canonical volume forms on M and N, respectively; then use Fubini's theorem to factor the integral into two parts.

15. Suppose $\{(V_\gamma, \psi_\gamma), \gamma \in J'\}$ is a positively oriented atlas on an oriented manifold M, and K and K' are disjoint measurable subsets of M whose union is contained in a compact subset of M. Prove that, for any n-form ρ on M,

$$\int_{K \cup K'} \rho = \int_K \rho + \int_{K'} \rho. \tag{8.34}$$

16. Prove the following variant of Stokes's Theorem in an open subset U of R^3: For every constant vector field W, with associated flux form $\phi_W \in \Omega^2 U$, and every smooth function h on U,

$$\int_{\partial P} h\phi_W = \int_P \langle W | \text{grad } h \rangle \sigma, \tag{8.35}$$

for every 3-dimensional submanifold-with-boundary $P \subseteq U$.

17. Suppose M is a compact orientable n-dimensional manifold, and $\omega \in \Omega^{n-1}M$. Prove that $d\omega$ must vanish somewhere on M.

18. Let P be a compact n-dimensional submanifold-with-boundary of an orientable n-dimensional manifold M, and let $\Phi: \partial P \to N$ be a smooth map into a manifold N.

(i) Show that if Φ can be smoothly extended to a map F on a neighborhood of P in M, then for all $\omega \in \Omega^{n-1}N$ such that $d\omega = 0$,

$$\int_{\partial P} \Phi^* \omega = 0. \tag{8.36}$$

(ii) In the special case where $M = R^n$, $N = \partial P$, and Φ is the identity, use this result to show that no such extension F of Φ can exist; in other words, a map from a submanifold-with-boundary onto its boundary which keeps the boundary points fixed necessarily involves "tearing."

Hint: Suppose there were such an F; take $\omega = F^1 (dF^2 \wedge \ldots \wedge dF^n)$, which satisfies $d\omega = 0$ since $N = \partial P$ is $(n-1)$-dimensional. Obtain a contradiction to (8.36) using Stokes's Theorem.

19. Let M be an orientable n-dimensional manifold M, oriented by a volume form μ. If X is a vector field on M, then the Lie derivative $L_X\mu \in \Omega^n M$, as discussed in the exercises in Chapter 2, must be some function multiplied by μ, since the nth exterior power of each cotangent space is one-dimensional; this function is called the **divergence** $\text{div}_\mu X$ of the vector field X. To summarize:

$$L_X\mu = (\operatorname{div}_\mu X)\,\mu. \tag{8.37}$$

(i) Derive from Stokes's Theorem the **Divergence Theorem**, which states that for every compact n-dimensional submanifold-with-boundary P of M,

$$\int_P (\operatorname{div}_\mu X)\,\mu = \int_{\partial P} \iota_X\mu, \tag{8.38}$$

where $\iota_X\mu \in \Omega^{n-1}M$ is the interior product, that is,

$$\iota_X\mu \cdot X_2 \wedge \dots \wedge X_n = \mu \cdot X \wedge X_2 \wedge \dots \wedge X_n.$$

Hint: Look in the exercises of Chapter 2 for a formula for $d\,(\iota_X\mu)$.

(ii) Find the explicit formula for the integrands on both sides of (8. 38) in the case where μ is the canonical volume form $\sinh s\,(ds \wedge d\theta)$ on the hyperbolic plane with (s, θ) parametrization (see Chapter 7), and

$$X = u\,(s, \theta)\frac{\partial}{\partial s} + v\,(s, \theta)\frac{\partial}{\partial \theta}.$$

8.11 History and Bibliography

Stokes's Theorem (in three dimensions) was used by George Stokes (1819–1903) in his Smith Prize examination at Cambridge University in 1854. Fuller treatments of the integration theory of forms, including de Rham cohomology, may be found in Abraham, Marsden, and Ratiu [1988], Berger and Gostiaux [1988], and Spivak [1979], among others.

8.12 Appendix: Proof of Stokes's Theorem

Proof is needed only for $n \geq 2$, in view of 8.8.1.1.

8.12.1 Step I

Take an atlas for M of the special kind that was used for the definition of a submanifold-with-boundary; in other words, for every chart (U, φ), either $\partial P \cap U = \varnothing$, or else

$$\varphi\,(P \cap U) = \{\,(x_1, \dots, x_n) \in \varphi\,(U) : x_1 \leq 0\}, \tag{8.39}$$

with $\varphi\,(\partial P \cap U) = \{\,(x_1, \dots, x_n) \in \varphi\,(U) : x_1 = 0\}$.

Given an orientation for M, we can turn this atlas into a positively oriented atlas without disturbing (8. 39), by simply replacing φ by $s_2 \bullet \varphi$ for any chart map φ which is not positively oriented, where $s_2 (x_1, ..., x_n) = (x_1, -x_2, x_3, ..., x_n)$. We will assume that this has been done. Next take an open cover $\{V_\alpha, \alpha \in I\}$ of M together with a collection of smooth functions $\{\upsilon_\alpha : M \to [0, 1], \alpha \in I\}$ that form a partition of unity subordinate to this atlas. P has an open cover each element of which intersects only finitely many of the $\{V_\alpha\}$ (see the definition of a partition of unity). Since P is compact, it has a finite subcover, and therefore $P \cap V_\alpha \neq \varnothing$ for only finitely many α. Thus we can write

$$\int_P d\omega = \int_P \sum_\alpha \upsilon_\alpha d\omega = \sum_\alpha \int_P \upsilon_\alpha d\omega, \tag{8. 40}$$

where the sum is finite, and of course each n-form $\upsilon_\alpha d\omega$ has its support inside the domain of a particular chart $U_{j(\alpha)}$. Thus the proof is complete as soon as we have verified (8. 27) for every summand on the right side of (8. 40); in other words, it suffices to check, for every chart (U, φ) in our special atlas, and for every $(n - 1)$-form ω with support contained in U, that

$$\partial P \cap U = \varnothing \Rightarrow \int_P d\omega = 0; \text{ also} \tag{8. 41}$$

$$\varphi (P \cap U) = \{ (x_1, ..., x_n) \in \varphi (U) : x_1 \leq 0\} \Rightarrow \int_{P \cap U} d\omega = \int_{\partial P \cap U} \iota^* \omega. \tag{8. 42}$$

8.12.2   Proof Step II: Verification of (8. 41)

If $\partial P \cap U = \varnothing$, then either $P \cap U = \varnothing$, in which case (8. 41) is vacuously true, or else $U \subseteq P$, in which case we have to prove that, for every $(n - 1)$-form ω with compact support contained in U, the integral of $d\omega$ over U is zero. Suppose that

$$(\varphi^{-1})^* \omega = \sum_{i = 1}^n h_i (dx^1 \wedge ... \wedge d\hat{x}^i \wedge ... \wedge dx^n), \tag{8. 43}$$

using the notation of (8. 2). By the rules for exterior differentiation, it follows that

$$(\varphi^{-1})^* d\omega = d((\varphi^{-1})^* \omega) = \sum_{i = 1}^n (-1)^{i-1} \frac{\partial h_i}{\partial x^i} (dx^1 \wedge ... \wedge dx^n), \tag{8. 44}$$

where the functions $\{h_i\}$ and their derivatives have their support in the interior of some closed finite rectangle C in R^n, and therefore are zero on the perimeter of C. The definition of the integral of $d\omega$ given in Section 8.5 gives the following equation; proof of (8. 41) amounts to showing the expression is zero:

$$\int\limits_{\varphi(U)} (\varphi^{-1})^* \, d\omega = \sum_{i=1}^{n} \int\limits_{C} (-1)^{i-1} \frac{\partial h_i}{\partial x_i} dx_1 \ldots dx_n. \tag{8.45}$$

Consider only the first of the summands in the integral on the right. Express C as $[a, b] \times C_{n-1}$, where C_{n-1} is a closed finite rectangle C in R^{n-1}. We want to show that

$$\int\limits_{a}^{b} dx_1 \int\limits_{C_{n-1}} \frac{\partial h_1}{\partial x_1} (x_1, \ldots, x_n) \, dx_2 \ldots dx_n = 0. \tag{8.46}$$

The situation is represented in the following diagram. Using Fubini's Theorem, which

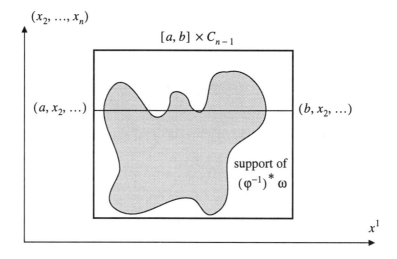

allows the interchange of the order of integration, we have

$$\int\limits_{a}^{b} dx_1 \int\limits_{C_{n-1}} \frac{\partial h_1}{\partial x_1} (x_1, \ldots, x_n) \, dx_2 \ldots dx_n = \int\limits_{C_{n-1}} \left\{ \int\limits_{a}^{b} \frac{\partial h_1}{\partial x_1} (x_1, \ldots, x_n) \, dx_1 \right\} dx_2 \ldots dx_n$$

$$= \int\limits_{C_{n-1}} (h_1 (b, x_2, \ldots, x_n) - h_1 (a, x_2, \ldots, x_n)) \, dx_2 \ldots dx_n = 0,$$

because the points $(b, x_2, \ldots, x_n)$ and $(a, x_2, \ldots, x_n)$ are outside the support of the functions $\{h_i\}$ (see picture). This verifies (8.46), and all the other integrals of the right side of (8.45) can be shown to be zero in the same way. This completes the verification of (8.41).

8.12.3 Proof Step III: Verification of (8. 42)

Now we suppose that $\varphi(P \cap U) = \{(x_1, ..., x_n) \in \varphi(U) : x_1 \leq 0\}$. Suppose that $C = [a, b] \times C_{n-1}$ is a closed rectangle in R^n, with $a < 0 < b$, containing the support of the $(n-1)$-form $(\varphi^{-1})^* \omega$; here $C_{n-1} = [a_2, b_2] \times ... \times [a_n, b_n] \subset R^{n-1}$. As in Step II, we have the formula

$$\int_{P \cap U} d\omega = \int_{\varphi(P \cap U)} (\varphi^{-1})^* d\omega = \sum_{i=1}^{n} \int_{C'} (-1)^{i-1} \frac{\partial h_i}{\partial x_i} dx_1 ... dx_n, \qquad (8.47)$$

where $C' = [a, 0] \times C_{n-1}$. The situation is shown in the following diagram. It is not

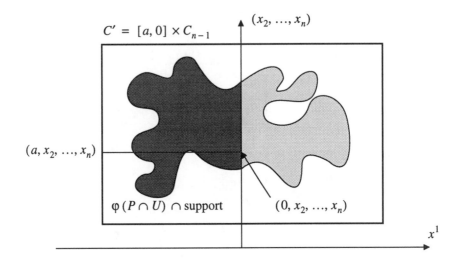

difficult to see that for $i = 2, ..., n$, the same reasoning as in Step II gives

$$\int_{a_i}^{b_i} dx_i \int_{C'_i} \frac{\partial h_i}{\partial x_i} (x_1, ..., x_n) \, dx_1 ... d\hat{x}_i ... dx_n = 0, \qquad (8.48)$$

where $C'_i = [a, 0] \times [a_2, b_2] \times ... \times [\hat{a}_i, \hat{b}_i] \times ... \times [a_n, b_n]$, since by choice of C,

$$h_i(x_1, ..., x_{i-1}, a_i, x_{i+1}, ..., x_n) = 0 = h_i(x_1, ..., x_{i-1}, b_i, x_{i+1}, ..., x_n).$$

However, this reasoning does not apply to the case $i = 1$, because here we are integrating not from a to b but from a to 0 only, and possibly $h_1(0, x_2, ..., x_n) \neq 0$ (see illustration). Using the fact that $(a, x_2, ..., x_n)$ is outside the support of the functions $\{h_i\}$, we obtain from (8. 47) and (8. 48)

$$\int_{P \cap U} d\omega = \int_a^0 dx_1 \int_{C_{n-1}} \frac{\partial h_1}{\partial x_1} (x_1, ..., x_n)\, dx_2...dx_n$$

$$= \int_{C_{n-1}} (h_1 (0, x_2, ..., x_n) - h_1 (a, x_2, ..., x_n))\, dx_2...dx_n$$

$$= \int_{C_{n-1}} h_1 (0, x_2, ..., x_n)\, dx_2...dx_n$$

$$= \int_{\varphi(\partial P \cap U)} h_1 (dx^2 \wedge ... \wedge dx^n).$$

Let $v: R^{n-1} \to R^n$ be the inclusion $(x_2, ..., x_n) \to (0, x_2, ..., x_n)$, which can be restricted to be the embedding of $\varphi (\partial P \cap U)$ into $\varphi (U)$; then we have

$$((\varphi^{-1} \bullet v)^* \omega) \cdot (\frac{\partial}{\partial x^1} \wedge ... \wedge \frac{\partial}{\partial \hat{x}^i} \wedge ... \wedge \frac{\partial}{\partial x^n})$$

$$= (\varphi^{-1})^* \omega \cdot (v_* \frac{\partial}{\partial x^1} \wedge ... \wedge v_* \frac{\partial}{\partial \hat{x}^i} \wedge ... \wedge v_* \frac{\partial}{\partial x^n}),$$

which is zero unless $i = 1$, since $v_* (\partial/\partial x^1) = 0$; in that case, by (8.43) the last expression is

$$= \sum_{i=1}^n h_i (dx^1 \wedge ... \wedge d\hat{x}^i \wedge ... \wedge dx^n) \cdot (v_* \frac{\partial}{\partial x^2} \wedge ... \wedge v_* \frac{\partial}{\partial x^n}) = h_1.$$

Moreover $\iota = \varphi^{-1} \bullet v \bullet \varphi$, and so $(\varphi^{-1})^* (\iota^* \omega) = (\varphi^{-1} \bullet v)^* \omega$. Thus it follows that

$$\int_{P \cap U} d\omega = \int_{\varphi(\partial P \cap U)} h_1 (dx^2 \wedge ... \wedge dx^n) = \int_{\varphi(\partial P \cap U)} (\varphi^{-1} \bullet v)^* \omega$$

$$= \int_{\partial P \cap U} \iota^* \omega,$$

as desired. ¤

9   Connections on Vector Bundles

The theory of connections can be presented in many equivalent ways. Here we present it as a theory of exterior differentiation of bundle-valued differential forms. First we will develop the theory for general vector bundles, and then specialize to the important example of the Levi-Civita connection on the tangent bundle of a Riemannian manifold.

9.1   Koszul Connections

If V is a vector space, with a basis $\{v_1, ..., v_m\}$, let $C^\infty(M \to V)$ denote the smooth maps from a differential manifold M to V. There is a natural way to differentiate a function $F \in C^\infty(M \to V)$ with respect to a vector field $X \in \mathfrak{I}(M)$; write F as $F^1 v_1 + ... + F^m v_m$, and take the vector-valued Lie derivative

$$L_X F = (dF^1 \cdot X) v_1 + ... + (dF^m \cdot X) v_m = (XF^1) v_1 + ... + (XF^m) v_m. \qquad \textbf{(9.1)}$$

The left-hand side is independent of the basis chosen for V. What can be said of the case where F is replaced by a section of a vector bundle over M, with fibers isomorphic to V? Now $\{v_1, ..., v_m\}$ is not a fixed basis, but a local frame field for the vector bundle, and $F^1 v_1 + ... + F^m v_m$ is the local expression for a section of the bundle.

9.1.1   Definition of a Koszul Connection

Suppose $\pi: E \to M$ is a smooth rank-m vector bundle, and $\sigma: M \to E$ is a section, expressed in terms of a local frame field $\{s_1, ..., s_m\}$ by

$$\sigma(r) = \sigma^1(r) s_1(r) + ... + \sigma^m(r) s_m(r) \in E_r,$$

where the $\{\sigma^i\}$ are smooth functions on some open set in the manifold. Our goal is as follows. Given a vector field X on M, we want to have some way (possibly not unique) to "differentiate a section σ with respect to X"; in other words to construct a new section, denoted $\nabla_X \sigma$, which will satisfy, $\forall h \in C^\infty(M)$, the "Leibniz rule"

$$\nabla_X(h\sigma) = h(\nabla_X \sigma) + (Xh)\sigma, \tag{9.2}$$

and the $C^\infty(M)$-linearity property

$$\nabla_{hX}\sigma = h(\nabla_X \sigma), \tag{9.3}$$

as well as the obvious additivity properties

$$\nabla_{X_1 + X_2}\sigma = \nabla_{X_1}\sigma + \nabla_{X_2}\sigma, \nabla_X(\sigma_1 + \sigma_2) = \nabla_X \sigma_1 + \nabla_X \sigma_2. \tag{9.4}$$

An operation ∇ satisfying (9.2), (9.3), and (9.4) is called a **Koszul connection** on E, and $\nabla_X \sigma$ is called the **covariant derivative** of σ along X. The "naturalness" of these conditions will become especially apparent when we present connections in terms of vector-bundle-valued forms in the next section.

9.1.2   Example: The Euclidean Connection on the Cotangent Bundle

The cotangent bundle of R^n has a natural connection $\hat{\nabla}$, which can be called the flat or **Euclidean connection**, and which the reader has been using for many years probably without knowing that it was a connection.[1] It is important here to take the standard coordinate system $\{x^1, ..., x^n\}$ on R^n. The sections of $T^* R^n \to R^n$ are simply the 1-forms, and so we can define $\hat{\nabla}_X \omega$ for an arbitrary 1-form $\omega = h_1 dx^1 + ... + h_n dx^n$ by taking $\hat{\nabla}_X(dx^i) = 0$ for all vector fields X, and using (9.2) and (9.4), namely,

$$\hat{\nabla}_X \omega = \sum_i \hat{\nabla}_X(h_i dx^i) = \sum_i h_i \hat{\nabla}_X(dx^i) + \sum_i (Xh_i) dx^i$$

$$= (Xh_1) dx^1 + ... + (Xh_n) dx^n, \tag{9.5}$$

which is the same as the situation described in (9.1), with only a change of notation. This is clearly consistent with (9.3) also. For more information about this connection, see Exercise 1 in Section 9.4.

9.1.3   Example: The Euclidean Connection on the Tangent Bundle

A general formula will be given in Exercise 3 in Section 9.4 for deriving a connection on a vector bundle from a connection on a dual bundle; as a special case of this we

---

[1] One recalls Molière's *bourgeois gentilhomme*, who discovered that all his life he had been speaking prose.

create from (9. 33) the **Euclidean connection on the tangent bundle of** R^n as follows; for vector fields

$$X = \sum \xi^i \partial/\partial x_i, \; Y = \sum \zeta^j \partial/\partial x_j$$

on R^n, $\nabla_X Y$ is the vector field

$$\nabla_X Y = \sum_j (X\zeta^j) \, \partial/\partial x_j = \sum_{i,j} \xi^i (\partial \zeta^j/\partial x_i) \, \partial/\partial x_j. \tag{9. 6}$$

The relationship of this connection to Example 9.1.2 is shown in Exercise 2 in Section 9.4.

9.1.4   The Obvious Generalization Is Not Intrinsic

Any attempt to generalize the (9. 1) construction of $\nabla_X \sigma$ to the case of a general vector bundle, that is, "$\nabla_X (\sigma^1 s_1 + \dots + \sigma^m s_m) = (X\sigma^1) s_1 + \dots + (X\sigma^m) s_m$," has limited usefulness, because it is not intrinsic; in other words, if we take another local frame field $\{t_1, \dots, t_m\}$, and write $\sigma = \tau^1 t_1 + \dots + \tau^m t_m$, then in general

$$(X\sigma^1) s_1 + \dots + (X\sigma^m) s_m \neq (X\tau^1) t_1 + \dots + (X\tau^m) t_m. \tag{9. 7}$$

For example, if $M = R^2 - \{0\}$, E is the tangent bundle, $\sigma = \partial/\partial r$ and $X = \partial/\partial\theta$ in the usual (r, θ) system of polar coordinates, $\{s_1, s_2\} = \{\partial/\partial r, \partial/\partial\theta\}$, and $\{t_1, t_2\} = \{\partial/\partial x, \partial/\partial y\}$, the reader can easily check that

$$\sigma = 1\frac{\partial}{\partial r} + 0\frac{\partial}{\partial\theta} = \cos\theta\frac{\partial}{\partial x} + \sin\theta\frac{\partial}{\partial y}$$

and so applying $X = \partial/\partial\theta$ to the component functions shows that the two sides of (9. 7) are indeed unequal. The heart of the matter is that we need to take into consideration the way the frame changes with position as well as the variation in the component functions. The extra ingredient can be understood by doing Exercise 5 in Section 9.4.

9.1.5   Origin of the Word "Connection"

The word "connection" has the following geometric origin. If ∇ is a connection on the tangent bundle, then $\nabla_X Y$ is a vector field which is supposed to describe the rate of change of the vector field Y along the "flow" of the vector field X. The difficulty is that the values of Y at different points along the flow belong to different tangent spaces, which are not in any natural relationship in general. The role of the "connection" is to connect these different tangent spaces. The Christoffel symbols (see Section 9.2.4.1) were devised as a way to describe how to do this, because they tell us how a given frame field (in terms of which X and Y may be defined) changes along the flow of each of its constituent vector fields.

9.2 Connections via Vector-Bundle-valued Forms

In order to develop the theory of connections in an efficient way, we shall now extend the theory of differential forms to cover vector-bundle-valued forms.

Suppose $\pi: E \to M$ is a rank-m vector bundle over an n-dimensional manifold M, and $1 \leq q \leq n$. Combining two constructions from Chapter 6, we obtain a vector bundle Hom $(\Lambda^q TM; E)$, whose fiber over $r \in M$, denoted Hom $(\Lambda^q TM; E)_r$, consists of the linear maps from $\Lambda^q T_r M$ to E_r. It may help the reader to recall the characterization of the exterior power of a vector space in terms of multilinear alternating maps (see Chapter 1), which identifies Hom $(\Lambda^q TM; E)_r$ with the q-multilinear alternating maps from $T_r M \times \ldots \times T_r M$ to E_r.

An **E-valued q-form** means a section of the vector bundle Hom $(\Lambda^q TM; E)$. The set of E-valued q-forms is denoted $\Omega^q (M; E)$, and q is called the **degree** of such a form, abbreviated to deg μ for a form μ. In keeping with the terminology for 0-forms on a manifold, the sections of $\pi: E \to M$ will be called **E-valued 0-forms**. For future reference, we make the formal statement:

$$\Omega^q (M; E) = \Gamma (\text{Hom} (\Lambda^q TM; E)), q = 1, 2, \ldots, n; \Omega^0 (M; E) = \Gamma E. \tag{9.8}$$

The union of the sets $\Omega^q (M; E)$ as q runs from 0 through n is called the set of **E-valued forms**. The action of an E-valued q-form μ on the wedge product of q vector fields on M will be denoted in the same way as the action of an ordinary differential form, namely,

$$\mu \cdot X_1 \wedge \ldots \wedge X_q \in \Omega^0 (M; E). \tag{9.9}$$

To find a local expression for such a form, take a chart (U, φ) for M, inducing a local frame field $\{dx^1, \ldots, dx^n\}$ for the cotangent bundle over U. If $\{s_1, \ldots, s_m\}$ is a local frame field for E over U, then $\mu \in \Omega^q (M; E)$ has the local expression

$$\mu|_U = \sum_{j=1}^{m} \sum_I s_j h_I^j (dx^{i(1)} \wedge \ldots \wedge dx^{i(q)}), \tag{9.10}$$

where $I = (1 \leq i(1) < \ldots < i(q) \leq n)$, and where the $\{h_I^j\}$ are smooth functions on U. Since

$$\sum_I h_I^j (dx^{i\,(1)} \wedge \ldots \wedge dx^{i\,(q)}) \in \Omega^q (M),$$

a suggestive way to write $\mu \in \Omega^q (M;E)$ is as

$$\mu|_U = \sum_{j=1}^m \eta^j \wedge s_j = \sum_{j=1}^m s_j \wedge \eta^j, \eta^j \in \Omega^q (M), \qquad \text{(9. 11)}$$

which is consistent with the notation $h\omega = h \wedge \omega$ when h is a 0-form and ω is a p-form, since $s_j \in \Gamma E|_U = \Omega^0 (M|_U;E)$ is an E-valued 0-form. However, the wedge symbols are actually superfluous, and we shall sometimes omit them.

9.2.1 Examples of Bundle-valued Forms

- $\Omega^q (M;R) = \Omega^q M$, the set of differential forms of degree q on M.

- If E is the trivial bundle $M \times R^m$, then $\Omega^q (M;M \times R^m)$, which is usually abbreviated to $\Omega^q (M;R^m)$, can be identified with the set of m-tuples $(\eta_1, \ldots, \eta_m)$ of q-forms on M; the vector-valued 1-forms and 2-forms encountered in Chapter 4 are good examples.

- If E is the tangent bundle TM, an obvious element of $\Omega^1 (M;TM)$ is the identity map ι from TM to TM, which can be regarded as an element of $L (T_rM;T_rM)$ for each $r \in M$.

- For any fixed vector field X on M, the Lie derivative L_X, acting on tangent vectors, is an element of $\Omega^1 (M;TM)$.

- We saw in Chapter 6 that the differential of a map from M to N can be viewed as a bundle-valued 1-form on M with values in the pullback of the tangent bundle of N.

- The primary motivation for this concept is that the all-important "Curvature tensor" (see below) for a connection on a vector bundle E is an element of $\Omega^2 (M;\mathrm{Hom}\ (E;E))$.

9.2.2 Exterior Products

There are natural mappings

$$\Omega^q (M;E) \times \Omega^p M \to \Omega^{p+q} (M;E) \leftarrow \Omega^p M \times \Omega^q (M;E),$$

which generalize the exterior product of differential forms; the map on the left is written

$$(\mu, \omega) \to \mu \wedge \omega = (-1)^{pq} \omega \wedge \mu,$$

and the map on the right is $\omega \wedge \mu \leftarrow (\omega, \mu)$. They can be defined locally in the notation of (9. 11), and the outcome is independent of the choice of local frame field for E; namely,

$$\omega \wedge \mu|_U = \sum_{j=1}^{m} (\omega \wedge \eta^j) \wedge s_j, \mu|_U \wedge \omega = \sum_{j=1}^{m} s_j \wedge (\eta^j \wedge \omega). \tag{9.12}$$

Obvious associativity and distributivity properties, which may be deduced from those of the ordinary exterior product, will be used without further comment.

9.2.3 A Connection as an Exterior Derivative of Bundle-valued Forms

In the notation of (9.8), a **covariant exterior derivative (of bundle-valued forms)** on a vector bundle $\pi: E \rightarrow M$ means a map

$$d^E: \Omega^p (M;E) \rightarrow \Omega^{p+1} (M;E), p = 0, 1, ..., n, \tag{9.13}$$

that satisfies the "Leibniz rule": That is, for every E-valued form μ and every differential form ω on M

$$d^E (\mu \wedge \omega) = d^E\mu \wedge \omega + (-1)^{\deg \mu}\mu \wedge d\omega. \tag{9.14}$$

Remark: Note how this resembles the rule "$d (\eta \wedge \omega) = d\eta \wedge \omega + (-1)^{\deg \eta}\eta \wedge d\omega$" for exterior differentiation. Observe also that the analog of the rule "$d (d\omega) = 0$" has been omitted; the reality is that $d^E (d^E\mu)$ need not vanish, and the obstruction to its vanishing will turn out to be the curvature of the connection.

How is a covariant exterior derivative related to the Koszul connection defined in the previous section? They are in one-to-one correspondence, and for this reason we shall often refer to d^E itself as a connection.

9.2.3.1 Equivalence between a Koszul Connection and a Covariant Exterior Derivative

Given a Koszul connection ∇ on a vector bundle $\pi: E \rightarrow M$, there is a unique covariant exterior derivative d^E such that the following formula holds:

$$d^E\sigma \cdot X = \nabla_X\sigma, \tag{9.15}$$

for every vector field X and every $\sigma \in \Gamma E = \Omega^0 (M;E)$.

Proof: Assume that we are given an operation ∇ satisfying (9.2), (9.3), and (9.4). Then for every vector field X and every $\sigma \in \Gamma E = \Omega^0 (M;E)$, the map which takes $X (r)$ to $d^E\sigma (r) (X (r)) = (d^E\sigma \cdot X) (r) = (\nabla_X\sigma) (r)$ belongs to $L (T_rM;E_r)$ by (9.3) and (9.4), and the image varies smoothly with r; this shows that $d^E\sigma$ indeed belongs to $\Omega^1 (M;E)$.

Next we shall verify that (9.14) holds for the case $p = 0$. Translated into the new notation, (9.2) says that, for every smooth function h on M and every $\sigma \in \Omega^0 (M;E)$,

$$d^E(\sigma \wedge h) = d^E(h\sigma) = h(d^E\sigma) + \sigma dh = d^E\sigma \wedge h + (-1)^0 \sigma \wedge dh,$$

where the latter version is designed to show concordance with (9. 14). So far we have proved that there is a unique map $d^E \colon \Omega^0(M;E) \to \Omega^1(M;E)$ that satisfies (9. 14). It remains to show that there exists a unique extension to E-valued forms of higher degrees consistent with (9. 14). We shall treat the uniqueness and existence questions in that order.

Uniqueness: If we use the local expression (9. 11), the action of d^E on $\mu \in \Omega^p(M;E)$ must be given by

$$d^E\mu \vert_U = \sum_{j=1}^{m} d^E(s_j \wedge \eta^j) \tag{9. 16}$$

$$= \sum_{j=1}^{m} (d^E s_j \wedge \eta^j + (-1)^0 s_j \wedge d\eta^j),$$

using (9. 14) and the fact that s_j has degree 0. Since $d^E s_j$ is uniquely specified by (9. 15), it follows that $d^E\mu \vert_U$ is uniquely specified by (9. 16).

Existence: One can use the local expression (9. 16) to show existence of $d^E\mu$ for all $\mu \in \Omega^p(M;E)$ such that (9. 14) and (9. 15) are satisfied; however, this requires that we check that the same value for $d^E\mu$ is obtained regardless of which local frame field is used for the vector bundle E, which is quite a tedious calculation. A more sophisticated way to prove existence is to find an intrinsic way to define $d^E\mu$ starting from ∇. This is suggested by the formula, given in the exercises to Chapter 2, for the exterior derivative of a differential form in terms of the Lie derivative of functions; namely, for any p-form ω and vector fields $X_0, \ldots, X_p$,

$$d\omega \cdot X_0 \wedge \ldots \wedge X_p = \sum_{i=0}^{p} (-1)^i L_{X_i}(\omega \cdot X_0 \wedge \ldots \wedge \hat{X}_i \wedge \ldots \wedge X_p)$$

$$+ \sum_{i<j} (-1)^{i+j} \omega \cdot [X_i, X_j] \wedge X_0 \wedge \ldots \wedge \hat{X}_i \wedge \ldots \wedge \hat{X}_j \wedge \ldots \wedge X_p,$$

where the superscript $^\wedge$ denotes a missing entry. Here we simply replace the Lie derivative L_X by ∇_X, and define $d^E\mu$ for all $\mu \in \Omega^p(M;E)$ by the formula

$$d^E\mu \cdot X_0 \wedge \ldots \wedge X_p = \sum_{i=0}^{p} (-1)^i \nabla_{X_i}(\mu \cdot X_0 \wedge \ldots \wedge \hat{X}_i \wedge \ldots \wedge X_p) \tag{9. 17}$$

$$+ \sum_{i<j} (-1)^{i+j} \mu \cdot [X_i, X_j] \wedge X_0 \wedge \ldots \wedge \hat{X}_i \wedge \ldots \wedge \hat{X}_j \wedge \ldots \wedge X_p.$$

The reader should note that the case $p = 0$ is precisely the formula (9. 15), so this is consistent with the earlier definition of $d^E \mu$ for $\mu \in \Omega^0 (M;E)$. A few (long!) lines of calculation will verify that (9. 14) is satisfied also.   ¤

The version of (9. 17) when $p = 1$ is so useful that we state it separately.

9.2.3.2 Covariant Exterior Derivative of an *E*-valued 1-Form
For every $\mu \in \Omega^1 (M;E)$ *and every pair X, Y of vector fields,*

$$d^E \mu \cdot X \wedge Y = \nabla_X (\mu \cdot Y) - \nabla_Y (\mu \cdot X) - \mu \cdot [X,Y] \tag{9. 18}$$

$$= d^E (\mu \cdot Y) \cdot X - d^E (\mu \cdot X) \cdot Y - \mu \cdot [X,Y]. \tag{9. 19}$$

9.2.4 Local Expression for a Covariant Exterior Derivative
Suppose $\{s_1, \ldots, s_m\}$ is a local frame field for the rank-m vector bundle $\pi: E \to M$ over some open set U in M. Since $d^E s_i \in \Omega^1 (M;E)$ for each i, there must exist an $m \times m$ matrix of 1-forms $\varpi_i^j \in \Omega^1 M$ such that

$$d^E s_i = \sum_{j=1}^m s_j \wedge \varpi_i^j. \tag{9. 20}$$

The 1-forms $\{\varpi_i^j\}$ constitute the **connection matrix** relative to the local frame field $\{s_1, \ldots, s_m\}$, and the $\{\varpi_i^j\}$ are traditionally called the **connection forms**. For any section $\sigma \in \Omega^0 (M;E)$, expressed in terms of the local frame field $\{s_1, \ldots, s_m\}$ by

$$\sigma (r) = \sigma^1 (r) s_1 (r) + \ldots + \sigma^m (r) s_m (r) = \sum_i s_i \wedge \sigma^i (r),$$

the rule (9. 14) gives the local form of the covariant exterior derivative of σ as

$$d^E \sigma = \sum_{i,j} s_j \wedge \sigma^i \varpi_i^j + \sum_i s_i \wedge d\sigma^i. \tag{9. 21}$$

9.2.4.1 Christoffel Symbols
Consider the specific case where E is the tangent bundle TM, and $\{e_1, \ldots, e_n\}$ is a local frame field for the tangent bundle, with a corresponding dual frame field $\{\varepsilon_1, \ldots, \varepsilon_n\}$ for the cotangent bundle. For example, a chart (U, φ) for M induces a local frame field $\{D_1, \ldots, D_n\}$, meaning $\{(\varphi^{-1})_* \partial/\partial x_1, \ldots, (\varphi^{-1})_* \partial/\partial x_n\}$, for the tangent bundle, with dual frame field $\{\varphi^* dx^1, \ldots, \varphi^* dx^n\}$. By expressing the connection 1-forms $\{\varpi_i^j\}$ in terms of the coframe field $\{\varepsilon_1, \ldots, \varepsilon_n\}$, we obtain a collection $\{\Gamma_{jk}^i\}$ of n^3 smooth

functions on U, called the **Christoffel symbols** of the connection in terms of this local frame field, determined by

$$\omega_i^j = \sum_k \Gamma_{ki}^j \varepsilon^k. \tag{9.22}$$

These are related to the Koszul connection ∇ as follows:

$$\nabla_{e_k} e_i = d^{TM} e_i \cdot e_k = \sum_j e_j \wedge \omega_i^j \cdot e_k = \sum_j \Gamma_{ki}^j e_j. \tag{9.23}$$

9.3   Curvature of a Connection

Let us see what happens to a section σ of a vector bundle when we apply the covariant exterior derivative twice. Unlike the case of the ordinary exterior derivative, for which $dd\omega = 0$ for all differential forms ω, it will usually happen that $d^E d^E \sigma \neq 0$. The nonvanishing of $d^E d^E \sigma$ will be described by a quantity called the curvature of the connection. Naturally the reader will be curious to know what this has got to do with the various kinds of curvature of a surface in R^3, described in Chapter 4; this will be explained in 9.3.5.

9.3.1   The Curvature as a Hom(E;E)-valued 2-Form

*Given a connection on a rank-m vector bundle $\pi: E \to M$, there exists a bundle-valued 2-form $\mathfrak{R} \in \Omega^2 (M; \mathrm{Hom} \, (E; E))$, called the **curvature of the connection**, such that*

$$d^E (d^E \sigma) = \mathfrak{R} \wedge \sigma, \, \sigma \in \Omega^0 (M; E). \tag{9.24}$$

*In terms of a local frame field $\{s_1, ..., s_m\}$ for E, $\mathfrak{R}$ is expressed by the $m \times m$ matrix of ordinary 2-forms (R_i^j), called **curvature forms**, defined by:*

$$d^E (d^E s_i) = \sum_{j=1}^m R_i^j \wedge s_j. \tag{9.25}$$

In terms of the connection matrix $\{\omega_i^j\}$ appearing in (9.20), we have

$$R_i^j = \sum_k \omega_k^j \wedge \omega_i^k + d\omega_i^j. \tag{9.26}$$

Proof: Note first that for any smooth function h,

$$d^E (d^E (h\sigma)) = d^E (hd^E \sigma + dh \wedge \sigma)$$

$$= hd^E(d^E\sigma) + dh \wedge d^E\sigma - dh \wedge d^E\sigma + d(dh) \wedge \sigma$$

$$= hd^E(d^E\sigma).$$

In other words, the map $\sigma \to d^E(d^E\sigma)$ is $C^\infty(M)$-linear. It follows that, if we take a local frame field and define 2-forms (R_i^j) by (9. 25), then for any section $\sigma = \sum \sigma^i s_i$,

$$d^E(d^E\sigma) = \sum_{i,j}(R_i^j\sigma^i) \wedge s_j,$$

and the right side is exactly what is meant by $\mathfrak{R} \wedge \sigma$ for $\mathfrak{R} \in \Omega^2(M; \text{Hom}(E;E))$; thus (9. 24) is established. As for (9. 26), (9. 20) implies

$$d^E(d^E s_i) = d^E(\sum_j s_j \wedge \varpi_i^j) = \sum_j \{d^E s_j \wedge \varpi_i^j + s_j \wedge d\varpi_i^j\};$$

consequently

$$\sum_k R_i^k \wedge s_k = \sum_{j,k} s_k \wedge \varpi_j^k \wedge \varpi_i^j + \sum_k s_k \wedge d\varpi_i^k,$$

from which (9. 26) immediately follows. ¤

9.3.2 The Curvature Tensor

Classical differential geometry usually deals with the set of m^2n^2 real-valued functions, called the curvature tensor, given by

$$R_{i\alpha\beta}^k = R_i^k \cdot D_\alpha \wedge D_\beta,$$

where $\{D_1, ..., D_n\}$ is a local frame field for the tangent bundle. Here are the gory details: If $X = \sum \xi^\alpha D_\alpha$, $Y = \sum \zeta^\beta D_\beta$, then

$$d^E(d^E s_i) \cdot X \wedge Y = \sum_{k,\alpha,\beta} \xi^\alpha \zeta^\beta R_{i\alpha\beta}^k s_k. \tag{9. 27}$$

9.3.3 Curvature in Terms of the ∇ Operator

For every pair X, Y of vector fields, and every section σ of the vector bundle $\pi: E \to M$,

$$\mathfrak{R} \wedge \sigma \cdot X \wedge Y = \nabla_X \nabla_Y \sigma - \nabla_Y \nabla_X \sigma - \nabla_{[X,Y]}\sigma. \tag{9. 28}$$

Proof: This is a routine application of formula (9. 18) on page 201 to the bundle-valued form $d^E\sigma \in \Omega^1(M;E)$. We have

$$\Re \wedge \sigma \cdot X \wedge Y = d^E (d^E \sigma) \cdot X \wedge Y = \nabla_X (d^E \sigma \cdot Y) - \nabla_Y (d^E \sigma \cdot X) - d^E \sigma \cdot [X,Y],$$

and this is exactly the right side of (9. 28) by (9. 15).   ¤

9.3.4 Examples

9.3.4.1 The Euclidean Connection Has Zero Curvature

Recall from 9.1.3 that, for $X = \sum \xi^i \partial/\partial x_i$, $Y = \sum \zeta^j \partial/\partial x_j$, the Euclidean connection on the tangent bundle of R^n is given by

$$\nabla_X Y = \sum_j (X\zeta^j) \, \partial/\partial x_j = \sum_{i,j} \xi^i (\partial\zeta^j/\partial x_i) \, \partial/\partial x_j.$$

It follows from (9. 28) that the vector field

$$\Re \wedge Z \cdot \partial/\partial x_i \wedge \partial/\partial x_j = 0, \ \forall Z, \ \forall i, j.$$

Thus the curvature of the Euclidean connection is zero.

9.3.4.2 A Case of Nonvanishing Curvature

Consider the 2-dimensional manifold $M = R^2 - \{0\}$, on which $\xi_1 = \partial/\partial r$, $\xi_2 = h(r)^{-1}\partial/\partial \phi$ form a frame field for the tangent bundle, where (r, ϕ) are polar coordinates, and h is a nowhere-vanishing smooth function. In Example 9.5.3, we shall see that a natural connection on TM is given by

$$\varpi = \begin{bmatrix} \varpi_1^1 & \varpi_2^1 \\ \varpi_1^2 & \varpi_2^2 \end{bmatrix} = \begin{bmatrix} 0 & -h'd\phi \\ h'd\phi & 0 \end{bmatrix}.$$

Using $R_i^j = \sum_k \varpi_k^j \wedge \varpi_i^k + d\varpi_i^j$, we see that the curvature is given by

$$\begin{bmatrix} R_1^1 & R_2^1 \\ R_1^2 & R_2^2 \end{bmatrix} = \begin{bmatrix} 0 & -h'd\phi \\ h'd\phi & 0 \end{bmatrix} \wedge \begin{bmatrix} 0 & -h'd\phi \\ h'd\phi & 0 \end{bmatrix} + \begin{bmatrix} 0 & -h''dr \wedge d\phi \\ h''dr \wedge d\phi & 0 \end{bmatrix},$$

$$= \begin{bmatrix} 0 & -h''dr \wedge d\phi \\ h''dr \wedge d\phi & 0 \end{bmatrix}. \tag{9. 29}$$

This means that, for example,

$$d^{TM} (d^{TM} \xi_2) = R_2^1 \wedge \xi_1 + R_2^2 \wedge \xi_2 = -h''dr \wedge d\phi \wedge \xi_1.$$

9.3.5 Why Is It Called "Curvature"?

In Chapter 4, we were dealing with a three-dimensional adapted moving frame for a parametrized surface $M \subset R^3$. The vector bundle involved there was

$$E = TR^3|_M \to M,$$

that is, the tangent bundle of R^3, restricted to M, which has rank 3. The connection was the Euclidean connection, suitably restricted. The matrix of connection forms $\{\varpi_i^j\}$, in terms of the orthonormal frame field[2] $\{\xi_1, \xi_2, \xi_3\}$ was precisely the matrix

$$\begin{bmatrix} \varpi_1^1 & \varpi_2^1 & \varpi_3^1 \\ \varpi_1^2 & \varpi_2^2 & \varpi_3^2 \\ \varpi_1^3 & \varpi_2^3 & \varpi_3^3 \end{bmatrix} = \begin{bmatrix} 0 & -\eta^0 & -\eta^1 \\ \eta^0 & 0 & -\eta^2 \\ \eta^1 & \eta^2 & 0 \end{bmatrix},$$

which we examined in Chapter 4, and is related to the orthonormal frame field by

$$d^E \xi_i = \sum_j \xi_j \wedge \varpi_i^j.$$

The 1-forms $\{\theta^1, \theta^2, \theta^3\}$ make up the orthonormal coframe field. Since the Euclidean connection is flat, the curvature $\Re = 0$, so (9. 31) implies

$$d\varpi_i^j = -\sum_q \varpi_q^j \wedge \varpi_i^q,$$

which is nothing other than the "second structure equation" of Chapter 4, from which Gauss's equation and the Codazzi-Mainardi equations are obtained. Since the Gaussian curvature K is characterized by $d\eta^0 = -K(\theta^1 \wedge \theta^2)$, we see that knowing the curvature $\Re$ tells us $d\varpi$, which in turn leads to knowledge of the Gaussian curvature. In summary, the computation of $\Re$ gives the data from which various concrete sorts of "curvature" can be obtained.

The next formula describes what happens when one takes an exterior derivative of a curvature form. A more concise and abstract version is derived in Exercise 14 in Section 9.4.

---

[2] A point of notation: Here we prefer to use the vector field ξ_i rather than the vector $\vec{\xi}_i$ of component functions.

9.3.6 Bianchi's Identity in Terms of the Connection Matrix

Exterior differentiation of the curvature forms (R_i^j) *appearing in (9. 25), in terms of the connection forms in (9. 20), gives* **Bianchi's identity:**

$$dR_i^j = \sum_k \{R_k^j \wedge \omega_i^k - \omega_k^j \wedge R_i^k\}. \tag{9.30}$$

Proof: Equation (9. 26) can be rewritten as

$$d\omega_i^j = R_i^j - \sum_q \omega_q^j \wedge \omega_i^q. \tag{9.31}$$

Exterior differentiation of (9. 26) gives

$$dR_i^j = \sum_k d\omega_k^j \wedge \omega_i^k - \sum_k \omega_k^j \wedge d\omega_i^k.$$

Substituting for $d\omega_k^j$, etc., using (9. 31) gives

$$dR_i^j = \sum_k \{R_k^j - \sum_q \omega_q^j \wedge \omega_k^q\} \wedge \omega_i^k - \sum_k \omega_k^j \wedge \{R_i^k - \sum_q \omega_q^k \wedge \omega_i^q\},$$

and two of the terms on the right cancel (just interchange indices k and q in the last product), leaving (9. 30).   ¤

The following table is intended to review the similarities and differences between ordinary exterior differentiation and covariant exterior differentiation.

	Exterior Derivative	**Covariant Exterior Derivative**
Domain and Range	$\Omega^p M \to \Omega^{p+1} M$	$\Omega^p (M;E) \to \Omega^{p+1} (M;E)$
Action on 0-forms	$dh \cdot X = Xh = L_X h$	$d^E \sigma \cdot X = \nabla_X \sigma$
Leibniz Rule	$d(\eta \wedge \omega) =$ $d\eta \wedge \omega + (-1)^{\deg \eta} \eta \wedge d\omega$	$d^E(\mu \wedge \omega) =$ $d^E \mu \wedge \omega + (-1)^{\deg \mu} \mu \wedge d\omega$
Iteration Rule	$d(d\omega) = 0$	$d^E(d^E \sigma) = \Re \wedge \sigma$

Table 9.1   Comparison between the exterior derivative and the covariant exterior derivative

9.4   Exercises

1. If $f \in C^\infty(R^n)$, and X and Y are the vector fields $X = \sum \xi^i \dfrac{\partial}{\partial x_i}$, $Y = \sum \zeta^i \dfrac{\partial}{\partial x_i}$, let

$$D^2 f(X, Y)(x) = D^2 f(x)(X(x), Y(x)) = \sum_{i,j} D_{ij} f(x) \xi^i(x) \zeta^j(x). \qquad \text{(9.32)}$$

(i) Show that if $\hat{\nabla}$ is the Euclidean connection on the cotangent bundle of R^n, as in Example 9.1.2, then

$$\hat{\nabla}_X df = \sum_i X(D_i f) dx^i, \text{ and } \hat{\nabla}_X df \cdot Y = D^2 f(X, Y). \qquad \text{(9.33)}$$

(ii) Verify that if $df = h \, dg$, then $\hat{\nabla}_X(h \, dg) \cdot Y = (Xh)(dg \cdot Y) + h \hat{\nabla}_X dg \cdot Y$.

Hint: $Df = hDg$, so $D^2 f(X, Y) = Dh(X) Dg(Y) + hD^2 g(X, Y)$.

2. Prove that (9. 5) and (9. 6) are related by

$$df \cdot \nabla_X Y = X(df \cdot Y) - \hat{\nabla}_X df \cdot Y, \qquad \text{(9.34)}$$

for all vector fields X and Y, and smooth functions f.

3. (Dual Connections) Prove that, to every connection ∇ on a vector bundle $E \to M$, there corresponds a unique connection $\hat{\nabla}$ on the dual bundle $E^* \to M$ such that, for $s \in \Gamma E, \lambda \in \Gamma E^*, X \in \mathfrak{I}(M)$, if $\lambda \bullet s$ denotes the action of λ on s, then

$$(\hat{\nabla}_X \lambda) \bullet s = X(\lambda \bullet s) - \lambda \bullet \nabla_X s, \qquad \text{(9.35)}$$

by taking the following steps:

(i) By canceling two terms of the type $(Xh) \lambda \bullet s$, show that the right side is $C^\infty(M)$-linear in s, that is, $(\hat{\nabla}_X \lambda) \bullet hs = h(\hat{\nabla}_X \lambda) \bullet s$ for smooth functions h; this shows that $\hat{\nabla}_X \lambda \in \Gamma E^*$.

(ii) Check conditions (9. 2), (9. 3), and (9. 4) for $\hat{\nabla}$.

(iii) Also show that (9. 34) is a special case of (9. 35) in the case where $E \to M$ is the tangent bundle and $E^* \to M$ is the cotangent bundle.

4. Consider the special case of constructing a connection for the cotangent bundle $\pi: T^* M \to M$. One wonders whether some judicious combination of the exterior derivative, the Lie derivative of forms, and the interior product would do the trick. If ω is a 1-form and X is a vector field, one candidate is $\nabla^1_X \omega = \iota_X(d\omega) \in \Omega^1 M$, where the right side uses the interior product described in the exercises to Chapter 2, so $\iota_X(d\omega) \cdot Y = d\omega \cdot X \wedge Y$. Another candidate is the Lie derivative $\nabla^2_X \omega = L_X \omega$, whose effect on 1-forms is given by $L_X(g \, dh) = (Xg) \, dh + g \, d(Xh)$. Prove that one candidate fails (9. 2), and the other fails (9. 3), and so *neither* is a connection.

5. Suppose $\Phi_U : \pi^{-1}(U) \to U \times R^k$ and $\Phi_V : \pi^{-1}(V) \to V \times R^k$ are local trivializations of a rank-m vector bundle $\pi : E \to M$, with $U \cap V \neq \varnothing$, giving local frame fields $r \to \{s_1(r), \ldots, s_m(r)\}$ and $r \to \{t_1(r), \ldots, t_m(r)\}$ over $U \cap V$, where

$$s_i(r) = \Phi_U^{-1}(r, e_i), \, t_i(r) = \Phi_V^{-1}(r, e_i), \, i = 1, 2, \ldots, m, \, r \in U \cap V.$$

Suppose the connection forms with respect to these two local frame fields for the same covariant exterior derivative d^E are given by

$$d^E s_i = \sum_{j=1}^{m} s_j \wedge \omega_i^j, \, d^E t_k = \sum_{j=1}^{m} t_j \wedge \hat{\omega}_k^j.$$

Derive the identity $\hat{\omega} = g^{-1} dg + g^{-1} \omega g$, meaning that

$$\sum_j g_j^i \hat{\omega}_k^j = dg_k^i + \sum_j \omega_j^i g_k^j, \tag{9.36}$$

where $g_j^i(r)$ denotes the (i, j) entry of the transition function $g_{UV}(r)$.

Hint: In the exercises to Chapter 7, we proved that $t_j = g_j^1 s_1 + \ldots + g_j^m s_m$.

6. (Continuation of Exercise 5) Suppose that the curvature forms with respect to these two local frame fields are given by (9.25), namely,

$$d^E(d^E s_i) = \sum_{j=1}^{m} R_i^j \wedge s_j, \, d^E(d^E t_i) = \sum_{j=1}^{m} \hat{R}_i^j \wedge t_j.$$

(i) Prove that $\hat{R} = g^{-1} R g$, meaning that

$$\sum_j g_j^k \hat{R}_i^j = \sum_j R_j^k g_i^j. \tag{9.37}$$

(ii) Give an alternative proof of (9.37), based on (9.36) and the formulas

$$R_i^j = \sum_k \omega_k^j \wedge \omega_i^k + d\omega_i^j, \, \hat{R}_i^j = \sum_k \hat{\omega}_k^j \wedge \hat{\omega}_i^k + d\hat{\omega}_i^j.$$

Remark: In classical language, the curvature forms transform as "tensors" whereas the connection forms do not.

7. Given a connection ∇ for TM, and a chart (U, φ) for M, let $\{\Gamma_{jk}^i\}$ be the Christoffel symbols associated with the local frame field $\{D_1, \ldots, D_n\}$ for the tangent bundle, where $D_i = (\varphi^{-1})_*(\partial/\partial x_i)$, as in 9.2.4.1.

The Christoffel symbols induce a section Γ of Hom $(TR^n \otimes TR^n; TR^n) \to \varphi(U)$, called the **local connector,** namely, the unique $C^\infty(R^n)$-bilinear map which sends a pair F, G of vector fields on $\varphi(U)$ to a vector field $\Gamma(F, G)$ on $\varphi(U)$ such that

$$\Gamma(\partial/\partial x_i, \partial/\partial x_j) = \sum_k \Gamma^k_{ij} \partial/\partial x_k = \varphi_*(\nabla_{D_i} D_j). \tag{9.38}$$

(i) Verify that the following formula holds for every pair X, Y of vector fields on U and every smooth function f on $\varphi(U)$:

$$df \cdot \Gamma(\varphi_* X, \varphi_* Y) = d(\varphi^* f) \cdot (\nabla_X Y) - X(d(\varphi^* f) \cdot Y) + D^2 f(\varphi_* X, \varphi_* Y).$$

(ii) Conversely, show that every section Γ of Hom $(TR^n \otimes TR^n; TR^n) \to \varphi(U)$ (i.e., a vector bundle over $\varphi(U)$) defines a connection ∇ on the tangent bundle, restricted to U, by the formula

$$(\nabla_X Y)(\varphi^* f) = X(d(\varphi^* f) \cdot Y) - D^2 f(\varphi_* X, \varphi_* Y) + df \cdot \Gamma(\varphi_* X, \varphi_* Y).$$

(iii) Suppose another chart (V, ψ) induces a local coordinate system $\{\hat{x}^1, ..., \hat{x}^n\}$ and a corresponding local frame field $\{\hat{D}_1, ..., \hat{D}_n\}$ for the tangent bundle, with corresponding Christoffel symbols $\{\hat{\Gamma}^\alpha_{\gamma\delta}\}$ and local connector $\hat{\Gamma}$. Prove that, on $U \cap V$, if $h = \varphi \bullet \psi^{-1}$, then for every pair F, G of vector fields on $\psi(U \cap V)$,

$$\hat{\Gamma}(F, G) = (h^{-1})_* \{\Gamma(h_* F, h_* G) + D^2 h(F, G)\} \tag{9.39}$$

in the notation of (9.32).

8. The setup is the same as for Exercise 7. Suppose $\hat{\nabla}$ is the "dual connection" on the cotangent bundle, given by (9.34). Prove that if D_{ij} denotes $D_i \bullet D_j$, then

$$\hat{\nabla}_{D_i} df \cdot D_j = D_{ij} f - \sum_k \Gamma^k_{ij} D_k f. \tag{9.40}$$

9. Suppose the image of a curve $\gamma = \gamma(t)$ is a one-dimensional submanifold N of a manifold M, and ∇ is a Koszul connection on the tangent bundle TM. We may restrict ∇ to the vector bundle $\pi: TM|_N \to N$ (the restriction of TM to N), of which $\dot{\gamma} = \partial\gamma/\partial t$ is a section; thus it makes sense to refer to $\nabla_{\dot\gamma} X$ as a "vector field along γ," for any section X of $\pi: TM|_N \to N$. We define γ to be a **geodesic** with respect to ∇ if

$$\nabla_{\dot\gamma} \dot\gamma = 0, \text{ for all } t. \tag{9.41}$$

(i) Suppose $\gamma(t) = (\gamma^1(t), ..., \gamma^n(t))$ in some coordinate system for which the Christoffel symbols are $\{\Gamma^i_{jk}\}$. Prove that γ satisfies the differential equation:

$$\frac{d^2\gamma^i}{dt^2} + \sum_{j,k} \Gamma^i_{jk}\frac{d\gamma^j}{dt}\frac{d\gamma^k}{dt} = 0. \qquad (9.\,42)$$

(ii) Write down the specific form of this equation for Example 9.3.4.2.

10. Work out all sixteen components of the curvature tensor for the curvature $\Re$ of 9.3.4.2 (most are zero).

11. Calculate the curvature of a covariant exterior derivative d^E on a rank-3 vector bundle $E \to R^2$ with connection matrix

$$\varpi = \begin{bmatrix} 0 & hdy & 0 \\ -hdy & 0 & fdx \\ 0 & -fdx & 0 \end{bmatrix},$$

where $f = f(x,y)$ and $h = h(x,y)$.

12. (Curvature of a line bundle) A connection ∇ on a complex line bundle $L \to M$ is specified by a complex-valued 1-form α such that

$$\nabla_X s = (\alpha \cdot X)\, s, \; s \in \Gamma L. \qquad (9.\,43)$$

Prove that the curvature of such a connection is given by $\Re = d\alpha$.

Hint: Calculate $\Re \wedge s \cdot X \wedge Y$ using formula (9.28).

13. (i) Suppose ∇ is a connection on a vector bundle $E \to M$. Prove that the following formula defines a connection ∇^{Hom} on the vector bundle $\mathrm{Hom}\,(E,E) \to M$:

$$(\nabla^{\mathrm{Hom}}_X F)\,\sigma = \nabla_X (F\sigma) - F(\nabla_X \sigma) \qquad (9.\,44)$$

for $X \in \mathfrak{S}(M)$, $F \in \Gamma(\mathrm{Hom}\,(E,E))$, and $\sigma \in \Gamma E$.

(ii) Show that the associated covariant exterior derivative d^{Hom} satisfies

$$d^E (F\sigma) = (d^{\mathrm{Hom}} F)\,\sigma + F(d^E \sigma). \qquad (9.\,45)$$

(iii) Extend (9.45) to $A \in \Omega^p (M;\mathrm{Hom}\,(E,E))$ and $\mu \in \Omega^q (M;E)$ as follows:

$$d^E (A \wedge \mu) = d^{\mathrm{Hom}} A \wedge \mu + (-1)^p A \wedge d^E \mu. \qquad (9.\,46)$$

Hint: Write

$$A = \sum F_k \wedge \varpi^k, \; \mu = \sum s_j \wedge \eta^j$$

for $F_k \in \Gamma (\text{Hom } (E,E))$ and $s_j \in \Gamma E$, as in (9. 11). Now apply (9. 14) and (9. 45).

14. Let $\mathfrak{R} \in \Omega^2 (M; \text{Hom } (E,E))$ be the curvature of a covariant exterior derivative d^E on a vector bundle $E \to M$. Using the associated covariant exterior derivative d^{Hom} defined in Exercise 13, prove the following concise form of Bianchi's identity:

$$d^{\text{Hom}} \mathfrak{R} = 0. \tag{9. 47}$$

Hint: Apply (9. 46) when $A = \mathfrak{R}$ and $\mu = \sigma \in \Gamma E$. Since $d^E (d^E d^E \sigma) = d^E d^E (d^E \sigma)$, the result follows.

15. (Continuation of Exercise 13) Let $\mathfrak{R}$ be the curvature of a covariant exterior derivative d^E on a vector bundle $E \to M$, and let $A \in \Omega^1 (M; \text{Hom } (E,E))$.

   (i) Show that the formula $d^{E, A} \mu = d^E \mu + A \wedge \mu$, for $\mu \in \Omega^p (M; E)$, defines another covariant exterior derivative $d^{E, A}$ on $E \to M$.

   (ii) Let $\mathfrak{R}^A$ be the curvature of $d^{E, A}$. By computing $\mathfrak{R}^A \wedge \sigma$ for $\sigma \in \Gamma E$, show that

$$\mathfrak{R}^A = \mathfrak{R} + d^{\text{Hom}} A + A \wedge A. \tag{9. 48}$$

Comment: (9. 48) plays a crucial role in deriving the Yang–Mills equations in Chapter 10.

16. The definition of the curvature $\mathfrak{R}$ of a covariant exterior derivative d^E says only that $d^E (d^E \sigma) = \mathfrak{R} \wedge \sigma$, $\sigma \in \Omega^0 (M; E)$. Show that

$$d^E (d^E \mu) = \mathfrak{R} \wedge \mu, \text{ for all } \mu \in \Omega^p (M; E), \tag{9. 49}$$

by writing μ in terms of a local frame field $\{s_1, ..., s_m\}$ as $\mu = \sum_i s_i \wedge \eta^i$, for some $\eta^i \in \Omega^p M$.

17. The setup is the same as for Exercise 5. The local frame field $\{D_1, ..., D_n\}$ for the tangent bundle, where $D_i = (\varphi^{-1})_* (\partial / \partial x_i)$, induces a coframe field $\{\varphi^* dx^1, ..., \varphi^* dx^n\}$, which will be abusively referred to as $\{dx^1, ..., dx^n\}$.

   (i) By exterior differentiation of formula (9. 22), show

$$d\omega_i^j = \sum_k d\Gamma_{ki}^j \wedge dx^k. \tag{9. 50}$$

   (ii) From the formula (9. 26) for the curvature forms $\{R_i^j\}$, deduce that

$$R_{imq}^j = R_i^j \cdot D_m \wedge D_q = D_m \Gamma_{qi}^j - D_q \Gamma_{mi}^j + \sum_k \{\Gamma_{mk}^j \Gamma_{qi}^k - \Gamma_{qk}^j \Gamma_{mi}^k\}. \tag{9. 51}$$

9.5 Torsion-free Connections

We shall be concerned here only with connections on the tangent bundle. To restrict this class somewhat, we shall introduce the notion of a connection with "zero torsion."

9.5.1 Torsion of a Connection on *TM*

There is a canonical TM-valued 1-form, namely, $\iota \in \Omega^1\,(M;TM)$ defined by $\iota\,(\xi)\,=\,\xi$ for all tangent vectors ξ. The **torsion** of a connection ∇ on TM, and of the associated covariant exterior derivative d^{TM}, is the TM-valued 2-form

$$\tau = d^{TM}\iota. \tag{9.52}$$

To evaluate $\tau \cdot X \wedge Y$ for any pair X, Y of vector fields, we use (9.19):

$$\tau \cdot X \wedge Y = d^{TM}\,(\iota \cdot Y)\,\cdot X - d^{TM}\,(\iota \cdot X)\,\cdot Y - \iota \cdot [X,Y]$$

$$= d^{TM}\,Y \cdot X - d^{TM}\,X \cdot Y - [X,Y]$$

$$= \nabla_X Y - \nabla_Y X - [X,Y]. \tag{9.53}$$

A connection is called **torsion-free** if $\tau\,=\,0$, that is, if $\nabla_X Y - \nabla_Y X - [X,Y]\,=\,0$ for every pair X, Y of vector fields. The condition of being torsion-free appears to be a very natural one, because one would expect a "derivative" of an identity map (in this case, ι) to vanish.

9.5.1.1 Example of a Torsion-free Connection

The Euclidean connection on the tangent bundle of R^n (Example 9.1.3) is torsion-free, because if $X\,=\,\sum \xi^i \partial/\partial x_i$, $Y\,=\,\sum \zeta^j \partial/\partial x_j$, then

$$\tau \cdot X \wedge Y = \sum_{i,j} \xi^i\,(\partial \zeta^j/\partial x_i)\,D_j - \sum_{i,j} \zeta^i\,(\partial \xi^j/\partial x_i)\,D_j - \left[\sum_i \xi^i \partial/\partial x_i, \sum_j \zeta^j \partial/\partial x_j\right]$$

and a routine calculation shows that all the terms on the right cancel out.

9.5.1.2 Interpretation in Terms of Christoffel Symbols

In terms of a local frame field $\{D_1, ..., D_n\}$ induced by a chart, the Christoffel symbols of a torsion-free connection ∇, characterized by

$$\nabla_{D_\beta} D_\gamma = \sum_\alpha \Gamma^\alpha_{\beta\gamma} D_\alpha, \tag{9.54}$$

satisfy $\Gamma^\alpha_{\beta\gamma} = \Gamma^\alpha_{\gamma\beta}$.

Proof: Simply take $X = D_\beta$ and $Y = D_\gamma$ in the equation $\nabla_X Y - \nabla_Y X - [X,Y] = 0$, and note that $[D_\beta, D_\gamma] = 0$. ¤

9.5.2 A Method of Constructing Torsion-free Connections

Suppose $\{e_1, ..., e_n\}$ is any local frame field for the tangent bundle, and $\{\theta^1, ..., \theta^n\}$ is the associated coframe field of 1-forms, that is, $\theta^k \cdot e_i = \delta_i^k$.

(i) If $\{\omega_i^k\}$ is a matrix of 1-forms such that

$$d\theta^k = \sum_i \theta^i \wedge \omega_i^k, \tag{9.55}$$

then the connection given by

$$d^{TM} e_i = \sum_k e_k \wedge \omega_i^k \tag{9.56}$$

(i.e., taking the $\{\omega_i^k\}$ as connection forms) is torsion-free.

(ii) There is a unique matrix of 1-forms $\{\omega_i^k\}$ that satisfy both (9.55) and (9.57):

$$\omega_i^k + \omega_k^i = 0. \tag{9.57}$$

Proof: (i) Observe that the canonical TM-valued 1-form, namely, $\iota \in \Omega^1(M;TM)$ defined by $\iota(\xi) = \xi$, can also be expressed as

$$\iota = \sum_k e_k \wedge \theta^k. \tag{9.58}$$

To verify this, take an arbitrary vector field $X = \sum \xi^i e_i$, and observe that

$$\sum_k e_k \wedge \theta^k \cdot X = \sum_{i,k} \xi^i e_k \wedge (\theta^k \cdot e_i) = \sum \xi^i e_i = X.$$

Now suppose d^{TM} is the covariant exterior derivative on TM given by (9.56). Using (9.55), (9.56), and (9.14), we see that

$$d^{TM} \iota = d^{TM} \{\sum_k e_k \wedge \theta^k\} = \sum_k (d^{TM} e_k) \wedge \theta^k + \sum_k e_k \wedge d\theta^k$$

$$= \sum_k \{\sum_j e_j \wedge \omega_k^j\} \wedge \theta^k + \sum_k e_k \wedge \{\sum_i \theta^i \wedge \omega_i^k\}$$

$$= \sum_{i,k} e_k \wedge \theta^i \wedge \{\omega_i^k - \omega_i^k\} = 0.$$

(ii)[3] **Uniqueness:** Suppose that $\{\omega_i^k\}$ satisfy both (9. 55) and (9. 57). There are unique functions $\{a_{ij}^k\}$ and $\{b_{ij}^k\}$ such that

$$\omega_i^k = \sum_j a_{ij}^k \theta^j; \tag{9.59}$$

$$d\theta^k = \frac{1}{2} \sum_{i,j} b_{ij}^k \theta^i \wedge \theta^j, \text{ with } b_{ji}^k = -b_{ij}^k. \tag{9.60}$$

Then (9. 55) and (9. 57) imply

$$a_{kj}^i = -a_{ij}^k; \tag{9.61}$$

$$\frac{1}{2} \sum_{i,j} b_{ij}^k \theta^i \wedge \theta^j = \sum_i \theta^i \wedge \omega_i^k = \sum_{i,j} a_{ij}^k \theta^i \wedge \theta^j \Rightarrow a_{ij}^k - a_{ji}^k = b_{ij}^k. \tag{9.62}$$

Therefore

$$\frac{1}{2}\{b_{ij}^k + b_{jk}^i - b_{ki}^j\} = \frac{1}{2}\{a_{ij}^k - a_{ji}^k + a_{jk}^i - a_{kj}^i - a_{ki}^j + a_{ik}^j\} = a_{ij}^k. \tag{9.63}$$

This proves that the $\{\omega_i^k\}$ are uniquely determined by (9. 55) and (9. 57).

Existence: Starting from (9. 60), let us define $\{a_{ij}^k\}$ by (9. 63), and define $\{\omega_i^k\}$ by (9. 59).[4] Then

$$a_{kj}^i = \frac{1}{2}\{b_{kj}^i + b_{ji}^k - b_{ik}^j\} = -a_{ij}^k,$$

since $b_{ji}^k = -b_{ij}^k$; thus (9. 61) and hence (9. 57) hold. To verify (9. 55), note that (9. 63) holds, and so, for any q and m,

$$\sum_i \theta^i \wedge \omega_i^k \cdot e_m \wedge e_q = \sum_i \{ (\theta^i \cdot e_m)(\omega_i^k \cdot e_q) - (\theta^i \cdot e_q)(\omega_i^k \cdot e_m) \}$$

$$= \sum_i \{\delta_m^i a_{iq}^k - \delta_q^i a_{im}^k\} = a_{mq}^k - a_{qm}^k$$

$$= b_{mq}^k = \frac{1}{2}\{b_{mq}^k - b_{qm}^k\}$$

---

[3] The remainder of this proof follows Spivak [1979], Volume 2.
[4] Note for future reference that the Christoffel symbols of the connection are $\Gamma_{ji}^k = a_{ij}^k$.

$$= \frac{1}{2} \sum_{i,j} b^k_{ij} \theta^i \wedge \theta^j \cdot e_m \wedge e_q$$

$$= d\theta^k \cdot e_m \wedge e_q,$$

which proves that $d\theta^k = \sum_i \theta^i \wedge \omega^k_i$. ¤

9.5.3 Example of Constructing a Torsion-free Connection

Consider the 2-dimensional manifold $M = R^2 - \{0\}$, on which $\theta^1 = dr$, $\theta^2 = h(r)\,d\phi$ form a frame field for the cotangent bundle, where (r, ϕ) are polar coordinates, and h is a nowhere-vanishing smooth function. Then $d\theta^1 = 0$ and $d\theta^2 = h'(r)\,dr \wedge d\phi = (h'/h)\,(\theta^1 \wedge \theta^2)$, so (9. 55) holds with

$$\omega^1_1 = 0, \omega^1_2 = 0, \omega^2_1 = (h'/h)\,\theta^2, \omega^2_2 = 0,$$

which gives a torsion-free connection, by part (i) of 9.5.2; however this connection does not satisfy (9. 57). The unique connection that satisfies (9. 57) is obtained using (9. 60); we see that

$$b^1_{ij} = 0, b^2_{12} = h'/h = -b^2_{21}, b^2_{ii} = 0;$$

$$\Gamma^k_{ji} = a^k_{ij} = \frac{1}{2}\{b^k_{ij} + b^i_{jk} - b^j_{ki}\};$$

it follows that

$$(\Gamma^1_{ij}) = \begin{bmatrix} 0 & 0 \\ 0 & -h'/h \end{bmatrix}, (\Gamma^2_{ij}) = \begin{bmatrix} 0 & 0 \\ h'/h & 0 \end{bmatrix}.$$

Even though the connection is torsion-free, the Christoffel symbols are not symmetric in (i, j) because the frame field is not induced by a chart.

Since $\omega^k_i = \sum_j \Gamma^k_{ji}\theta^j$, the desired connection is given by

$$\omega^1_1 = 0, \omega^1_2 = -(h'/h)\,\theta^2, \omega^2_1 = (h'/h)\,\theta^2, \omega^2_2 = 0, \text{ or}$$

$$\omega = \begin{bmatrix} \omega^1_1 & \omega^1_2 \\ \omega^2_1 & \omega^2_2 \end{bmatrix} = \begin{bmatrix} 0 & -h'd\phi \\ h'd\phi & 0 \end{bmatrix}. \tag{9. 64}$$

9.6   Metric Connections

The "product rule" $(uv)' = u'v + uv'$ is familiar to every calculus student, and leads easily to the rule for differentiating the Euclidean inner product of two R^n-valued functions $\vec{u}(t)$ and $\vec{v}(t)$, namely, $\langle \vec{u}|\vec{v}\rangle' = \langle \vec{u}'|\vec{v}\rangle + \langle \vec{u}|\vec{v}'\rangle$. We are now going to generalize this idea in the obvious way to a connection ∇ on TM, where the manifold M has an indefinite metric $\langle \cdot|\cdot \rangle$, as described in Chapter 7. We say that ∇ is **compatible with** $\langle \cdot|\cdot \rangle$, or is a **metric connection** with respect to $\langle \cdot|\cdot \rangle$, if

$$X(\langle Y|Z\rangle) = \langle \nabla_X Y|Z\rangle + \langle Y|\nabla_X Z\rangle \tag{9.65}$$

for all vector fields X, Y, Z. Note that the left side denotes the action of the vector field X on the function formed by taking the inner product of Y and Z.

9.6.1   Example of a Metric Connection

The Euclidean connection on the tangent bundle of R^n (Example 9.1.3) is compatible with the Euclidean metric, because if $Y = \sum \psi^j \partial/\partial x_j$, $Z = \sum \zeta^k \partial/\partial x_k$, then

$$X(\langle Y|Z\rangle) = X\left(\sum \psi^i \zeta^i\right)$$

$$= \langle \sum_j (X\psi^j)\, \partial/\partial x_j | Z\rangle + \langle Y| \sum_j (X\zeta^j)\, \partial/\partial x_j \rangle,$$

$$= (\langle \nabla_X Y|Z\rangle + \langle Y|\nabla_X Z\rangle).$$

9.6.2   Fundamental Theorem of Riemannian Geometry

9.6.2.1   Short Version
*Given an indefinite metric on M, there is a unique torsion-free metric connection on TM, called the **Levi-Civita connection**.*

9.6.2.2   Elaborated Version
Suppose that M has an indefinite metric, and that $\{e_1, \ldots, e_m\}$ is an orthonormal frame field for the tangent bundle, with orthonormal coframe field $\{\theta^1, \ldots, \theta^n\}$ of 1-forms, that is, $\theta^k \cdot e_i = \delta^k_i$ (see Chapter 7). Then the unique torsion-free connection on TM whose connection forms satisfy (9.55) and (9.57) (see 9.5.2) is compatible with the metric, and is the only torsion-free metric connection.

9.6.2.3   Corollary: Formula for the Christoffel Symbols of the Levi-Civita Connection
In terms of a local frame field $\{D_1, \ldots, D_n\}$ induced by a chart, the Christoffel symbols, characterized by

$$d^{\text{TM}} D_\beta \cdot D_\gamma = \sum_\alpha \Gamma^\alpha_{\beta\gamma} D_\alpha, \tag{9.66}$$

take the form

$$\Gamma^{\alpha}_{\beta\gamma} = \frac{1}{2}\sum_{\delta} g^{\alpha\delta}\{D_{\gamma}g_{\delta\beta} + D_{\beta}g_{\delta\gamma} - D_{\delta}g_{\beta\gamma}\}, \tag{9.67}$$

where $(g_{\alpha\gamma})$ *is the metric tensor, with inverse* $(g^{\alpha\delta})$.

Proof of uniqueness: Given a torsion-free metric connection ∇, the lack of torsion implies

$$\langle \nabla_X Y - \nabla_Y X | Z \rangle = \langle [X,Y] | Z \rangle, \tag{9.68}$$

and (9.65) implies

$$X\langle Y|Z\rangle + Y\langle Z|X\rangle - Z\langle X|Y\rangle$$

$$= \langle \nabla_X Y|Z\rangle + \langle Y|\nabla_X Z\rangle + \langle \nabla_Y Z|X\rangle + \langle Z|\nabla_Y X\rangle - \langle \nabla_Z X|Y\rangle - \langle X|\nabla_Z Y\rangle$$

$$= \langle \nabla_X Y + \nabla_Y X|Z\rangle + \langle \nabla_Y Z - \nabla_Z Y|X\rangle - \langle \nabla_Z X - \nabla_X Z|Y\rangle.$$

It follows that

$$\langle \nabla_X Y + \nabla_Y X|Z\rangle = X\langle Y|Z\rangle + Y\langle Z|X\rangle - Z\langle X|Y\rangle - \langle [Y,Z]|X\rangle + \langle [Z,X]|Y\rangle. \tag{9.69}$$

Adding (9.68) and (9.69) gives

$$2\langle \nabla_X Y|Z\rangle = X\langle Y|Z\rangle + Y\langle Z|X\rangle - Z\langle X|Y\rangle - \langle [Y,Z]|X\rangle + \langle [Z,X]|Y\rangle + \langle [X,Y]|Z\rangle, \tag{9.70}$$

which proves that $\langle \nabla_X Y|Z\rangle$, and hence $\nabla_X Y$, are uniquely defined by the metric.

As for the existence part of 9.6.2.1, the truly industrious can verify that if $\nabla_X Y$ is defined by (9.70), then the axioms for a Koszul connection are satisfied. However, existence follows easily from the approach taken in 9.5.2, as we shall now see.

Proof of existence: We need only prove that the connection ∇ whose connection matrix $\{\varpi^k_i\}$ is characterized by

$$d\theta^k = \sum_i \theta^i \wedge \varpi^k_i, \; \varpi^k_i + \varpi^i_k = 0, \tag{9.71}$$

is a metric connection. By (9.23), $\nabla_X e_i = \sum_j e_j \wedge \varpi^j_i \cdot X$, and so for any vector field X,

$$X\langle e_i|e_k\rangle = X(\pm\delta_{ik}) = 0.$$

On the other hand,

$$\langle \nabla_X e_i | e_k \rangle + \langle e_i | \nabla_X e_k \rangle = \sum_j (\varpi_i^j \cdot X) \langle e_j | e_k \rangle + \sum_j (\varpi_k^j \cdot X) \langle e_j | e_i \rangle,$$

$$= (\varpi_i^k + \varpi_k^i) \cdot X,$$

which is zero by (9. 71). Thus for an orthonormal frame field, we have shown that

$$X \langle e_i | e_k \rangle = \langle \nabla_X e_i | e_k \rangle + \langle e_i | \nabla_X e_k \rangle, \forall i, k. \qquad (9.\,72)$$

By writing arbitrary vector fields Y and Z as $Y = \sum \psi^j e_j$, $Z = \sum \zeta^m e_m$, (9. 2), (9. 4), and (9. 72) imply

$$X \langle Y | Z \rangle = \langle \nabla_X Y | Z \rangle + \langle Y | \nabla_X Z \rangle. \qquad (9.\,73)$$

Proof of 9.6.2.3: Taking $X = D_\beta$, $Y = D_\gamma$, and $Z = D_\delta$ in (9. 70), and noting that for these vector fields all the Lie brackets vanish, gives

$$2 \sum_\alpha \Gamma_{\beta\gamma}^\alpha g_{\alpha\delta} = D_\beta g_{\gamma\delta} + D_\gamma g_{\beta\delta} - D_\delta g_{\beta\gamma},$$

from which (9. 67) immediately follows.   ¤

9.6.3   Calculating a Levi-Civita Connection

Instead of using (9. 67), it is often better to write down an orthonormal coframe field, and then calculate the matrix of connection 1-forms as in Example 9.5.3. For example, consider, as in 9.5.3, the 2-dimensional manifold $M = R^2 - \{0\}$, on which $\theta^1 = dr$, $\theta^2 = h(r)\,d\phi$ form a frame field for the cotangent bundle, where (r, ϕ) are polar coordinates, and h is a nowhere-vanishing smooth function. Evidently $\{\theta^1, \theta^2\}$ is an orthonormal coframe field for the metric

$$\langle \cdot | \cdot \rangle = \theta^1 \otimes \theta^1 + \theta^2 \otimes \theta^2 = dr \otimes dr + h^2 (d\phi \otimes d\phi). \qquad (9.\,74)$$

Hence the Levi-Civita connection is precisely the one obtained in (9. 64), namely,

$$\varpi = \begin{bmatrix} \varpi_1^1 & \varpi_2^1 \\ \varpi_1^2 & \varpi_2^2 \end{bmatrix} = \begin{bmatrix} 0 & -(h'/h)\,\theta^2 \\ (h'/h)\,\theta^2 & 0 \end{bmatrix} = \begin{bmatrix} 0 & -h'd\phi \\ h'd\phi & 0 \end{bmatrix}. \qquad (9.\,75)$$

Consider two special cases: Equation (9. 74) gives the Euclidean metric if $h(r) = r$, and then it follows that $\mathfrak{R} = \varpi \wedge \varpi + d\varpi = 0$, as we know already. As we saw in Chapter 7, on the hyperbolic plane the metric is given by taking $h(r) = \sinh r$, so

$$\varpi = \begin{bmatrix} 0 & -(\cosh r)\, d\phi \\ (\cosh r)\, d\phi & 0 \end{bmatrix};$$

$$\Re = \varpi \wedge \varpi + d\varpi = \begin{bmatrix} 0 & -\sinh r\,(dr \wedge d\phi) \\ \sinh r\,(dr \wedge d\phi) & 0 \end{bmatrix} = \begin{bmatrix} 0 & -\theta^1 \wedge \theta^2 \\ \theta^1 \wedge \theta^2 & 0 \end{bmatrix}.$$

9.6.4 Sectional Curvature

Suppose $(M, \langle \cdot | \cdot \rangle)$ is a Riemannian manifold of dimension at least 2. If $\{e_1, \ldots, e_n\}$ is an orthonormal frame field with coframe field $\{\theta^1, \ldots, \theta^n\}$, then the **sectional curvature** of the 2-dimensional subbundle of $TM \to M$ spanned by the pair of vector fields $\{e_i, e_j\}$ is defined to be

$$K(e_i, e_j) = \langle \Re \wedge e_i \cdot e_j \wedge e_i | e_j \rangle = \sum_k \langle R_{iji}^k e_k | e_j \rangle = R_{iji}^j. \tag{9.76}$$

It can be proved that this depends only on the subbundle, not on the specific choice of frame field. In the case of a 2-dimensional Riemannian manifold, there is only one such subbundle, and hence only one sectional curvature

$$K(e_1, e_2) = R_{121}^2 = R_{212}^1. \tag{9.77}$$

For example, in the case of the hyperbolic plane, whose curvature was obtained in 9.6.3,

$$R_1^2 = \theta^1 \wedge \theta^2 \Rightarrow R_{121}^2 = -1;$$

thus the hyperbolic plane has constant sectional curvature -1. As the notation suggests, if M is a 2-dimensional submanifold of R^3 with the embedded metric, then the Gaussian curvature K defined in Chapter 4 is the same as the sectional curvature; see Exercise 21 below for the steps of the proof.

9.7 Exercises

18. Calculate the connection matrix of the Levi-Civita connection for R^3 with a metric of the type

$$\langle \cdot | \cdot \rangle = F\, dx \otimes dx + G\, dy \otimes dy + dz \otimes dz,$$

where $F = F(z)$ and $G = G(z)$ are smooth functions, and calculate the curvature $\Re$.

19. Calculate the Levi-Civita connection for a 2-dimensional Riemannian manifold with a
     local coordinate system (r, ϕ) and the metric

$$\langle \cdot | \cdot \rangle = dr \otimes dr + u^2 (d\phi \otimes d\phi),$$

where $u = u(r, \phi)$ is a nowhere-vanishing smooth function, and calculate the
curvature $\mathfrak{R}$.

20. We saw in Example 9.6.3 that if a 2-dimensional Riemannian manifold has a local coor-
     dinate system (r, ϕ) and the metric

$$\langle \cdot | \cdot \rangle = dr \otimes dr + h^2 (d\phi \otimes d\phi),$$

where $h = h(r)$ is a nowhere-vanishing smooth function, then the Levi-Civita
connection with respect to the orthonormal frame field $\theta^1 = dr$, $\theta^2 = h(r)\, d\phi$ is
given by (9.75). Derive the general formula for the sectional curvature of this manifold,
and check that for a portion of the 2-sphere (here $h(r) = \sin r$) we obtain $K = 1$.

21. Suppose M is a 2-dimensional submanifold of R^3 with the embedded metric. Prove that
     the Gaussian curvature K defined in Chapter 4 is the same as the sectional curvature
     defined in 9.6.4, using the following steps:

     (i) Using the orthonormal coframe $\{\theta^1, \theta^2\}$ induced from a parametrization Ψ, and
     using the first structure equation, show that the matrix of connection forms is

$$\varpi = \begin{bmatrix} 0 & \eta^0 \\ \eta^0 & 0 \end{bmatrix}, \tag{9.78}$$

where η^0 is the 1-form appearing in Chapter 4.

Hint: No computations with Ψ are needed!

(ii) Using Gauss's equation $d\eta^0 = -\eta^1 \wedge \eta^2 = -K\theta^1 \wedge \theta^2$ and formula (9.26) for $\mathfrak{R}$,
show that $R^2_{121} = K$.

22. Suppose M is a manifold, U is an open subset of R^n, and $f: U \times M \to (0, \infty)$ is a strictly
     positive function, integrable in the first variable and C^∞ in the second variable, with
     sufficient regularity such that all the integrals which arise in this problem exist and are
     finite, and that differentiation on M can be interchanged with integration over U. Also
     assume that f is sufficiently nondegenerate in p so that if $l_x(p) = \log f(x, p)$, then the
     following formula induces a Riemannian metric on M:

$$\langle W | Y \rangle_p = \int_U W l_x (p)\, Y l_x (p) f(x, p)\, dx, \quad W, Y \in \mathfrak{J}(M). \tag{9.79}$$

Abbreviate the last expression to $\int_U (Wl)\,(Yl)\, f dx$. Consider the formula, for real α,

$$\langle {}^{\alpha}\nabla_Y W|Z\rangle \;=\; \int\limits_{U} \{\, Y(Wl) + \frac{1-\alpha}{2}\,(Yl)\,(Wl)\,\}\,(Zl)\,f dx, \; Y, W, Z \in \mathfrak{I}\,(M). \qquad \textbf{(9. 80)}$$

(i) Prove that ${}^{\alpha}\nabla$ is a torsion-free connection on TM for all real α.

Hint: For any vector field Y, $Yf = f(Yl)$.

(ii) Prove that ${}^0\nabla$ is the Levi-Civita connection for this metric.

Hint: You only need to prove that the connection ${}^0\nabla$ is compatible with the metric.

Remark: This family of connections, whose curvature is related to information loss in statistics, is discussed in the article by Amari, in Amari et al. [1987], and in Murray and Rice [1993]. The function f is the likelihood associated with the observation x and the parameter p.

23. Let $\hat{\nabla}$ be the connection on the cotangent bundle corresponding, via (9. 35), to the Levi-Civita connection ∇ on the tangent bundle of a pseudo-Riemannian manifold M. Prove that $\hat{\nabla}$ satisfies the formula, for every smooth function h,

$$\hat{\nabla}_X dh \cdot Y \;=\; \frac{1}{2}\,\{\, \mathrm{grad}\,h\,(\langle X|Y\rangle) - \langle\,[\,\mathrm{grad}\,h, X]\,|Y\rangle - \langle X|\,[\,\mathrm{grad}\,h, Y]\,\rangle\}. \qquad \textbf{(9. 81)}$$

Hint: See Chapter 7 for the definition of grad, and use (9. 70).

24. (Continuation) A formula for the Laplacian for functions on a pseudo-Riemannian manifold M is given by

$$\Delta h \;=\; \sum_i \hat{\nabla}_{e_i} dh \cdot e_i, \qquad \textbf{(9. 82)}$$

where $\{e_1, \ldots, e_n\}$ is any orthonormal frame field, and $\hat{\nabla}$ is dual to the Levi-Civita connection.

(i) Show that this agrees with the usual formula $\sum \partial^2 h/\partial x_i^2$ in the case of the Euclidean connection.

(ii) Show that the formula

$$\Delta\,(uv) \;=\; u\Delta v + v\Delta u + 2\,\mathrm{grad}\,u \cdot \mathrm{grad}\,v \qquad \textbf{(9. 83)}$$

holds on any pseudo-Riemannian manifold M; grad was defined in Chapter 7.

(iii) Show that $\Delta h = \mathrm{div}_\mu\,(\mathrm{grad}\,h)$, where μ is the Riemannian volume form, and div was defined in the exercises to Chapter 8.

25. A connection Γ on the tangent bundle of an m-dimensional manifold M will be called **pyramidic** if, at each point p in M, there exists a coordinate system $\{x^1, \ldots, x^m\}$ in which the Christoffel symbols satisfy:

$$\Gamma^k_{ij} \;=\; 0 \text{ for } k \le \max\,\{i, j\}, \qquad \textbf{(9. 84)}$$

$$\Gamma_{ij}^{k} = \Gamma_{ij}^{k} (x^1, ..., x^k). \tag{9.85}$$

For example, in three dimensions, the only nonzero Christoffel symbols of a pyramidic connection are Γ_{11}^{2}, Γ_{11}^{3}, Γ_{12}^{3}, Γ_{21}^{3}, Γ_{22}^{3}, and Γ_{11}^{2} is a function of x^1 and x^2 only. Prove that, in three dimensions, a metric pyramidic connection is "flat," meaning that the curvature forms are identically zero.

Hint: Write down expressions for $D_i g$, where g is the 3×3 metric tensor, and use the fact that $D_i D_j g = D_j D_i g$, for $i, j \in \{1, 2, 3\}$.

9.8   History and Bibliography

The theory of connections has many versions, most of which are summarized and compared in Spivak [1979]. The parts we have studied here are mainly due to E. Christoffel (1829–1900), E. Cartan (1869–1951), T. Levi-Civita (1873–1941), J.L. Koszul (1921–), and S. Chern (1911–). For a much deeper treatment, consult Spivak [1979] and Kobayashi and Nomizu [1963]. For a stimulating and readable survey of some applications, see the article by Chern in Chern [1989].

10 Applications to Gauge Field Theory

The quantization of classical electromagnetic theory (see the appendix to Chapter 2) leads to the description of the photon, the quantum of electromagnetic radiation. One of the aims of quantum field theory is to explain all elementary particles as quanta of appropriate classical field theories. In the search for such field theories, non-Abelian gauge theories have become leading candidates since their introduction in 1954 by Yang and Mills. Only in the 1970s did it become generally recognized that the mathematical theory of connections on vector bundles is the proper context for gauge theory.

In this chapter we shall start out by giving a geometric description of the Yang–Mills Lagrangian, which must be minimized in order to construct the desired classical field. The minimization is actually carried out here using a special class of connections which are called "self-dual." We reformulate electromagnetism as an Abelian gauge field theory, and give a detailed account of a non-Abelian $SU(2)$ gauge theory over S^4, including the formula for "instantons," the connections which minimize the Yang–Mills Lagrangian. Much of this chapter is based on Lawson [1985].

10.1   The Role of Connections in Field Theory

Imagine a structured particle, that is, a particle located at some point p in a four-dimensional manifold M ("space-time"), and with an internal structure, or set of states ("spin", etc.) labeled by elements of a complex Lie group G (e.g., $SU(2)$). In practice we cannot observe this internal structure but only the action of G on some complex vector space V. Thus the total space of all states of such a particle is represented by a vector bundle E over M with fibers isomorphic to V.

An external field will cause the internal state of the particle to change as the particle flows along a vector field X in space-time. The internal state of the particle at position p in M is reflected by the value of a section σ of E at the point p. Therefore the external field is characterized by a method of "differentiating" σ along the flow of X, denoted $\nabla_X \sigma$, which to be geometrically meaningful must be given by a connection ∇ on the vector bundle, as discussed in Chapter 9. In any local frame field for E, the connection is specified by a collection of connection forms, here called the **gauge potential**.

The curvature of this connection tells us the second covariant exterior derivative of the section σ, and therefore can be said to represent the **field strength** of the external field, just as in mechanics the size of the acceleration (the second derivative of position) measures the strength of the applied force.

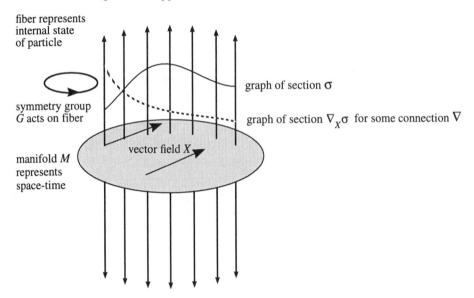

Figure 10. 1   Vector bundle formalism for field theories

The importance of the geometric formulation of the problem is that physicists are interested in properties which are invariant under:

- changes of coordinate system, that is, changes in local trivialization of the vector bundle, which are represented in the fibers by transition functions (see Chapter 6); in many cases the vector bundle is not trivial, and so more than one local trivialization must be considered.

- **gauge transformations**, in which the fibers are transformed by an element of G that varies smoothly with the fiber.

To ensure invariance, geometric formalism is preferable both for definitions and for calculations.

In the past physicists debated whether there is a physical interpretation of the gauge potential (i.e., the connection), or whether it is merely an unobservable mathematical artifact. However, an experiment proposed in 1959 by Y. Aharonov and D. Bohm, and performed in 1960 by Chambers, proved that the electromagnetic potential does play a role, even in the absence of a field. A coherent beam of electrons is reflected around a closed path encircling a solenoid, which is considered as a perfectly insulated tube. Although the field outside the tube is zero, the phase shift caused by self-interaction of the beam is found to vary with the intensity of current in the solenoid.

The following table may be a useful reminder:

Differential Geometry	Physics
connection forms	gauge potential
curvature forms	field strength

Table 10.1   Comparison of mathematics and physics terminology

10.2   Geometric Formulation of Gauge Field Theory

The technical preliminaries involve some more material on complex Lie groups, and some constructions of metrics on vector bundles.

10.2.1   Constructions with Complex Lie Groups

Suppose G is a Lie subgroup of the complex general linear group $GL_m(C)$ discussed in Chapter 3; in other words G is a collection of invertible complex $m \times m$ matrices, closed under matrix multiplication, and with a smooth manifold structure. The tangent space at the identity of G, under the operation of bracket of vector fields discussed in Chapter 2, is known as the **Lie algebra** of G, which we will denote by $L(G)$. In this chapter, we shall only be performing calculations with $U(1)$ and $SU(2)$, already mentioned in Chapter 3, whose Lie algebras are described below:

10.2.1.1   The Circle Group $U(1)$

The unitary group $U(1)$ is simply the set of complex numbers z with modulus 1, that is, the circle in the complex plane, under ordinary multiplication of complex numbers. In electromagnetic theory, this group models the polarization of a photon. As a manifold, it may be identified with the circle S^1, and the tangent space at 1, which is of course one-dimensional, is obtained by differentiating the curves $t \to e^{iat}$ at $t = 0$ for all real numbers a, giving as the Lie algebra the complex line iR (here i is the complex number $\sqrt{-1}$); thus $L(U(1)) = iR$.

10.2.1.2   The Special Unitary Group $SU(2)$

This is a 3-dimensional manifold (over the reals) consisting of the 2×2 complex unitary matrices of determinant 1. Its importance in physics is that it is used to model the "isotopic spin" of particles related to the **strong nuclear force**. The reader may check using the hint in Exercise 6 in Section 10.5 that the Lie algebra $L(SU(2))$ consists of the span of the **Pauli matrices**[1]:

$$\begin{bmatrix} i & 0 \\ 0 & -i \end{bmatrix}, \begin{bmatrix} 0 & 1 \\ -1 & 0 \end{bmatrix}, \begin{bmatrix} 0 & i \\ i & 0 \end{bmatrix}. \tag{10.1}$$

In Section 10.3 we shall see how the algebra of quaternions provides concise and efficient machinery for performing calculations involving $SU(2)$.

10.2.2   *G*-Bundles and *G*-Connections

Suppose G is represented as a set of $m \times m$ complex matrices, acting on a complex m-dimensional vector space V; this means that, for every $g \in G$, there is a linear mapping $v \to gv$ from V to V, with the property that $g_1(g_2 v) = (g_1 g_2) v$.

A complex vector bundle $\pi: E \to M$, whose fibers are isomorphic to V, is called a **G-vector bundle** if its transition functions (see Chapter 6) all belong to G. A Koszul connection ∇ on a G-vector bundle $\pi: E \to M$ is called a **G-connection** if the associated connection matrix (with respect to any local frame field), when viewed as a 1-form on M with values in $C^{m \times m}$, takes values in $L(G)$. For example, the connection matrix of an $SU(m)$-connection must satisfy $\omega_i^k + \overline{\omega}_k^i = 0$ (here the bar denotes complex conjugation) and $\omega_1^1 + \ldots + \omega_m^m = 0$, since the Lie algebra of $SU(m)$ consists of the skew-Hermitian matrices whose trace is zero.

10.2.3   Action of *G* on the Fibers, Preserving a Metric

Given a local trivialization Φ_α for the vector bundle at $r \in M$, G can be made to act on the fiber $E_r \cong V$ in the obvious way, namely,

$$g\Phi_\alpha^{-1}(r, v) = \Phi_\alpha^{-1}(r, gv). \tag{10.2}$$

Of course, this action depends on the trivialization; for example, in another local trivialization Φ_β, with transition function $g_{\beta\alpha}(r)$ as in Chapter 6, (10.2) becomes

$$g\Phi_\beta^{-1}(r, w) = \Phi_\beta^{-1}(r, g_{\beta\alpha}(r) gg_{\alpha\beta}(r) w). \tag{10.3}$$

Nevertheless since $\{g_{\beta\alpha}(r) gg_{\alpha\beta}(r) : g \in G\} = G$, the resulting group G_r of linear transformations of E_r does not depend on the trivialization. We are interested here in the case where $\pi: E \to M$ has a metric $r \to \langle \cdot | \cdot \rangle_r^E$ such that, in every fiber, G_r is identical with the set of **isometries** of E_r, that is, the nonsingular linear transformations that preserve the metric; in other words G is the set of $g \in L(V \to V)$ such that, for any r,

---

[1] These are $\sqrt{-1}$ times the Pauli matrices as often given in the physics literature.

$$\langle g\xi | g\xi \rangle_r^E = \langle \xi | \xi \rangle_r^E, \ \forall \xi \in E_r, \tag{10.4}$$

where of course the meaning of "$g\xi$" depends on the trivialization.

10.2.4 The Gauge Group

Suppose $\pi: E \to M$ is a G-vector bundle with a metric $r \to \langle \cdot | \cdot \rangle_r^E$. The **gauge group** G_E consists of the set of sections of Hom (E,E) which are vector bundle isomorphisms (see Chapter 6), preserving the metric; the latter means that, for $\sigma \in \Gamma E$ and $\phi \in G_E$,

$$\langle \phi \sigma | \phi \sigma \rangle^E = \langle \sigma | \sigma \rangle^E. \tag{10.5}$$

In view of (10.4), ϕ locally takes the form of a smooth mapping from an open set in M to G. $\phi \in G_E$ takes a G-connection ∇ to a G-connection ∇^ϕ, defined by

$$\nabla_X^\phi \sigma = \phi (\nabla_X (\phi^{-1} \sigma)). \tag{10.6}$$

In Exercise 1 in Section 10.5 the reader may check that if ∇ is a G-connection, then so is ∇^ϕ.

In terms of the covariant exterior derivatives d^E and $d^{E,\phi}$ for ∇ and ∇^ϕ, respectively,

$$d^{E,\phi}(d^{E,\phi}\sigma) = d^{E,\phi}(\phi d^E(\phi^{-1}\sigma)) = \phi d^E(d^E(\phi^{-1}\sigma))$$

$$\Rightarrow \mathfrak{R}^{\nabla^\phi} \wedge \sigma = \phi(\mathfrak{R}^\nabla \wedge \phi^{-1}\sigma). \tag{10.7}$$

The elements of the gauge group are called **gauge transformations**; formula (10.7) describes how the curvature of a G-connection changes under a gauge transformation.

10.2.5 Metrics on Various Vector Bundles

Suppose $\pi: E \to M$ is a G-vector bundle with a metric $r \to \langle \cdot | \cdot \rangle_r^E$, over an n-dimensional oriented Riemannian manifold $(M, \langle \cdot | \cdot \rangle)$.

10.2.5.1 A Metric on the Vector Bundle Hom $(\Lambda^q TM, E)$
A metric on the vector bundle Hom (TM, E), whose sections are the E-valued 1-forms, is given by the formula

$$\langle \mu | \lambda \rangle_r^{(1)} = \sum_{1 \le \alpha \le n} \langle \mu \cdot e_\alpha | \lambda \cdot e_\alpha \rangle_r^E, \tag{10.8}$$

for μ and λ in $\Omega^1(M;E)$, where $\{e_1, ..., e_n\}$ is an orthonormal frame field for TM in a neighborhood of r. The fact that this does not depend on the choice of $\{e_1, ..., e_n\}$ is a special case of the exercise in Chapter 7 on constructing a metric on Hom (E,E').

Next we seek a metric on the vector bundles Hom $(\Lambda^q TM, E)$ for $q = 2, \ldots, n$. Suppose that μ and λ are in $\Omega^q (M; E)$ for some $2 \le q \le n$; for $r \in M$, $\mu(r)$ and $\lambda(r)$ take values in $L(\Lambda^q T_r M \to E_r)$, which by a result in Chapter 1 can be viewed as the qth exterior power of $L(T_r M \to E_r)$. We learned in Chapter 1 how to extend an inner product on a vector space, in this case $L(T_r M \to E_r)$, to one on all of its exterior powers. Applying this construction here allows us to define a metric $\langle \mu | \lambda \rangle_r^{(q)}$ for all μ and λ in $\Omega^q (M; E)$. To be specific: If $\{s_i\}$ constitutes an orthonormal frame field for $\pi: E \to M$ in a neighborhood of r, and if $\{\varepsilon^1, \ldots, \varepsilon^n\}$ is an orthonormal coframe field for the cotangent bundle in a neighborhood of r, then the set of E-valued q-forms

$$\{s_i \wedge \omega^\alpha : i \ge 1, \alpha \in I_q\}, \tag{10.9}$$

where $\omega^\alpha = \varepsilon^{\alpha(1)} \wedge \ldots \wedge \varepsilon^{\alpha(q)}$, and I_q is the set of ascending multi-indices α with $1 \le \alpha(1) < \ldots < \alpha(q) \le n$, constitutes an orthonormal frame field for the vector bundle Hom $(\Lambda^q TM, E)$ under this new metric.

Finally a "global" inner product of E-valued q-forms may be defined by the formula

$$(\lambda, \mu) = \int_M \langle \lambda | \mu \rangle^{(q)} \rho, \tag{10.10}$$

where ρ is the canonical volume form, at least when λ and μ are zero outside a compact set.

10.2.6   The Norm of the Curvature

In the exercises to Chapter 7, we saw that a metric on Hom $(E, E) \to M$ is given by

$$\langle F | F' \rangle_r^{\mathrm{Hom}} = \sum_i \langle F(s_i) | F'(s_i) \rangle_r^E, \tag{10.11}$$

where $\{s_i\}$ is any orthonormal frame field for $\pi: E \to M$ at r; here F and F' are any two sections of Hom $(E, E) \to M$ (this metric does not depend on the choice of $\{s_i\}$).

With the construction given in 10.2.5.1, a metric on Hom $(\Lambda^q TM; \mathrm{Hom}(E, E))$, denoted $r \to \langle \cdot | \cdot \rangle_r^{\mathrm{Hom}, q}$, may be defined for all $1 \le q \le n$. For example, the curvature $\mathfrak{R} = \mathfrak{R}^\nabla$ of a connection ∇ is a 2-form on M with values in Hom (E, E), and this construction allows us to define

$$\| \mathfrak{R}^\nabla \|_r^2 = \langle \mathfrak{R}^\nabla | \mathfrak{R}^\nabla \rangle_r^{\mathrm{Hom}, 2} \tag{10.12}$$

$$= \sum_{1 \le \alpha < \beta \le n} \sum_i \langle \mathfrak{R}^\nabla \wedge s_i \cdot e_\alpha \wedge e_\beta | \mathfrak{R}^\nabla \wedge s_i \cdot e_\alpha \wedge e_\beta \rangle_r^E.$$

The reader has the opportunity to check the last identity in Exercise 2 in Section 10.5.

10.2.6.1 The Norm of the Curvature Is Gauge Invariant

For any gauge transformation ϕ (see 10.2.4) and any G-connection ∇,

$$\left\| \Re^{\nabla^\phi} \right\|_r^2 = \left\| \Re^\nabla \right\|_r^2. \tag{10.13}$$

Proof: Since ϕ preserves the metric, $\{\phi^{-1} s_i\}$ is an orthonormal frame field for $\pi: E \to M$, and for each α, β

$$\sum_i \left\| \phi \left(\Re^\nabla \wedge \phi^{-1} s_i \right) \cdot e_\alpha \wedge e_\beta \right\|^2 = \sum_i \left\| \Re^\nabla \wedge s_i \cdot e_\alpha \wedge e_\beta \right\|^2$$

However, $\Re^{\nabla^\phi} \wedge s_i = \phi \left(\Re^\nabla \wedge \phi^{-1} s_i \right)$ by (10.7). ¤

10.2.7 The Yang–Mills Lagrangian

We want to find out which connections, and hence which field strengths, are consistent with the Principle of Least Action. For this we will assume that, in the terminology of Chapter 7:

- M is a compact, oriented, Riemannian manifold (for example, $M = S^4$ with some metric);

- $\pi: E \to M$ is a G-vector bundle with a metric $r \to \langle \cdot | \cdot \rangle_r^E$.

The **Yang–Mills Lagrangian**, or **Yang–Mills action**, is the mapping from the set of G-connections on $\pi: E \to M$ to R^+ given by

$$L(\nabla) = \frac{1}{2} \int_M \left\| \Re^\nabla \right\|^2 \rho, \tag{10.14}$$

where ρ is the canonical volume form on the compact, Riemannian manifold M, and the norm under the integral is as in (10.12). It is important to note that, by 10.2.6.1, $L(\nabla^\phi) = L(\nabla)$ for every gauge transformation ϕ (see Section 10.2.4); in other words, the Lagrangian is gauge invariant. Do not be put off by the abstractness of this integral; more concrete versions will follow.

We are interested in finding connections that minimize the Yang–Mills action. To this end, a G-connection ∇ will be called a **Yang–Mills connection** if it is a stationary point of $L(\nabla)$. In that case its curvature $\Re^\nabla$ will be called a **Yang–Mills field**.

10.2.8 Example: Maxwell's Equations as a Gauge Theory

Let us to try to rephrase Maxwell's equations (see the appendix to Chapter 2) in terms of gauge theory. Recall that we encoded both the electric and the magnetic field in a 2-form $\eta = \varpi_E \wedge dt + \phi_B$ on $M = R^4$ with the Lorentz metric (so in this case, M is pseudo-Riemannian rather than Riemannian), and showed that solving Maxwell's equations amounts to finding a 1-form α, called the electromagnetic potential, such that $d\alpha = \eta$, $*d(*\eta) = 4\pi\gamma$, where γ is the 4-current; of course if α solves the equation, then so does $\alpha + db$, for any smooth function b.

Suppose $\pi: E \to M$ is a complex line bundle, with transition functions in the group $U(1)$ (see Section 10.2.1.1); thus $\pi: E \to M$ is a $U(1)$-bundle. Take the metric on $\pi: E \to M$ defined by

$$\langle \sigma | \tau \rangle_p^E = \mathrm{Re}\,\{\sigma(p)\,\bar{\tau}(p)\,\}, \tag{10.15}$$

referring to the real part of a complex product (the right side does not depend on the trivialization). The complex Lie group that preserves this metric is $U(1)$; thus a gauge transformation consists locally of multiplication in the fibers by a function of the form $\phi(p) = \exp(-ib(p)) \in U(1)$, for some smooth real-valued function b on an open set in M.

Take a local trivialization Φ_U for the complex line bundle. Since every section is of the form $\sigma = sf$, for some complex-valued function f, with respect to the frame field $s(p) = \Phi_U^{-1}(p, i)$, we may define a connection ∇ by

$$d^E(sf) = s \wedge i\alpha f + s \wedge df. \tag{10.16}$$

The reason for using $i\alpha$ instead of α is that we want the "gauge potential" (i.e., connection form) to take values in the Lie algebra iR of $U(1)$; in other words, ∇ is a $U(1)$-connection. Notice that under the gauge transformation induced by $\phi = e^{-ib}$, (10.6) gives

$$d^{E,\phi}(sf) = e^{-ib}(d^E(e^{ib}sf)) = s \wedge (i\alpha f + ifdb) + s \wedge df. \tag{10.17}$$

Thus the new gauge potential is $i\alpha^\phi = i\alpha + idb$.

The formula for the curvature form ("field strength") is $F = -\alpha \wedge \alpha + id\alpha$ from Chapter 9, but the $\alpha \wedge \alpha$ term is zero because α is an ordinary 1-form on M, and we are left with a field strength of

$$i\eta = id\alpha = i(\varpi_E \wedge dt + \phi_B). \tag{10.18}$$

The Yang–Mills Lagrangian is now simply the action of the electromagnetic field, namely (in the absence of a current),

$$L(\nabla) = -\frac{1}{2}\int_M \|d\alpha\|^2 \rho, \tag{10.19}$$

where $\rho = dx \wedge dy \wedge dz \wedge cdt$ (the minus sign can be cancelled out by suitable choice of volume form). If $\alpha_t = \alpha + t\beta$ is a family of potentials, corresponding to connections ∇_t, then

$$\| d\alpha_t \|^2 = \| d\alpha \|^2 + 2t\langle d\alpha | d\beta \rangle^{\text{Hom}, 2} + t^2 \| d\beta \|^2$$

$$\Rightarrow \frac{d}{dt}L(\nabla_t) \Big|_{t=0} = \int_M \langle d\alpha | d\beta \rangle^{\text{Hom}, 2} \rho = \int_M \langle * d(*d\alpha) | \beta \rangle^{\text{Hom}, 1} \rho.$$

Here the inner product inside the integral turns out to be the Lorentz inner product of 1-forms, and the last identity, which the reader is invited to check in Exercise 4 in Section 10.5, assumes that β is nonzero only inside a compact region. If ∇_0 is a Yang–Mills connection, then the last integral must be zero for every possible variation β; therefore *in vacuo* its field strength η satisfies

$$* d(* \eta) = 0, \tag{10.20}$$

which is precisely Maxwell's equations in the absence of a current. To summarize: Minimizing the Yang–Mills Lagrangian leads to a set of differential equations for the field, that is, for the curvature of the connection.

10.3 Special Unitary Groups and Quaternions

The purpose of this section is to introduce the formalism of quaternions, in order to carry out calculations on $SU(2)$-bundles in a streamlined fashion, without manipulating a lot of 2×2 complex matrices. The following account follows Atiyah [1979].

10.3.1 Quaternions and the Group $Sp(1)$

Just as the complex numbers C are formed from the real numbers R by adjoining a symbol i with $i^2 = -1$, so the **quaternions** H are formed from R by adjoining three symbols i, j, k satisfying the identities:

$$i^2 = j^2 = k^2 = -1, \tag{10.21}$$

$$ij = -ji = k, \; jk = -kj = i, \; ki = -ik = j. \tag{10.22}$$

Thus a general quaternion is of the form

$$z = z_1 + z_2 i + z_3 j + z_4 k, \tag{10.23}$$

where z_1, z_2, z_3, z_4 are real numbers. Addition of quaternions is accomplished by simply adding up the respective coefficients of $1, i, j, k$. Multiplication, which is associative (see Exercise 8 in Section 10.5) but not commutative, requires the use of (10. 21) and (10. 22). The conjugate quaternion $\bar{z}$ is defined by

$$\bar{z} = z_1 - z_2 i - z_3 j - z_4 k, \qquad (10.24)$$

and by virtue of (10. 22), conjugation satisfies $\overline{yz} = (\bar{z})(\bar{y})$. The identities (10. 21) and (10. 22) imply that

$$z\bar{z} = \bar{z}z = \sum_{1 \le i \le 4} z_i^2, \qquad (10.25)$$

and this quantity, denoted $|z|^2$, is zero only for $z = 0$. It makes sense to define the inverse of a nonzero quaternion by

$$z^{-1} = \bar{z}/|z|^2. \qquad (10.26)$$

If two quaternions y and z have unit norm, that is, $|y| = |z| = 1$, then so does their product, because

$$|yz|^2 = (yz)(\overline{yz}) = yz\bar{z}\bar{y} = y|z|^2\bar{y} = y\bar{y} = 1.$$

Thus the set of quaternions of unit norms forms a multiplicative group, denoted $Sp(1)$, in which $z^{-1} = \bar{z}$ by (10. 26). Moreover $Sp(1)$ may be identified with the differentiable manifold S^3 by virtue of (10. 25).

10.3.2 Identification of $Sp(1)$ and $SU(2)$

The groups $Sp(1)$ and $SU(2)$ are isomorphic; moreover the Lie algebra $L(SU(2))$ may be identified with the purely imaginary quaternions

$$\mathrm{Im}(H) = \{z_1 + z_2 i + z_3 j + z_4 k : z_1 = 0\}. \qquad (10.27)$$

Proof: Every quaternion has a unique expression of the form:

$$z = z_1 + z_2 i + z_3 j + z_4 k = u_1 + u_2 j,$$

where $u_1 = z_1 + z_2 i$, and $u_2 = z_3 + z_4 i$. This identifies each $z \in H$ with the pair of complex numbers $(u_1, u_2) \in C^2$. Consider the quaternion multiplication $z \to zv$ where $v = v_1 + v_2 j$, with $(v_1, v_2) \in C^2$:

$$zv = (u_1 + u_2 j)(v_1 + v_2 j) = u_1 v_1 - u_2 \bar{v}_2 + (u_1 v_2 + u_2 \bar{v}_1) j.$$

In other words, the multiplication $z \to zv$ amounts to multiplying the vector (u_1, u_2) on the right by the 2×2 complex matrix

$$\begin{bmatrix} v_1 & v_2 \\ -\bar{v}_2 & \bar{v}_1 \end{bmatrix}. \tag{10.28}$$

For $v \in Sp(1)$, $v_1\bar{v}_1 + v_2\bar{v}_2 = 1$, so this matrix has determinant 1; moreover matrices of this form make up the whole of $SU(2)$ by Exercise 5 in Section 10.5. Thus the map from H to $C^{2 \times 2}$ which sends (v_1, v_2) to the matrix (10.28) is one-to-one and onto from $Sp(1)$ to $SU(2)$, and preserves the group multiplication.

In particular, when v takes the values i, j, and k, respectively, then the corresponding forms of the matrix (10.28) are precisely the Pauli matrices (10.1). The reader may verify that the three Pauli matrices behave under multiplication in exactly the same way as the quaternions i, j, and k do in (10.21) and (10.22); this shows that we may identify the Lie algebra of $SU(2)$ with the purely imaginary quaternions.   ¤

To summarize: We are now in the position where calculations involving $SU(2)$, and its Lie algebra (spanned by the Pauli matrices), can be mimicked by calculations involving unit quaternions, and the purely imaginary quaternions, respectively.

10.4   Quaternion Line Bundles

Recall from Chapter 6 that a complex line bundle over a differentiable manifold M means a complex vector bundle whose fibers are copies of the complex line. Likewise we may define $\pi: E \to M$ to be a **quaternion line bundle** if the fibers are copies of H, the algebra of quaternions defined in Section 10.3.1. In checking the differentiability conditions in the definition of a vector bundle, as in Chapter 6, we may treat the fibers simply as 4-dimensional real vector spaces. However, the transition functions $\{g_{UU'}\}$ take values in the set of nonzero quaternions, acting on the right by quaternion multiplication; in other words, for $p \in M$,

$$g_{UU'}(p)(v) = vg_{UU'}(p), v \in H. \tag{10.29}$$

Evidently each transition function $g_{UU'}$ is H-linear on the left, in the sense that

$$g_{UU'}(p)(zv) = zvg_{UU'}(p), z \in H. \tag{10.30}$$

Note that the group $Sp(1)$ of unit quaternions, defined in Section 10.3.1, acts on H by simple quaternion multiplication. We shall call a quaternion line bundle an $Sp(1)$-bundle if its transition functions are all unit quaternions.

10.4.1   Example: a Quaternion Line Bundle over the 4-Sphere

This comes from Lawson [1985]. Let $M = S^4$ (we really have in mind a four-dimensional space-time which has been "compactified" by adding a "point at infinity," which is more convenient to work with than R^4 because integrals will now be over a compact manifold). Following along the lines of one of the exercises to Chapter 5, we see that it is possible to pick an atlas consisting of the pair of charts (U, φ) and (V, ψ), where

$$U = S^4 - \{ (1, 0, ..., 0) \} , \varphi (x_0, ..., x_4) = \frac{(x_1, ..., x_4)}{1 - x_0} ;$$

$$V = S^4 - \{ (-1, 0, ..., 0) \} , \psi (x_0, ..., x_4) = \frac{(x_1, -x_2, -x_3, -x_4)}{1 + x_0} .$$

The change-of-chart map $\varphi \bullet \psi^{-1} : R^4 - \{0\} \to R^4 - \{0\}$ has the property that if we identify $z \in R^4$ with $z^1 + z^2 i + z^3 j + z^4 k \in H$ as in (10. 23), then

$$\varphi \bullet \psi^{-1} (z^1, z^2, z^3, z^4) = \frac{(z^1, -z^2, -z^3, -z^4)}{|z|^2} = \frac{\bar{z}}{|z|^2}, \tag{10.31}$$

where $\bar{z}$ is the conjugate quaternion as in (10. 24).

We may now construct a quaternion line bundle $\pi : E \to M$ "of instanton number 1" (this is a topological classification defined in Section 10.7.3) using the vector bundle construction theorem given in Chapter 6. To apply this theorem, it suffices to specify that the fiber of this vector bundle is H, the open cover of S^4 is $\{ U, V \}$, and the transition function is

$$g_{UV}(\psi^{-1}(z)) v = vz/|z|, \tag{10.32}$$

where $z \in \psi(V)$ and v are both regarded as elements of H. In the other coordinate system $\{y^1, y^2, y^3, y^4\}$ induced by φ, where $y = \bar{z}/|z|^2 = z^{-1}$ by (10. 31), (10. 32) becomes

$$g_{UV}(\varphi^{-1}(y)) v = v\bar{y}/|y|. \tag{10.33}$$

Since $z/|z|$ is a unit quaternion, this is an $Sp(1)$-bundle.

10.4.2   Connections on Quaternion Line Bundles

Suppose $\pi : E \to M$ is an $Sp(1)$-bundle with the metric $r \to \langle \cdot | \cdot \rangle_r^E$ induced by quaternion multiplication in the fibers, namely,

$$\langle \sigma | \tau \rangle_p^E = \text{Re} \{ \sigma(p) \bar{\tau}(p) \} \tag{10.34}$$

(the right side does not depend on the trivialization). The group of transformations that preserve the metric in each fiber is the group $Sp(1)$ of unit quaternions (which would correspond to $SU(2)$ if we viewed the bundle instead as having 2-dimensional complex vector spaces as fibers). The gauge group consists of transformations which take a section σ to $\phi\sigma$, where ϕ is locally a smooth mapping from M into the unit quaternions, and $(\phi\sigma)(p)$ is simply the quaternion product $\phi(p)\sigma(p)$ computed in the fiber over p, in some trivialization.

We are interested in $Sp(1)$-**connections** on $\pi: E \to M$ in the sense of Section 10.2.2; in other words, the connection form (see below) is a 1-form on M taking values in $\mathrm{Im}(H)$, the purely imaginary quaternions defined in (10.27), since $\mathrm{Im}(H)$ corresponds to the Lie algebra of $SU(2)$ by 10.3.2. Moreover we require that the connection is linear with respect to right multiplication by constant quaternions, meaning that the Leibniz rule for the covariant exterior derivative, namely,

$$d^E(\mu \wedge \omega) = d^E\mu \wedge \omega + (-1)^{\deg \mu}\mu \wedge d\omega, \tag{10.35}$$

now holds for every E-valued p-form μ, and every differential form ω on M with values in H (replacing R by H is the only extension of what was presented in Chapter 9).

Consider the case where μ is a nonvanishing section s of E, that is, an element of $\Omega^0(M;E)$, which by itself constitutes a local frame field for E. If ω is an H-valued 0-form, that is, a function $f = f_1 + if_2 + jf_3 + kf_4$ from M to H, then we obtain

$$d^E(s \wedge f) = d^E(sf) = (d^Es)f + s \wedge df. \tag{10.36}$$

Since the fibers of the bundle are only "1-dimensional" in the quaternion sense (although 4-dimensional in the real sense), the E-valued 1-form d^Es is expressible as $s \wedge A$ for some $\mathrm{Im}(H)$-valued 1-form A on M, which corresponds to the "connection matrix" described in Chapter 9, and which will henceforward be called the **gauge potential** with respect to the local frame field $\{s\}$. Thus the formula for differentiating a general section $\sigma = sf$ of a quaternion line bundle is

$$d^E(sf) = s \wedge Af + s \wedge df. \tag{10.37}$$

Note that if the gauge potential A is expressed in local coordinates, then the product Af is the H-valued 1-form:

$$(A_1dx^1 + A_2dx^2 + A_3dx^3 + A_4dx^4)\,(f_1 + if_2 + jf_3 + kf_4). \tag{10.38}$$

10.4.3   Guidelines for Exterior Calculus with *H*-valued Forms

The iteration rule and Leibniz rule for exterior differentiation of *H*-valued forms are the same as for real-valued forms. However, since quaternion multiplication is not commutative, we have

$$\lambda \wedge \mu \neq (-1)^{pq} (\mu \wedge \lambda)$$

in general, for an *H*-valued *p*-form λ, and an *H*-valued *q*-form μ. For example, possibly $Af \neq fA$ in (10. 38). Ways to skirt around this difficulty will be apparent in Exercise 15 in Section 10.5.

10.4.4   Curvature as "Field Strength"

Given an $Sp(1)$-connection on a quaternion line bundle $\pi: E \to M$, the curvature $\mathfrak{R}$, according to Chapter 9, is an *H*-valued 2-form characterized by

$$d^E(d^E \sigma) = \mathfrak{R} \wedge \sigma, \ \sigma \in \Omega^0(M; E). \tag{10. 39}$$

With respect to the local frame field $\{s\}$, the curvature form is the *H*-valued 2-form *F* which satisfies:

$$d^E(d^E s) = F \wedge s. \tag{10. 40}$$

Actually it follows from (10. 42) below that *F* is an Im(*H*)-valued 2-form. In terms of a local coordinate system $\{x^1, ..., x^n\}$ for *M*,

$$F = \sum_{1 \leq \alpha < \beta \leq n} F_{\alpha\beta} (dx^\alpha \wedge dx^\beta). \tag{10. 41}$$

F, or the collection of Im(*H*)-valued functions $\{F_{\alpha\beta}\}$, is referred to as the **field strength** associated with the gauge potential $A_1 dx^1 + ... + A_n dx^n$. The reasoning behind this physical interpretation of curvature was sketched in Section 10.1.

The reader may check in Exercise 10 in Section 10.5 that the formula in Chapter 9 for the curvature forms in terms of the connection forms is now expressible as

$$F = A \wedge A + dA. \tag{10. 42}$$

The fact that *A* is Im(*H*)-valued means that $A \wedge A$ is not zero in general, unlike the case of real-valued 1-forms; see Exercise 18 in Section 10.9 for an example.

10.4.5 Example: A Quaternion Line Bundle over the 4-Sphere, Continued

The notation continues from Section 10.4.1. The local trivialization
$\Phi_U : \pi^{-1}(U) \to U \times H$ induces a local frame field $\{s\}$ for E over U, where
$s = \Phi_U^{-1}(p, 1)$. Thus an arbitrary section has a local expression

$$\sigma = s\,(f_1 + if_2 + jf_3 + kf_4), \tag{10.43}$$

where the $\{f_\alpha\}$ are real-valued. We would like to compare the local expressions for the
gauge potential of an $Sp\,(1)$-connection in two different trivializations of the bundle.
This will also tell us how A behaves under "gauge transformation." Suppose that over U
we have

$$d^E(\sigma) = d^E(sf) = s \wedge Af + s \wedge df, \tag{10.44}$$

while over V, if $t = \Phi_V^{-1}(p, 1)$, we have

$$d^E(\sigma) = d^E(tg) = t \wedge Bg + t \wedge dg. \tag{10.45}$$

10.4.5.1 Transformation Rule for Gauge Potentials
The formula for the gauge potential in the other trivialization is

$$B = \bar{u}Au + \bar{u}du, \tag{10.46}$$

where $u = z/|z| = \bar{y}/|y|$, and $z = \psi(p)$, $y = \varphi(p)$.

10.4.5.2 Remarks

- The formula (10.46) is a quaternion version of the identity $\hat{\omega} = g^{-1}dg + g^{-1}\omega g$,
  derived in the exercises to Chapter 9, relating two sets of connection forms under a
  change of local trivialization, where g denotes the transition function.

- B is an Im(H)-valued 1-form, just as A is; check this in Exercise 11 in Section 10.5.

- In terms of the respective coordinate systems on U and V, we could write the gauge
  potentials as

$$A = A_1 dy^1 + A_2 dy^2 + A_3 dy^3 + A_4 dy^4, \ B = B_1 dz^1 + B_2 dz^2 + B_3 dz^3 + B_4 dz^4.$$

Here each A_α is an Im(H)-valued function of $\{y^1, y^2, y^3, y^4\}$, etc. In Exercise 14
below, the reader can practice using the formula (10.46) in going from y- to
z-coordinates, where $z = z^1 + z^2 i + z^3 j + z^4 k = y^{-1} = (y^1 - y^2 i - y^3 j - y^4 k)/|y|^2$.

Proof: By expressing $\sigma(p)$ in both trivializations, and using the definition of the
transition function, we see that

$$\Phi_U(\sigma(p)) = \Phi_U \bullet \Phi_V^{-1}(\Phi_V(\sigma(p)))$$

$$\Rightarrow (p, f(p)) = (p, g(p)u),$$

where $u = z/|z| = \bar{y}/|y|$, by (10. 32). Thus $\sigma = tg = sf = sgu$, and taking $g \equiv 1$ shows that $t = su$. Since $s \wedge Af + s \wedge df = t \wedge Bg + t \wedge dg$ by (10. 44) and (10. 45), we have on taking $g \equiv 1$, $f = u$, that

$$s \wedge Au + s \wedge du = s \wedge uB$$

and therefore $uB = Au + du$; now the result follows on premultiplication by $\bar{u}$, the element of $Sp(1)$ inverse to u.   ¤

10.4.6   The Yang–Mills Lagrangian on a Quaternion Line Bundle

As the reader has guessed by now from our propaganda for quaternions, the case where $G = SU(2)$, acting on its Lie algebra, is going to be handled using a quaternion line bundle E, where the group is $Sp(1) \cong SU(2)$. Forgetting about compactness for a moment, take M to be R^4 with a metric

$$\langle \cdot | \cdot \rangle = \sum_{1 \le \alpha, \beta \le 4} g_{\alpha\beta} dx^\alpha \otimes dx^\beta. \tag{10. 47}$$

Let us take a local frame field $\{s\}$ for E such that $|s(p)|^2 = 1$; now the connection and its curvature are represented by equations (10. 37) and (10. 40). A special case of Exercise 2 in Section 10.5 shows that, in terms of the field strength defined in (10. 41), and the coordinates $\{x^1, x^2, x^3, x^4\}$,

$$\|\mathfrak{R}^\nabla\|_p^2 = \sum_{1 \le \alpha < \beta \le 4} F^{\alpha\beta} \bar{F}_{\alpha\beta}; \tag{10. 48}$$

here the $\{F^{\alpha\beta}\}$ are the Im(H)-valued functions given by

$$F^{\alpha\beta} = \sum_{\gamma, \delta} g^{\alpha\gamma} F_{\gamma\delta} g^{\delta\beta}, \tag{10. 49}$$

where $(g^{\gamma\delta}) = (g_{\alpha\beta})^{-1}$. Consequently

$$L(\nabla) = \frac{1}{2} \int_M \left(\sum_{1 \le \alpha < \beta \le 4} F^{\alpha\beta} \bar{F}_{\alpha\beta} \right) |g|^{1/2} (dx^1 \wedge dx^2 \wedge dx^3 \wedge dx^4). \tag{10. 50}$$

10.5   Exercises

1.   Verify that, in the setting of Section 10.2.4:

(i) If ∇ is a metric connection, then ∇^ϕ defined by (10. 6) is a metric connection for every gauge transformation ϕ.

(ii) If ∇ has connection matrix ω with respect to some local frame field $\{s_1, \ldots, s_m\}$, and if the matrix $g = (g_{jk})$ represents the gauge transformation ϕ with respect to this frame field, then $\omega^\phi = g\,d\,(g^{-1}) + g\omega g^{-1}$ is the connection matrix of ∇^ϕ.

Hint: Compare with one of the exercises for Chapter 9.

(iii) Deduce from (ii) that if ω is an $L\,(G)$-valued 1-form, then so is ω^ϕ; in other words, if ∇ is a G-connection, then so is ∇^ϕ.

2. (i) Check the last identity in formula (10. 12), namely,

$$\|\mathfrak{R}^\nabla\|_r^2 = \sum_{\alpha < \beta} \sum_i \|\mathfrak{R} \wedge s_i \cdot e_\alpha \wedge e_\beta\|^2. \tag{10. 51}$$

(ii) Suppose that $\{D_1, \ldots, D_n\}$ is a local frame field for $TM \to M$, and $\{t_i\}$ is a local frame field for $\pi: E \to M$, with

$$\langle D_\alpha | D_\beta \rangle = g_{\alpha\beta}, \quad \langle t_i | t_j \rangle^E = h_{ij}.$$

Prove that if $(g^{\gamma\delta}) = (g_{\alpha\beta})^{-1}$ and $(h^{km}) = (h_{ij})^{-1}$, then in terms of the curvature tensor defined in Chapter 9,

$$\|\mathfrak{R}^\nabla\|^2 = \frac{1}{2} \sum_{\alpha, \beta, \gamma, \delta} g^{\alpha\beta} g^{\gamma\delta} \sum_{i, j, k, m} h^{km} R^i_{k\alpha\gamma} R^j_{m\beta\delta} h_{ij}. \tag{10. 52}$$

Note: The comparison between (10. 51) and (10. 52) may serve to remind the reader of the advantages of coordinate-free notation!

3. Suppose the Riemannian metric $\langle \cdot | \cdot \rangle$ on TM is replaced by $\phi^2 \langle \cdot | \cdot \rangle$ for some smooth function ϕ. Show that the integrand in the Yang–Mills Lagrangian (10. 14) is multiplied by ϕ^{n-4}, where n is the dimension of M, and hence is unchanged when M is 4-dimensional.

Remark: The last property is called the **conformal invariance** of the Yang–Mills Lagrangian.

4. Suppose α and β are real-valued functions on R^4 with the Lorentz inner product, where the canonical volume form is $\rho = dx \wedge dy \wedge dz \wedge c\,dt$.

(i) Using a formula at the end of Chapter 1, and the rules for exterior differentiation given in Chapter 2, prove that

$$\langle d\alpha | d\beta \rangle \rho = \langle * d\,(* d\alpha) | \beta \rangle \rho - d\,(\beta \wedge * d\alpha). \tag{10. 53}$$

(ii) Suppose that P is a compact 4-dimensional submanifold-with-boundary of R^4, such that β vanishes on the boundary ∂P and on the complement of P. Using Stokes's Theorem, prove that

$$\int_{R^4} \langle d\alpha | d\beta \rangle \rho \ = \ \int_{R^4} \langle * \, d \, (* \, d\alpha) \, | \beta \rangle \rho \, . \tag{10.54}$$

5. Prove that $SU(2)$ consists of the complex matrices of the form

$$\begin{bmatrix} u & v \\ -\bar{v} & \bar{u} \end{bmatrix}, \tag{10.55}$$

where u and v are complex numbers with $|u|^2 + |v|^2 = 1$.

Hint: A general matrix in $C^{2 \times 2}$ may be written in the form $\begin{bmatrix} u & v \\ -y\bar{v} & z\bar{u} \end{bmatrix}$ for $y, z \in C$; prove that $y = z = 1$ when this matrix is in $SU(2)$.

6. Prove that the tangent space at the identity of the Lie group $SU(2)$ (i.e., the Lie algebra of $SU(2)$) has as its basis the Pauli matrices listed in (10. 1).

Hint: In the light of Exercise 5, consider the following three curves in $C^{2 \times 2}$ and their derivatives at 0:

$$t \to \begin{bmatrix} e^{it} & 0 \\ 0 & e^{-it} \end{bmatrix}, \ t \to \begin{bmatrix} 0 & e^t \\ -e^t & 0 \end{bmatrix}, \ t \to \begin{bmatrix} 0 & e^{it} \\ e^{it} & 0 \end{bmatrix}. \tag{10.56}$$

Note that the derivatives of these curves at 0 are linearly independent, and so they must span a space of dimension 3; also each tangent space to $SU(2)$ is 3-dimensional.

7. Define the **adjoint action** of a Lie group G on its Lie algebra $L(G)$ by the formula

$$\text{Ad}_g (\xi) \ = \ \frac{d}{dt} (g \varphi (t) \, g^{-1}) \Big|_{t=0}, \ g \in G, \tag{10.57}$$

where $\xi \in T_e G$ (i.e., the tangent space at the identity of G), and $\varphi(t)$ is a curve in G with $\varphi(0) = e$, $\dot{\varphi}(0) = \xi$. Calculate the adjoint action of a general element (10. 55) of $SU(2)$ on each of the Pauli matrices listed in (10. 1), using the curves described in (10. 56).

8. Check that multiplication of quaternions is associative; that is, check using (10. 21) and (10. 22) that for any quaternions x, y, z, we have $x(yz) = (xy)z$.

9. Check the formula (10. 31) for the change-of-chart map for the 4-sphere
    $\varphi \bullet \psi^{-1} : R^4 - \{0\} \to R^4 - \{0\}$.

10. Derive the formula $F = A \wedge A + dA$, that is, (10. 42), for the field strength in terms of the gauge potential on a quaternion line bundle, by using formula (10. 35) and the fact that $d^E s = s \wedge A$.

11. Suppose that two H-valued 1-forms A and B are related by the formula (10. 46). Prove that if A is purely imaginary, that is, $(1/2)\,(A - \bar{A}) = A$, then so is B.

12. Suppose z denotes the quaternion $z^1 + z^2 i + z^3 j + z^4 k$. Compute $dz \wedge dz$ and show that it is not zero, where $dz = dz^1 + dz^2 i + dz^3 j + dz^4 k$. What is the field strength associated with the gauge potential $A = (dz - d\bar{z})\,/2\,?$

13. Let z and dz be as in Exercise 12.

(i) Use the formula $d|z|^2 = d\,(\bar{z}z)$ to show that

$$d|z| = \frac{\bar{z}dz + (d\bar{z})\,z}{2|z|}. \tag{10. 58}$$

(ii) Show that if $u = z/|z|$, then

$$du = \frac{1}{2}\left[\frac{dz}{|z|} - \frac{z\,(d\bar{z})\,z}{|z|^3}\right]. \tag{10. 59}$$

(iii) Show that if $y = z^{-1}$, for $z \neq 0$, then $dy = -z^{-1}\,(dz)\,z^{-1}$.

14. Consider the formula (10. 46) in the case where the gauge potential A describes an "anti-instanton" (see Atiyah [1979], p. 21):

$$A = \frac{1}{2}\left[\frac{\bar{y}dy - (d\bar{y})\,y}{1 + |y|^2}\right]. \tag{10. 60}$$

Using the results of Exercise 13, show that the corresponding gauge potential B is given in terms of the quaternion variable $z = y^{-1}$ by

$$B = \frac{1}{2}\left[\frac{\bar{z}dz - (d\bar{z})\,z}{1 + |z|^2}\right]. \tag{10. 61}$$

15. In the example described in Section 10.4.5, calculate the field strength of the gauge potential $C = yd\bar{y} - (dy)\,\bar{y}$, using the following steps:

(i) By exterior differentiation of $|y|^2$, show that the real 1-form $\omega = d\,(|y|^2)$ satisfies $(dy)\,\bar{y} = \omega - yd\bar{y}$.

(ii) Using (i), show that

$$dC = 2dy \wedge d\bar{y}, \tag{10. 62}$$

$$C \wedge C = 2\omega \wedge C - 4|y|^2 dy \wedge d\bar{y}. \tag{10. 63}$$

16. In 10.4.5.1, we compared the gauge potentials A and B computed on two different local trivializations of a quaternion line bundle. Show that the field strengths F and $\hat{F}$ with respect to these local trivializations are related by the formula

$$\hat{F} = \bar{u}Fu. \tag{10.64}$$

Hint: Compute $B \wedge B + dB$ using (10. 46); $u\bar{u} = 1 \Rightarrow ud\bar{u} = -(du)\,\bar{u}$.

Remark: This is a quaternion version of the result "$\hat{R} = g^{-1}Rg$" in the exercises to Chapter 9, concerning the way that curvature forms transform under a transition function g.

10.6   The Yang–Mills Equations

The Yang–Mills equations are the conditions on the curvature of a connection for it to be a stationary point of the general Yang–Mills Lagrangian; thus in the special case of the Yang–Mills Lagrangian for the electromagnetic potential, discussed above, the Yang–Mills equations are simply Maxwell's equations. To state these equations we need the notion of codifferential.

10.6.1 The Codifferential

The **codifferential** δ^E associated with a covariant exterior derivative d^E on $\pi : E \to M$ is the mapping $\Omega^{q+1}(M;E) \to \Omega^q(M;E)$, for $q = 0, 1, ..., n$, defined by the formula

$$(d^E\lambda,\mu) = (\lambda,\delta^E\mu) \tag{10.65}$$

for $\mu \in \Omega^{q+1}(M;E)$ and $\lambda \in \Omega^q(M;E)$, where the inner product is the one defined in (10. 10). As we shall see in Exercise 17 in Section 10.9, the codifferential stands in similar relation to d^E as $*\,d*$ does to the exterior derivative d, where $*$ is the Hodge star operator.

10.6.1.1 The Codifferential on $\mathrm{Hom}\,(E,E)$

Recall from the exercises of Chapter 9 that a connection ∇ on $\pi : E \to M$ induces one on $\mathrm{Hom}\,(E,E) \to M$, whose covariant exterior derivative d^{Hom} is characterized by

$$(d^{\mathrm{Hom}}A) \wedge \mu = d^E(A \wedge \mu) - (-1)^{\deg A}A \wedge d^E\mu \tag{10.66}$$

for every $\mathrm{Hom}\,(E,E)$-valued form A and E-valued form μ. Its codifferential is denoted δ^{Hom}, and is calculated with respect to the inner product on $\mathrm{Hom}\,(E,E)$-valued q-forms given by

$$(A,B) = \int_M \langle A|B\rangle^{\mathrm{Hom},\,q}\rho, \tag{10.67}$$

in the notation of 10.2.6.

10.6.2 Formulation of the Yang–Mills Equations

Either of following conditions is necessary and sufficient for ∇ to be a Yang–Mills connection, that is, a stationary point of the Yang–Mills Lagrangian:

$$\delta^{\text{Hom}} \mathfrak{R}^{\nabla} = 0, \tag{10.68}$$

$$\Delta\, \mathfrak{R}^{\nabla} = 0, \tag{10.69}$$

where $\Delta = d^{\text{Hom}} \delta^{\text{Hom}} + \delta^{\text{Hom}} d^{\text{Hom}}$.

Remarks: The condition $\delta^{\text{Hom}} \mathfrak{R}^{\nabla} = 0$ is called the **Yang–Mills equations**. As we noted already, Maxwell's equations *in vacuo*, $*d\,(*\eta) = 0$, are a particular case. The condition $\Delta\, \mathfrak{R}^{\nabla} = 0$ says that the curvature of a Yang–Mills connection is "harmonic," with respect to the natural "Laplacian" on Hom (E,E)-valued 2-forms.

Proof: Fix a connection ∇ on a complex vector bundle $\pi: E \to M$ (in the setting described in Section 10.2.7), and select $A \in \Omega^1(M; \text{Hom}\,(E,E))$. There exists a family of connections ∇_t, parametrized by $t \in R$, whose covariant exterior derivatives are given by

$$d^{E,\,t}\mu = d^E\mu + tA \wedge \mu. \tag{10.70}$$

In the exercises to Chapter 9, it was proved that the Curvature of ∇_t is given by the formula

$$\mathfrak{R}^{\nabla,\,t} = \mathfrak{R}^{\nabla} + t d^{\text{Hom}} A + t^2 A \wedge A, \tag{10.71}$$

where $d^{\text{Hom}} A$ refers to (10.66). Therefore, to first order in t, we have the formula

$$\| \mathfrak{R}^{\nabla,\,t} \|^2 = \| \mathfrak{R}^{\nabla} \|^2 + 2t\langle d^{\text{Hom}} A | \mathfrak{R}^{\nabla} \rangle^{\text{Hom},\,2} + O\,(t^2). \tag{10.72}$$

Referring to (10.14), and the definition of δ^{Hom} on $\Omega^2(M; \text{Hom}\,(E,E))$ given in Section 10.6.1.1, we see that

$$\frac{d}{dt} L\,(\nabla_t)\,\Big|_{t\,=\,0} = \int_M \langle d^{\text{Hom}} A | \mathfrak{R}^{\nabla} \rangle^{\text{Hom},\,2} \rho = \int_M \langle A | \delta^{\text{Hom}} \mathfrak{R}^{\nabla} \rangle^{\text{Hom},\,1} \rho.$$

For ∇ to be a Yang–Mills connection amounts to saying that this integral is zero for all choices of A, which is precisely the condition $\delta^{\text{Hom}} \mathfrak{R}^{\nabla} = 0$.

It only remains to check the equivalence of (10.68) and (10.69). Recall that in the exercises to Chapter 9, we formulated Bianchi's identity as

$$d^{\mathrm{Hom}}\mathfrak{R}^{\nabla} = 0, \tag{10.73}$$

which holds for any connection. Thus (10.68) immediately implies (10.69). To see the converse, note

$$\int_M \langle \Delta\mathfrak{R}^{\nabla} | \mathfrak{R}^{\nabla} \rangle^{\mathrm{Hom},\,2} \rho = \int_M \langle d^{\mathrm{Hom}}(\delta^{\mathrm{Hom}}\mathfrak{R}^{\nabla}) + \delta^{\mathrm{Hom}}(d^{\mathrm{Hom}}\mathfrak{R}^{\nabla}) | \mathfrak{R}^{\nabla} \rangle^{\mathrm{Hom},\,2} \rho$$

$$= \int_M \| d^{\mathrm{Hom}}\mathfrak{R}^{\nabla} \|^2 \rho + \int_M \| \delta^{\mathrm{Hom}}\mathfrak{R}^{\nabla} \|^2 \rho$$

$$= \int_M \| \delta^{\mathrm{Hom}}\mathfrak{R}^{\nabla} \|^2 \rho,$$

and so $\Delta\mathfrak{R}^{\nabla} = 0$ implies $\delta^{\mathrm{Hom}}\mathfrak{R}^{\nabla} = 0$.   ¤

10.7   Self-duality

Since the curvature involves one order of differentiation of the connection forms (i.e., of the gauge potential), the Yang–Mills equations $\delta^{\mathrm{Hom}}\mathfrak{R}^{\nabla} = 0$ are equations of second order in the gauge potential. However, there is a stronger, first-order system of equations whose solutions automatically satisfy the Yang–Mills equations, and which is much easier to solve. The situation is analogous to replacing the study of Laplace's equation in the plane by the study of the Cauchy-Riemann equations, which are much stronger.

10.7.1 Self-dual 2-Forms

We assume that $\pi\colon E \to M$ is a real or complex vector bundle with a metric $r \to \langle \cdot | \cdot \rangle_r^E$, over a compact, oriented, 4-dimensional Riemannian manifold $(M, \langle \cdot | \cdot \rangle)$. Recall that a metric on $\mathrm{Hom}\,(\Lambda^q TM, E)$ was constructed in Section 10.2.5.1, for $q = 1, 2, 3, 4$. The discussion of the Hodge star operator in Chapter 1 applies immediately to each vector space $L(\Lambda^q T_r M \to E_r)$, for $q = 1, 2, 3, 4$, in the case where E is a line bundle; moreover it extends, with only minor notational changes, to the case of an arbitrary vector bundle E. If $\{s_i\}$ is an orthonormal frame field for $\pi\colon E \to M$, and $\{\varepsilon^1, \varepsilon^2, \varepsilon^3, \varepsilon^4\}$ is an orthonormal coframe field for $T^*M \to M$, then the effect of the Hodge star operator on E-valued 2-forms is given by:

$$*(s_i(\varepsilon^\alpha \wedge \varepsilon^\beta)) = s_i(\varepsilon^\gamma \wedge \varepsilon^\delta), \tag{10.74}$$

where $(\alpha, \beta, \gamma, \delta)$ is an even permutation of $(1, 2, 3, 4)$; for reasons of space, we omit the wedge symbol after s_j. For example,

$$* \left(s_j (\varepsilon^1 \wedge \varepsilon^3 + \varepsilon^4 \wedge \varepsilon^2) \right) = s_j (\varepsilon^4 \wedge \varepsilon^2 + \varepsilon^1 \wedge \varepsilon^3). \tag{10.75}$$

Define $\lambda \in \Omega^2(M;E)$ to be **self-dual** if $*\lambda = \lambda$, and **anti–self-dual** if $*\lambda = -\lambda$, and denote the self-dual and the anti–self-dual E-valued 2-forms by $\Omega_+^2(M;E)$ and $\Omega_-^2(M;E)$, respectively. For example, (10.75) illustrates a self-dual E-valued 2-form.

It follows from (10.74) that an orthonormal frame field for $\Omega_+^2(M;E)$ is given by

$$\{ s_i (\varepsilon^1 \wedge \varepsilon^2 + \varepsilon^3 \wedge \varepsilon^4), s_j (\varepsilon^1 \wedge \varepsilon^3 + \varepsilon^4 \wedge \varepsilon^2), s_k (\varepsilon^1 \wedge \varepsilon^4 + \varepsilon^2 \wedge \varepsilon^3) : i, j, k \geq 1 \}, \tag{10.76}$$

and an orthonormal frame field for $\Omega_-^2(M;E)$ is given by

$$\{ s_i (\varepsilon^1 \wedge \varepsilon^2 - \varepsilon^3 \wedge \varepsilon^4), s_j (\varepsilon^1 \wedge \varepsilon^3 - \varepsilon^4 \wedge \varepsilon^2), s_k (\varepsilon^1 \wedge \varepsilon^4 - \varepsilon^2 \wedge \varepsilon^3) : i, j, k \geq 1 \}. \tag{10.77}$$

Moreover it is clear that the E-valued 2-forms listed in (10.76) are orthogonal to those listed in (10.77). We could summarize by saying that $\Omega^2(M;E)$ admits an orthogonal decomposition

$$\Omega^2(M;E) = \Omega_+^2(M;E) \oplus \Omega_-^2(M;E) \tag{10.78}$$

into the self-dual and the anti–self-dual E-valued 2-forms.

10.7.2 Self-dual Connections

Replace E by $\mathrm{Hom}(E,E)$ in the preceding discussion. For any connection ∇, the associated curvature $\mathfrak{R}^\nabla \in \Omega^2(M;\mathrm{Hom}(E,E))$ decomposes according to (10.78) into self-dual and anti–self-dual parts:

$$\mathfrak{R}^\nabla = \mathfrak{R}_+^\nabla + \mathfrak{R}_-^\nabla. \tag{10.79}$$

We say that ∇ is a **self-dual connection** if $*\mathfrak{R}^\nabla = \mathfrak{R}^\nabla$, that is, if $\mathfrak{R}_-^\nabla = 0$.

10.7.2.1 Self-dual Connections Satisfy the Yang–Mills Equations

Suppose M has dimension four. If $*\mathfrak{R}^\nabla = \mathfrak{R}^\nabla$, then $\delta^{\mathrm{Hom}} \mathfrak{R}^\nabla = 0$.

Proof: Recall the following formula from Chapter 1, for the case of a four-dimensional M, and an inner product of signature 0:

$$\lambda \wedge *\mu = \langle \lambda | \mu \rangle^{\mathrm{Hom}, q} \rho, \text{ for } \lambda, \mu \in \Omega^q(M;\mathrm{Hom}(E,E)), \tag{10.80}$$

where ρ is the canonical volume form. It follows that for $\lambda \in \Omega^1(M;\mathrm{Hom}(E,E))$ and $\mu \in \Omega^2(M;\mathrm{Hom}(E,E))$

$$\int_M \langle d^{\mathrm{Hom}} \lambda | \mu \rangle^{\mathrm{Hom},\,2} \rho \ = \ \int_M d^{\mathrm{Hom}} \lambda \wedge *\mu$$

$$= \ \int_M d^{\mathrm{Hom}} \left(\lambda \wedge *\mu \right) + \int_M \lambda \wedge d^{\mathrm{Hom}} \left(*\mu \right)$$

by the Leibniz rule. Since the boundary of M is empty, the first integral is zero by Stokes's Theorem. On the other hand $* \left(*\phi \right) \ = \ -\phi$ for $\phi \in \Omega^3 \left(M; \mathrm{Hom}\left(E, E \right) \right)$ (see Chapter 1), so we obtain

$$\int_M \langle d^{\mathrm{Hom}} \lambda | \mu \rangle^{\mathrm{Hom},\,2} \rho \ = \ -\int_M \lambda \wedge * \left(* \left(d^{\mathrm{Hom}} *\mu \right) \right) \ = \ -\int_M \langle \lambda | * \left(d^{\mathrm{Hom}} *\mu \right) \rangle^{\mathrm{Hom},\,1} \rho .$$

Now it follows from (10. 10) and (10. 65) that $\delta^{\mathrm{Hom}} \mathfrak{R}^\nabla \ = \ -* \left(d^{\mathrm{Hom}} * \mathfrak{R}^\nabla \right)$. If ∇ is self-dual then $d^{\mathrm{Hom}} * \mathfrak{R}^\nabla = d^{\mathrm{Hom}} \mathfrak{R}^\nabla = 0$, by the Bianchi identity (10. 73). Therefore $\delta^{\mathrm{Hom}} \mathfrak{R}^\nabla = 0$.   ¤

10.7.2.2 Anti–self-dual Connections

If the connection is anti–self-dual, that is, if $*\mathfrak{R}^\nabla = -\mathfrak{R}^\nabla$, then simply reversing the orientation of M makes it self-dual, and so the preceding result applies.

10.7.3 Self-dual Connections Minimize the Yang–Mills Lagrangian

It follows from the Chern–Weyl theory of characteristic classes, which cannot be covered here (see Kobayashi and Nomizu [1963]), that

$$i \left(E \right) \ = \ \frac{1}{8\pi^2} \int_M \left(\left\| \mathfrak{R}_+^\nabla \right\|^2 - \left\| \mathfrak{R}_-^\nabla \right\|^2 \right) \rho \tag{10. 81}$$

does not depend upon the choice of connection ∇; in other words, it is an invariant of the bundle. In the case where $\pi : E \to M$ is an $Sp \left(1 \right)$-quaternion line bundle, as described in Section 10.4, with the metric (10. 34), we call $i \left(E \right)$ the **instanton number.**

This yields considerable insight into the minimization of the Yang–Mills Lagrangian. By the orthogonality aspect of decomposition (10. 78),

$$L \left(\nabla \right) \ = \ \frac{1}{2} \int_M \left(\left\| \mathfrak{R}_+^\nabla \right\|^2 + \left\| \mathfrak{R}_-^\nabla \right\|^2 \right) \rho \ge \frac{1}{2} \int_M \left(\left\| \mathfrak{R}_+^\nabla \right\|^2 - \left\| \mathfrak{R}_-^\nabla \right\|^2 \right) \rho \ = \ 4\pi^2 i \left(E \right),$$

and the right side does not depend upon the connection; thus $L \left(\nabla \right) \ge 4\pi^2 i \left(E \right)$, with equality if and only if ∇ is self-dual. Thus, among the solutions to the Yang–Mills equations, self-dual connections are precisely the ones which absolutely minimize the Yang–Mills Lagrangian.

10.8 Instantons

Let us return to the study of the quaternion line bundle discussed in Section 10.4.1 and Section 10.4.5, and whose Yang–Mills Lagrangian was described in Section 10.4.6. We shall take as the Riemannian metric on S^4:

$$\langle \cdot | \cdot \rangle = \sum_{1 \le \alpha \le 4} \frac{4}{(1+|y|^2)^2} dy^\alpha \otimes dy^\alpha = \sum_{1 \le \alpha \le 4} \frac{4}{(1+|z|^2)^2} dz^\alpha \otimes dz^\alpha \qquad \textbf{(10. 82)}$$

in terms of the two coordinate systems presented previously. Take a local frame field for E over U consisting of a nonvanishing section s of unit norm. Since the metric on S^4 is just a function multiplied by the Euclidean metric, it is clear from (10. 76) that an orthonormal frame field for $\Omega_+^2(M;E)$ over U is given by

$$\{ s \, (dy^1 \wedge dy^2 + dy^3 \wedge dy^4), s \, (dy^1 \wedge dy^3 + dy^4 \wedge dy^2), s \, (dy^1 \wedge dy^4 + dy^2 \wedge dy^3) \}.$$

10.8.1 Self-dual Connections on a Quaternion Line Bundle

Self-dual solutions to the Yang–Mills equations over S^4, with the metric (10. 82), are called **instantons**. The following instantons were reported by Belavin, Polyakov, Schwartz, and Tyupkin in 1975. Define a family of $Sp\,(1)$-connections ∇_t for $t > 0$ by taking their gauge potentials, with respect to this frame field $\{s\}$ (see (10. 37)), to be

$$A^t = \frac{1}{2} \left[\frac{y d\bar{y} - (dy)\bar{y}}{t^2 + |y|^2} \right]. \qquad \textbf{(10. 83)}$$

Comparing this expression to Exercise 15, we see that, for $C = y d\bar{y} - (dy)\bar{y}$ and $\omega = d|y|^2$, (10. 62) and (10. 63) give

$$A^t \wedge A^t = \frac{C \wedge C}{4(t^2 + |y|^2)^2} = \frac{\omega \wedge C - 2|y|^2 dy \wedge d\bar{y}}{2(t^2 + |y|^2)^2},$$

$$dA^t = \frac{dy \wedge d\bar{y}}{t^2 + |y|^2} - \frac{\omega \wedge C}{2(t^2 + |y|^2)^2}.$$

According to (10. 42), the corresponding curvature 2-form is $F^t = dA^t + A^t \wedge A^t$, which simplifies to

$$F^t = \frac{t^2 dy \wedge d\bar{y}}{(t^2 + |y|^2)^2}, \qquad \textbf{(10. 84)}$$

which is self-dual, since by Exercise 18 in Section 10.9, $dy \wedge d\bar{y}$ is a self-dual E-valued 2-form. Changing to the other local trivialization as in Exercise 14, we find the gauge potential transforms to

$$B^{1/t} = \frac{1}{2} \left[\frac{zd\bar{z} - (dz)\,\bar{z}}{t^{-2} + |z|^2} \right] \tag{10.85}$$

so clearly we have a family of self-dual connections defined over the whole vector bundle. Using the metric (10. 82) and the formula (10. 48), we see that the norm of the curvature is

$$\| \mathfrak{R}^{\nabla, t} \|^2 = \sum_{\alpha < \beta} |F^t_{\alpha, \beta}|^2 \frac{(1 + |y|^2)^4}{16} = \frac{3t^4}{2} \left(\frac{1 + |y|^2}{t^2 + |y|^2} \right)^4, \tag{10.86}$$

using (10. 89) below.

Let us say that two G-connections are **gauge equivalent** if they are related by a gauge transformation, as in (10. 6); this is indeed an equivalence relation. It follows from 10.2.6.1 that two G-connections cannot be gauge equivalent if they lead to different values for the normed curvature. In this light, formula (10. 86) tells us several things about the equivalence classes of self-dual connections that we have constructed:

- When $t = 1$, the norm of the curvature is a constant, and this is the same for any choice of the "pole" omitted from the chart U in 10.4.1.

- $t = 0$ is not permitted, because in that case the gauge potential cannot be extended to the other local trivialization, by (10. 85).

- Changing from y to its antipode $z = y^{-1}$ and from t to $1/t$ would give the same values for the normed curvature; thus we should only consider $t \in (0, 1)$, if we are going to allow all choices of the pole.

- For $0 < t < 1$, (10. 86) achieves its unique maximum when $y = (0, 0, 0, 0)$, that is, at the "pole" omitted from the chart U. Thus different choices of $t \in (0, 1)$ and different choices of the "pole" $p \in S^4$ omitted from the chart map lead to different values for the normed curvature, and hence to different equivalence classes of connections, by the observation above.

In other words, we have constructed a family of distinct equivalence classes of self-dual connections on a quaternion line bundle "of instanton number 1," parametrized by $\{1\} \cup (S^4 \times (0, 1))$. Using more advanced methods, Atiyah, Hitchin and Singer [1978] proved that these are the **only** classes of self-dual connections on this bundle, and gave this set the geometric structure of a 5-dimensional Riemannian manifold:

10.8.2 Theorem

The set of connections described above represents all the equivalence classes of self-dual connections on this quaternion line bundle; these equivalence classes can be identified with 5-dimensional hyperbolic space.

To summarize: In view of 10.7.3, the connections defined by (10. 83), for $t = 1$, and for $t \in (0, 1)$ with different choices of the "pole" $p \in S^4$ omitted from the chart map, comprise a complete set of those connections that minimize the Yang–Mills Lagrangian for this $Sp\,(1)$-bundle over S^4.

10.9   Exercises

17. Suppose $\pi : E \to M$ is a real or complex local $U\,(1)$-line bundle over $M = R^4$ with the Lorentz metric, and has the metric

$$\langle \sigma | \tau \rangle_p^E = \mathrm{Re}\,\{\sigma\,(p)\,\bar{\tau}\,(p)\,\}.$$

The E-valued q-forms may be regarded as ordinary (real or complex) q-forms under the global trivialization, and an example of a covariant exterior derivative on this bundle is the ordinary exterior derivative, $d^E = d$. Prove that the codifferential δ^E defined in Section 10.6.1 is here given by the formula

$$\delta^E \mu = *\,(d*\mu), \qquad\qquad\qquad (10.\,87)$$

for $\mu \in \Omega^q\,(M)$, $q = 1, 2, 3, 4$.

Hint: Use the methods of Exercise 4.

18. Suppose z is the quaternion $z^1 + z^2 i + z^3 j + z^4 k$, and $dz = dz^1 + dz^2 i + dz^3 j + dz^4 k$.
    (i) Show that the H-valued 2-form $dz \wedge d\bar{z}$ is self-dual, by showing that

$$-\frac{1}{2}\,(dz \wedge d\bar{z}) = (dz^1 \wedge dz^2 + dz^3 \wedge dz^4)\,i + (\,(dz^1 \wedge dz^3 + dz^4 \wedge dz^2)\,j) \qquad (10.\,88)$$

$$+\,(dz^1 \wedge dz^4 + dz^2 \wedge dz^3)\,k.$$

    (ii) Suppose $F_{\alpha\beta} = (dz \wedge d\bar{z}) \cdot D_\alpha \wedge D_\beta$, where $D_\alpha = \partial/\partial z^\alpha$. Prove that

$$\sum_{\alpha < \beta} |F_{\alpha\beta}|^2 = 24. \qquad\qquad\qquad (10.\,89)$$

19. In the situation described in Section 10.8, show that for any $t > 0$ the following gauge potential represents an anti–self-dual connection ("anti-instanton"):

$$A^t = \frac{1}{2}\left[\frac{\bar{y}dy - (d\bar{y})\,y}{t^2 + |y|^2}\right].$$

(10. 90)

20.   Check by direct integration over R^4 that, for the curvatures of the self-dual connections described in (10. 86),

$$\int_{S^4} \|\,\mathfrak{R}^{\nabla,\,t}\|^2 \rho$$

(10. 91)

is the same for all $t > 0$, and explain how this fact is related to the formula (10. 81).

10.10   History and Bibliography

The unification of the electromagnetic and weak interactions by Weinberg and Salam, using Yang–Mills theory, received experimental confirmation in the discovery of massive particles called intermediate bosons, which were predicted by the theory. In recent years much work has been done on a $U(1) \times SU(2) \times SU(3)$ Yang–Mills theory known as the "standard model," which seeks to provide the basic framework for unifying the electromagnetic, weak, and strong interactions; a complete description of the solution of the Yang–Mills equations for this case is available. Presently physicists are postulating "grand unified theories" for the unification of the electromagnetic, weak, strong, and gravitational fields, but the sort of particles predicted by these theories seem to be beyond the range of present experimental capabilities.

Many physics books give accounts of gauge field theories. For a mathematician's account of the geometrization of physics in the twentieth century, see Bourguignon [1987]. The lecture series of Atiyah [1979] and of Lawson [1985] present gauge theories in greater depth.

Bibliography

Abraham, R.; Mardsen, J. E.; and Ratiu, T. (1988). *Manifolds, Tensor Analysis, and Applications*. Springer, New York.

Amarsi, S.-I.; Barndorff-Nielsen, O. E.; Kass, R. E.; Lauritzen, S. L; and Rao, C. R. (1987). *Differential Geometry in Statistical Inference*. Institute of Mathematical Statistics Lecture Notes, Vol. 10. Institute of Mathematical Statistics, Hayward,CA.

Atiyah, M. F. (1979). *Geometry of Yang–Mills fields*. Lezione Fermiane, Accademia Nazionale dei Lincei & Scuola Normale Superiore, Pisa.

Atiyah, M. F.; Hitchin, N. J.; Singer, I. M. (1978). Self-Duality in Four-Dimensional Riemannian Geometry. *Proceedings of the Royal Society, London, A* **362**, 425–61.

Berger, M., and Gostiaux, B. (1988*). Differential Geometry: Manifolds, Curves, and Surfaces*. Springer, New York.

Bourguignon, J. P. (1987). Yang–Mills Theory: The Differential Geometric Side. Lecture Notes in Mathematics 1263, 13–54. Springer, New York.

Cartan, H. (1970). *Differential Forms*. Hermann, Paris.

Chern, S. S. (1989), ed. *Global Differential Geometry*. Mathematical Association of America, Studies in Mathematics, Vol. 27. Prentice-Hall, Englewood Cliffs, NJ.

Curtis, W. D. and Miller, F. R. (1985). *Differential Manifolds and Theoretical Physics*. Academic Press, New York..

do Carmo, M. P. (1992). *Riemannian Geometry*. Birkhäuser, Boston.

Edelen, D. G. B. (1985). *Applied Exterior Calculus*. Wiley, New York.

Eisenhart, L. P. (1940). *An Introduction to Differential Geometry with Use of the Tensor Calculus*. Princeton University Press, Princeton, NJ.

Flanders, H. (1989). *Differential Forms with Applications to the Physical Sciences.* Dover, New York.

Gallot, S.; Hulin, D.; and Lafontaine, J. (1990). *Riemannian Geometry.* Springer, Berlin.

Greub, W. (1978). *Multilinear Algebra.* Springer, Berlin.

Helgason, S. (1978). *Differential Geometry, Lie Groups, and Symmetric Spaces.* Academic Press, New York.

Husemoller, D. (1975). *Fiber Bundles.* Springer, New York.

Klingenberg (1982). *Riemannian Geometry.* De Gruyter, Berlin.

Kobayashi, N., and Nomizu, K. (1963). *Foundations of Differential Geometry.* Wiley-Interscience, New York.

Lancaster, P., and Tismenetsky, M. (1985). *The Theory of Matrices*. Academic Press, New York.

Lang, S. (1972). *Differential Manifolds*. Addison-Wesley, Reading, MA.

Lawson, H. B. (1985). *The Theory of Gauge Fields in Four Dimensions*. CBMS Regional Conference Series 58, American Mathematical Society, Providence, RI.

Murray, M. K., and Rice, J. W. (1993). *Differential Geometry and Statistics*. Chapman and Hall, London.

Okubo, T. (1987). *Differential Geometry.* Dekker, New York.

Sattinger, D. H., and Weaver, O. L. (1986). *Lie Groups and Lie Algebras with Applications to Physics, Geometry, and Mechanics.* Springer, New York.

Spivak, M. (1979). *A Comprehensive Introduction to Differential Geometry, Vols. I–V.* Publish or Perish, Berkeley.

Struik, D. (1961). *Lectures on Classical Differential Geometry.* Addison-Wesley, Reading, MA; republished by Dover, New York (1988).

Warner, F. W. (1983). *Foundations of Differentiable Manifolds and Lie Groups.* Springer, New York.

Index

adapted moving frame 82
adjoint action 240
alternating 5
Ampère's Law 50
analytic function 163
anti-instanton 241, 249
antisymmetric 5
area form 86
Atiyah–Hitchin–Singer Theorem 249
atlas 99

base manifold 120
Bianchi's identity 206, 211, 243
bilinear 5
binormal 80
boundary 112
  orientation of 169
bracket of vector fields 27
bump function 118
bundle-valued forms 197

canonical parametrization 87
canonical volume form 168, 182
Cartan, Élie 49
Cayley transform 74
chain rule 42
charge density 50
chart 98
  compatible 98
  positively oriented 165
Christoffel symbols 202, 216
closed set
  in a manifold 101
Codazzi–Mainardi equations 84, 205
codifferential 242
codimension 105

compact set
  in a manifold 101
Compactness Lemma 109
complex
  general linear group 71
  line bundle 128
  special linear group 71
  vector bundle 128
component function 25
connection 195
  curvature 202
  dual 207
  Euclidean 195, 196
  forms 78, 201
  G- 226
  Levi–Civita 216
  matrix 201
  metric 216
  pyramidic 221
  self-dual 245
  $Sp(1)$- 235
  torsion of 212
  Yang–Mills 229
continuity equation 50
contraction 49
coordinate functions 25
cotangent bundle 138
  Euclidean 121
cotangent space
  Euclidean 28
  to manifold 138
cotangent vector 138
  Euclidean 28
covariant derivative 195
covariant exterior derivative 199
cross product 7, 19

curl 37
current 51
curvature
  forms 202
  Gaussian 89, 205, 220
  mean 89
  negative 92
  norm 228
  of a connection 202
  of a curve 80
  positive 92
  principal 91
  sectional 219
  tensor 203
curve in a manifold 103, 147
cylinder 87, 126

density form 36
derivation 26, 147
determinant 9
diffeomorphism 102
differentiable structure 99, 104
differential form
  anti-self-dual 245
  bundle-valued 124
  Euclidean 30
  matrix-valued 78
  on a manifold 145, 148
  1-form 145
  self-dual 245
  vector-bundle valued 197
differential manifold 99
differential of a map 35, 41
differential operator 147
div 37, 188, 221
Divergence Theorem 184, 189
dot product 14
dual
  basis 28
  bundle 124
  frame 157
  space 11, 28

electric current density 50
electric field 50
electromagnetic potential 52
electromagnetic theory 50, 223
ellipsoid 96, 161
embedded submanifold 108
embedding 108
equivalent atlas 99
Euclidean metric 153
exterior calculus 149, 236
exterior derivative 36
  bundle-valued form 199
  geometric meaning 186

exterior power 1
  of a linear transformation 10
  vector bundle 125
exterior product 6
  bundle-valued forms 198
  differential forms 31

Faraday's Law 50
fiber 120
field strength 224, 236
first fundamental form 95
first structure equation 79
flux form 36, 177
frame field 144
  positively oriented 181
Frenet frame 80
fundamental forms 95
Fundamental Theorem of Calculus 184

gauge
  group 227
  potential 224, 237
  transformations 224, 227
Gauss's equation 84, 205
general linear group 69
geodesic 209
grad 37
  on a manifold 160
Grassmann, Hermann 23
Green's Theorem 184

hairy ball 129
Hodge star operator 17, 32
homomorphism bundle 123
hyperbolic paraboloid 97, 161
hyperbolic plane 155
hyperbolic space 155, 161
hyperboloid 58, 95, 165

immersion 53, 102
implicit function 63
  parametrization 63
  theorem 64
inclusion map 103
indefinite metric 152
indefinite Riemannian metric 152
induced metric 154
inner product 13
  on exterior powers 15
instanton 247
  number 234, 246
integral of a form 174, 179
interior 112
interior product 49
intermediate bosons 250
Inverse Function Theorem 64

Iteration Rule 36

Jacobi identity 27

Koszul connection 195, 199

Laplacian 221
Leibniz rule 195, 206
level set, orientability of 165
Levi-Civita connection 216
Lie algebra 225
Lie derivative 27, 48, 149
  of differential forms 40
  using flows 48
Lie subgroup 69, 167
Lie, Sophus 69
likelihood 221
line bundle 121
line integral 36, 176
linear forms 11
local connector 209
local trivialization 126
local vector bundle 120
  isomorphism 123
  morphism 122
  section 120
locally finite 117
Lorentz group 74
Lorentz inner product 14, 50, 155, 182

magnetic field 50
matrix groups 69
Maxwell's equations 50, 229
metric 152
  connection 216
  Riemannian 152
Möbius strip 127, 169
Molière 195
moving orthonormal frame 76
multilinear 5

neighborhood 25
non-degeneracy 13
non-orientable manifold 165
normal 80
normal bundle 140

open set
  Euclidean 24
  in a manifold 101
  in a submanifold 109
orientable manifold 165
orientation 165
orientation of basis 17
oriented 165
orthogonal group 70

orthonormal
  basis 14
  coframe 157
  frame 156

parametrization 61
parametrized curve 61, 176
parametrized surface 61, 159, 177
partition of unity 117
Pauli matrices 226
permutations 3
Poincaré, Henri 49
projective plane 136
pseudo-Riemannian manifold 152
pullback 43, 122, 149
  tangent bundle 124
push-forward 44
p-vectors 1

quadric surface 95
quaternion 231
  imaginary 232
  line bundle 233

regular mapping 60
repère mobile 76
Riemannian
  manifold 152
  metric tensor 96, 153, 163
  volume 183

second fundamental form 96
second structure equation 79, 205
section 130
Serret–Frenet formulas 81
sharp # 160
signature
  inner product space 15
  permutation 3
smooth map 24, 102
special
  linear group 70
  orthogonal group 70
  unitary group 71, 225
sphere 57, 154, 165
spherical coordinates 45
stereographic projection 104
Stokes's Theorem 183
strong nuclear force 225
structure equations 78
submanifold 105
  Euclidean 56
submanifold-with-boundary 112, 169
submersion 54, 102
surface integral 36, 177
surface of revolution 54, 87, 94

symmetry 13
symplectic group 74

tangent bundle 136, 169
  atlas 136
  Euclidean 121
  projective plane 137
tangent map 138, 151
  Euclidean 122
tangent plane 81
tangent space
  Euclidean 24
  to manifold 136
  to submanifold 66
tangent vector 136
  as class of curves 147
  Euclidean 25
tensor
  (0,2)-tensor 33
tensor product 32
  of 1-forms 33
  vector bundles 124
Theorema Egregium 89, 90
torsion 212
  of curve 80
torsion-free 212
torus 57, 165
transition function 127
transposition 3
triple product 7, 19
trivial bundle 129
trivializing cover 126

unitary group 71
universal mapping property 5

vector 227
vector algebra 19
vector bundle 125
  base 126
  construction 130, 141
  equivalent 129
  fiber 125
  metric 227
  morphism 129
  projection 125
  pullback 133
  section 130
  sub-bundle 129
  tensor product 124
  transition function 127
vector field 25, 144
volume
  form 164
  Riemannian 183

work form 36, 176

Yang–Mills
  conformal invariance 239
  equations 243
  field 229
  Lagrangian 229, 238

zero section 120